T3-BNR-803

SUBJECT-SPECIFIC INSTRUCTIONAL METHODS AND ACTIVITIES

ADVANCES IN RESEARCH ON TEACHING

Series Editor: Jere Brophy

ADVANCES IN RESEARCH ON TEACHING VOLUME 8

SUBJECT-SPECIFIC INSTRUCTIONAL METHODS AND ACTIVITIES

EDITED BY

JERE BROPHY

College of Education, Michigan State University, USA

2001

JAI
An Imprint of Elsevier Science

Amsterdam – London – New York – Oxford – Paris – Shannon – Tokyo

ELSEVIER SCIENCE Ltd
The Boulevard, Langford Lane
Kidlington, Oxford OX5 1GB, UK

First edition 2001

Library of Congress Cataloging in Publication Data
A catalog record from the Library of Congress has been applied for.

British Library Cataloguing in Publication Data
A catalogue record from the British Library has been applied for.

ISBN: 0-7623-0615-7

♾ The paper used in this publication meets the requirements of ANSI/NISO Z39.48-1992 (Permanence of Paper).
Printed in The Netherlands.

CONTENTS

LIST OF CONTRIBUTORS

Donna E. Alvermann College of Education, University of Georgia, USA

Jo-Ann M. Amadeo Department of Human Development (Education), University of Maryland at College Park, USA

Michael T. Battista College of Education, Kent State University, USA

Jere Brophy College of Education, Michigan State University, USA

Gail Chittleborough National Key Centre for School Science & Mathematics, Curtin University of Technology, USA

Stephen F. Cunha Department of Geography, Humboldt State University, USA

Colette Daiute The Graduate Center, City University of New York, USA

Gerald G. Duffy College of Education (Emeritus), Michigan State University, USA

Richard A. Duschl School of Education, Kings College, London, UK

Rolland Fraser Department of Geography, Western Michigan University, USA

Sarah Warshauer Freedman School of Education, University of California, Berkeley, USA

Carole L. Hahn Division of Educational Studies, Emory University, USA

James V. Hoffman	College of Education, University of Texas at Austin, USA
George G. Hruby	College of Education, University of Georgia, USA
Pamela A. Kraus	Talaria, Inc., Seattle, USA
James D. Laney	College of Education, University of North Texas, USA
Jim Minstrell	Talaria, Inc., Seattle, USA
Michael J. Smith	Director of Education, American Geological Institute, USA
Mary Kay Stein	College of Education, University of Pittsburgh, USA
Joseph P. Stoltman	Department of Geography, Western Michigan University, USA
Stephen J. Thornton	Teachers College, Columbia University, USA
Judith Torney-Purta	Department of Human Development (Education), University of Maryland at College Park, USA
David F. Treagust	National Key Centre for School Science & Mathematics, Curtin University of Technology, USA
James H. Wandersee	College of Education, Louisiana State University, USA

INTRODUCTION

Jere Brophy

The *Advances in Research on Teaching* series is designed to bring focus and visibility to significant bodies of research on teaching, especially research on emerging new topics or on older topics that are being revived and studied from a fresh perspective. Volume 8 focuses on the very old topic of instructional methods and learning activities (e.g. lecture, discussion, simulation, etc.). However, instead of construing these broadly and assessing them as generic approaches to instruction, the chapters in this volume embed analyses of methods and activities within the contexts of subject matter-specific purposes and goals of instruction.

The idea for the volume was suggested by Steve Thornton, one of the chapter authors. Steve had been thinking about issues of generic vs. subject- and situation-specific applicability of instructional methods raised in a chapter on professional and subject-matter knowledge for teacher education written by Lauren Sosniak (1999). In that chapter, Sosniak had cast these issues as a dilemma currently facing the field. On one hand, Lee Shulman (1986) and others have argued convincingly that we need to shift from generic to more subject matter-specific approaches to conceptualizing and studying teaching (and ultimately, educating teachers). Elaborating on this point, Shulman has called for attention to "pedagogical content knowledge," which subsumes knowledge about how to teach particular subject matter to particular students:

> Within the category of pedagogical content knowledge I include, for the most regularly taught topics in one's subject area, the most useful forms of representation of those ideas, the most powerful analogies, illustrations, examples, explanations, and demonstrations – in a word, the ways of representing and formulating the subject that make it most

Subject-Specific Instructional Methods and Activities, Volume 8, pages 1–23.
2001 by Elsevier Science Ltd.
ISBN: 0-7623-0615-7

> comprehensible to others. Since there are no single most powerful forms of representation, the teacher must have at hand a veritable armamentarium of alternative forms of representation, some of which derive from research whereas others originate in the wisdom of practice (Shulman, 1986, p. 5).

Sosniak noted that although this notion appears sound in theory, attempts to follow up on it in practice have been challenging because the potential combinations of specific subject matter with particular types of students and other situational factors are seemingly infinite. Pondering this, educators may feel caught between traditional ideas about instructional methods that seem too generic to take them very far and more contemporary ideas that carry daunting implications concerning the need for a myriad of studies of topic-specific pedagogy:

> Work on pedagogical content knowledge or subject-specific instruction thus seems likely to find itself at the same dead end as efforts to become increasingly more precise and specific about educational objectives or aptitude treatment interactions. The research can proceed forever, of course, as scholars continue to identify slices of content knowledge to investigate. Its meaning for practice, however, likely will become less significant as the studies become overly specific and reach levels of detail that are not responsive to the daily work of teaching.
>
> Perhaps what might help teachers and teacher educators much more than increasingly detailed work on subject-specific pedagogy would be a small number of big subject-specific ideas and concepts, well articulated and well elaborated, that have broad consequences for teaching and learning (Sosniak, 1999, pp. 197–198).

This volume addresses the dilemma highlighted by Sosniak, which can be described somewhat more broadly as follows. Teacher education textbooks and much research on teaching have focused on relatively generic methods (lecture, discussion, projects, cooperative learning, transmission vs. social construction of knowledge, etc.), often with little consideration of subject matter or instructional goals. However, commonly endorsed ideas about good teaching typically include features such as alignment (i.e. curriculum, instruction, and assessment all need to be aligned to the instructional goals and intended student outcomes), pedagogical content knowledge (wisdom of practice concerning optimal examples, analogies, activities, etc. for accomplishing the goals), and connecting with students' prior knowledge and teaching within their zones of proximal development (which vary with their ages and prior experiences). All of these notions imply that good teaching will vary with the situation. With respect to instructional methods and activities, they imply that the relevance and value of particular methods and activities will vary with the nature of the students, the instructional goals, and the curricular content.

If taken literally, this suggests the need for a separate scholarly literature on each major topic and skill taught at each grade in each subject. Ultimately, this may in fact be needed, or at least helpful. In the meantime, however, there should be value in identifying core instructional principles and prototype instructional methods and activities that reflect best practices in various subject areas. These would be subject-specific, yet generalizable across many if not all of the subject's major goals and content clusters. Their levels of generality and spheres of application would lie in between those of generic principles and methods that appear to apply across all subjects and specific, detailed lesson plans designed to develop highly specific content. This volume is designed to synthesize what is currently known (or least believed to be true) about such mid-range principles, methods, and activities that represent best practices in teaching various subjects.

The volume addresses the four major academic subjects (language arts, mathematics, science, and social studies). However, to make it easier for authors to focus on topic-specific pedagogy, the chapters are organized around fourteen curricular strands that lie within these four subjects. Language arts is represented by chapters on beginning reading, content area reading and literature studies, and writing; mathematics by chapters on number and geometry; science by chapters on biology, physics, chemistry, and earth science; and social studies by chapters on history, physical geography, cultural studies, citizenship education, and economics.

The chapters have been prepared by authors who are scholarly leaders in these respective fields. They were invited to participate because they have been involved for some years in research and teacher education in their subject and have focused on instructional method (not just curricular) issues in ways that reflect emphasis on teaching the subjects for understanding.

In contrast to what was done in previous volumes in this series, the contributors were asked not to present their own work in detail but instead to synthesize all relevant work in their subject to produce a "state-of-the-field" contribution. Thus, their task was more similar to that of an author of a handbook chapter than to that of an author of a typical contribution to an edited volume.

Authors were asked to focus on instructional methods and learning activities emphasized in their subject area. Some of these are unique to the subject. For example, dissections are unique to biology and proofs are unique to mathematics (although at more abstract levels of analysis, parallels might be drawn between these respective activities and activities used in other subjects). Additional subject-specific methods and activities are adaptations of more general techniques tailored to fit the unique content of the subject. For example, most subjects include teacher-student discourse, but the particular forms and rhythms that this discourse commonly takes differ across subjects because of differences in the nature and affordances of the content.

I encouraged the authors to take a broad view in identifying information that might be included in their chapters. Along with scholarly literature featuring relatively formal studies of methods and activities, I encouraged them to consider the wisdom of practice as represented in case studies of good teaching and in the methods and activities commonly featured in subject-specific teacher education texts and in K-12 textbook series and other instructional materials. This is especially relevant in the case of learning activities (as contrasted to general instructional principles or methods). Even where systematic research on a given activity is limited or absent, there is often a consensus within the field that certain activities are especially powerful for developing certain content clusters. It seemed likely that for each major topic commonly addressed in each subject, there would be certain classic learning activities that are commonly recommended because they engage students in working with big ideas in ways that promote progress toward important understanding, appreciation, and application goals. The chapter might highlight these classic methods, identify questions that might be used to guide discourse or activities that might be used to foster learning, and discuss why these are known or believed to be effective.

The authors were asked to focus on the classic big ideas and fundamental skills emphasized in teaching their subjects and consider what research and the wisdom of practice suggest about ideal methods and learning activities to use for accomplishing major goals. Where relevant, they could comment about the trade-offs to be expected in using recognized alternative methods that are relevant to a given content cluster. They might also comment on what kinds of research on teaching methods and activities have or have not been useful, and on what their chapter suggests for teacher educators and for writers of instructional materials.

GENERIC GUIDELINES FOR GOOD TEACHING

To set a context and provide a common take-off point for the subject-specific chapters, I have included in this introduction to the volume a brief summary of what is currently known (or at least, widely believed to be true) about generic aspects of effective teaching. There is broad agreement among educators associated with all school subjects that students should learn each subject with understanding of its big ideas, appreciation of its value, and the capability and disposition to apply it in their lives outside of school. Analyses of research done in the different subject areas have identified some commonalities in conclusions drawn about curricular, instructional, and assessment practices that foster this kind of learning. If phrased as general principles rather than specific behavioral rules, these emerging guidelines can be seen as mutually supportive components of a coherent approach to teaching that applies across subjects and situations. Thus, it is possible to identify generic features of good teaching, although not to outline a specific instructional model to be implemented step-by-step.

I recently synthesized these generic features of good teaching in a booklet developed for the Educational Practices Series sponsored by the International Academy of Education (Brophy, 1999). My charge was to focus on generic aspects of teaching, so I excluded principles that apply only to particular school subjects, grade levels, or instructional settings. The principles reflect aspects of classrooms that are much more similar than different across countries and cultures: Most subject-matter teaching involves whole-class lessons in which content is developed through teacher explanation and teacher-student interaction, followed by practice and application activities that students work on individually or in pairs or small groups. Instruction is directed primarily to the class as a whole, with the teacher seeking to individualize around the margins.

In preparing the booklet, I construed teaching broadly, addressing all of the major tasks that teachers need to address: instructional goals, content selection and representation, classroom discourse, learning activities, assessment, adjusting instruction to meet the needs of individual students, and managing classrooms in ways that support the instructional program. Adopting this broad view of teaching meant adopting a broad view of relevant research and scholarship to draw from in identifying guidelines.

Much of the research support for these principles comes from studies of relationships between classroom processes and student outcomes. However, some principles are rooted in the logic of instructional design (e.g. the need for alignment among a curriculum's goals, content, instructional methods, and assessment measures). In addition, attention was paid to emergent theories of teaching and learning (e.g. sociocultural, social constructivist) and to the standards statements circulated by organizations representing the major school subjects. Priority was given to principles that have been shown to be applicable under ordinary classroom conditions and associated with progress toward desired student outcomes.

These principles rest on a few fundamental assumptions about optimizing curriculum and instruction. First, school curricula subsume different types of learning that call for somewhat different types of teaching, so no single teaching method (e.g. direct instruction, social construction of meaning) can be the method of choice for all occasions. An optimal program will feature a mixture of instructional methods and learning activities.

Second, within any school subject or learning domain, students' instructional needs change as their expertise develops. Consequently, what constitutes an optimal mixture of instructional methods and learning activities will evolve as school years, instructional units, and even individual lessons progress.

Third, students should learn at high levels of mastery yet progress through the curriculum steadily. This implies that, at any given time, curriculum content and learning activities need to be difficult enough to provide some challenge and extend learning, but not so difficult as to leave many students confused or frustrated. Instruction should focus on the zone of proximal development, which is the range of knowledge and skills that students are not yet ready to learn on their own but can learn with help from the teacher.

Finally, although 12 principles are highlighted for emphasis and discussed individually, each principle should be applied within the context of its relationships with the others. That is, the principles are meant to be understood as mutually supportive components of a coherent approach to teaching in which the teacher's plans and expectations, the classroom learning environment and management system, the curriculum content and instructional materials, and the learning activities and assessment methods are all aligned as means to help students attain intended outcomes.

1. Supportive Classroom Climate: Students Learn Best Within Cohesive and Caring Learning Communities

Research findings. Productive contexts for learning feature an ethic of caring that pervades teacher-student and student-student interactions and transcends gender, race, ethnicity, culture, socioeconomic status, handicapping conditions, or other individual differences. Students are expected to assume individual and group responsibilities for managing instructional materials and activities and for supporting the personal, social, and academic well being of all members of the classroom community (Good & Brophy, 2000; Sergiovanni, 1994).

In the classroom. To create a climate for molding their students into a cohesive and supportive learning community, teachers need to display personal attributes that will make them effective as models and socializers: a cheerful disposition, friendliness, emotional maturity, sincerity, and caring about students as individuals as well as learners. The teacher displays concern and affection for students, is attentive to their needs and emotions, and socializes them to display these same characteristics in their interactions with one another.

In creating classroom displays and in developing content during lessons, the teacher connects with and builds on the students' prior knowledge and experiences, including their home cultures. Extending the learning community from the school to the home, the teacher establishes and maintains collaborative relationships with parents and encourages their active involvement in their children's learning.

The teacher promotes a learning orientation by introducing activities with emphasis on what students will learn from them, treating mistakes as natural parts of the learning process, and encouraging students to work collaboratively and help one another. Students are taught to ask questions without embarrassment, to contribute to lessons without fear of ridicule of their ideas, and to collaborate in pairs or small groups on many of their learning activities.

2. Opportunity to Learn: Students Learn More when Most of the Available Time is Allocated to Curriculum-Related Activities and the Classroom Management System Emphasizes Maintaining Students' Engagement in Those Activities

Research findings. A major determinant of students' learning in any academic domain is their degree of exposure to the domain at school through participation in lessons and learning activities. The lengths of the school day and the school year create upper limits on these opportunities to learn. Within these limits, the learning opportunities actually experienced by students depend on how much of the available time they spend participating in lessons and learning activities. Effective teachers allocate most of the available time to activities designed to accomplish instructional goals.

Research indicates that teachers who approach management as a process of establishing an effective learning environment tend to be more successful than teachers who emphasize their roles as disciplinarians. Effective teachers do not need to spend much time responding to behavior problems because they use management techniques that elicit student cooperation and engagement in activities and thus minimize the frequency of such problems. Working within the positive classroom climate implied by the principle of learning community, the teacher articulates clear expectations concerning classroom behavior in general and participation in lessons and learning activities in particular, follows through with any needed cues or reminders, and ensures that students learn procedures and routines that foster productive engagement during activities and smooth transitions between them (Brophy, 1983; Denham & Lieberman, 1980; Doyle, 1986).

In the classroom. There are more things worth learning than there is time available to teach them, so it is essential that limited classroom time be used efficiently. Effective teachers allocate most of this time to lessons and learning activities rather than to nonacademic pastimes that serve little or no curricular purpose. Their students spend many more hours each year on curriculum-related activities than do students of teachers who are less focused on instructional goals.

Effective teachers convey a sense of the purposefulness of schooling and the importance of getting the most out of the available time. They begin and end lessons on time, keep transitions short, and teach their students how to get started quickly and maintain focus when working on assignments. Good planning and preparation enable them to proceed through lessons smoothly without having to stop to consult a manual or locate an item needed for display or demonstration. Their activities and assignments feature stimulating variety and optimal challenge, which helps students to sustain their task engagement and minimizes disruptions due to boredom or distraction.

Successful teachers are clear and consistent in articulating their expectations. At the beginning of the year they model or provide direct instruction in desired procedures if necessary, and subsequently they cue or remind their students when these procedures are needed. They monitor the classroom continually, which enables them to respond to emerging problems before they become disruptive. When possible, they intervene in ways that do not disrupt lesson momentum or distract students who are working on assignments. They teach students strategies and procedures for carrying out recurring activities such as participating in whole-class lessons, engaging in productive discourse with classmates, making smooth transitions between activities, collaborating in pairs or small groups, storing and handling equipment and personal belongings, managing learning and completing assignments on time, and knowing when and how to get help. The teachers' emphasis is not on imposing situational control but on building students' capacity for managing their own learning, so that expectations are adjusted and cues, reminders, and other managerial moves are faded out as the school year progresses.

These teachers do not merely maximize "time on task," but spend a great deal of time actively instructing their students during interactive lessons, in which the teachers elaborate the content for students and help them to interpret and respond to it. Their classrooms feature more time spent in interactive discourse and less time spent in independent seatwork. Most of their instruction occurs during interactive discourse with students rather than during extended lecture-presentations.

Note: The principle of maximizing opportunity to learn is not meant to imply maximizing the scope of the curriculum (i.e. emphasizing broad coverage at the expense of depth of development of powerful ideas). The breadth/depth dilemma must be addressed in curriculum planning. The point of the opportunity-to-learn principle is that, however the breadth/depth dilemma is addressed and whatever the resultant curriculum may be, students will make the most progress toward intended outcomes if most of the available classroom time is allocated to curriculum-related activities.

Note: Opportunity to learn is sometimes defined as the degree of overlap between what is taught and what is tested. This definition can be useful if both the curriculum content and the test content reflect the major goals of the instructional program. Where this is not the case, achieving an optimal alignment may require making changes in the curriculum content, the test content, or both (see next principle).

3. Curricular Alignment: All Components of the Curriculum are Aligned to Create a Cohesive Program for Accomplishing Instructional Purposes and Goals

Research findings. Research indicates that educational policymakers, textbook publishers, and teachers often become so focused on content coverage or learning activities that they lose sight of the larger purposes and goals that are supposed to guide curriculum planning. Teachers typically plan by concentrating on the content they intend to cover and the steps involved in the activities their students will do, without giving much thought to the goals or intended outcomes of the instruction. Textbook publishers, in response to pressure from special interest groups, tend to keep expanding their content coverage. As a result, too many topics are covered in not enough depth; content exposition often lacks coherence and is cluttered with insertions; skills are taught separately from knowledge content rather than integrated with it; and in general, neither the students' texts nor the questions and activities suggested in the teachers' manuals are structured around powerful ideas connected to important goals.

Students taught using such textbooks may be asked to memorize parades of disconnected facts or to practice disconnected subskills in isolation instead of learning coherent networks of connected content structured around powerful ideas. These problems are often exacerbated by externally imposed assessment programs that emphasize recognition of isolated bits of knowledge or performance of isolated subskills. Such problems can be minimized through goal-oriented curriculum development, in which the overall purposes and goals of the instruction, not miscellaneous content coverage pressures or test items, guide curricular planning and decision making (Beck & McKeown, 1988; Clark & Peterson, 1986; Wang, Haertel & Walberg, 1993).

In the classroom. A curriculum is not an end in itself but a means, a tool for helping students to learn what is considered essential as preparation for fulfilling adult roles in society and realizing their potential as individuals. Its goals are learner outcomes – the knowledge, skills, attitudes, values, and dispositions to action that the society wishes to develop in its citizens. The goals are the reason for the existence of the curriculum, so that beliefs about what is needed to accomplish them should guide each step in curriculum planning and implementation. Goals are most likely to be attained if all of the curriculum's components (content clusters, instructional methods, learning activities, and assessment tools) are selected because they are believed to be needed as means for helping students to reach the goals.

This involves planning curriculum and instruction not just to cover content but to accomplish important student outcomes – capabilities and dispositions to be developed in students and used in their lives inside and outside of school, both now and in the future. In this regard, it is important to emphasize goals of understanding, appreciation, and life application. Understanding means that students learn both the individual elements in a network of related content and the connections among them, so that they can explain the content in their own words and connect it to their prior knowledge. Appreciation means that students value what they are learning because they understand that there are good reasons for learning it. Life application means that students retain their learning in a form that makes it useable when needed in other contexts.

Content developed with these goals in mind is likely to be retained as meaningful learning that is internally coherent, well connected with other meaningful learning, and accessible for application. This is most likely to occur when the content itself is structured around powerful ideas and the development of this content through classroom lessons and learning activities focuses on these ideas and their connections.

4. Establishing Learning Orientations: Teachers Can Prepare Students for Learning by Providing an Initial Structure to Clarify Intended Outcomes and Cue Desired Learning Strategies

Research findings. Research indicates the value of establishing a learning orientation by beginning lessons and activities with advance organizers or previews. These introductions facilitate students' learning by communicating the nature and purpose of the activity, connecting it to prior knowledge, and cueing the kinds of student responses that the activity requires. This helps students to remain goal oriented and strategic as they process information and respond to the questions or tasks embodied in the activity. Good lesson orientations also stimulate students' motivation to learn by communicating enthusiasm for the learning or helping students to appreciate its value or application potential (Ausubel, 1968; Brophy, 1998; Meichenbaum & Biemiller, 1998).

In the classroom. Advance organizers tell students what they will be learning before the instruction begins. They characterize the general nature of the activity and give students a structure within which to understand and connect the specifics presented by a teacher or text. Such knowledge of the nature of the activity and the structure of the content will help students to focus on the main ideas and order their thoughts effectively. Therefore, before beginning any lesson or activity, the teacher should see that students know what they will be learning and why it is important for them to learn it.

Other ways to help students learn with a sense of purpose and direction include calling attention to the activity's goals, overviewing main ideas or major steps to be elaborated, pretests that sensitize students to main points to learn, and prequestions that stimulate student thinking about the topic.

5. Coherent Content: To Facilitate Meaningful Learning and Retention, Content is Explained Clearly and Developed with Emphasis on its Structure and Connections

Research findings. Research indicates that networks of connected knowledge structured around powerful ideas can be learned with understanding and retained in forms that make them accessible for application. In contrast, disconnected bits of information are likely to be learned only through low-level processes such as rote memorizing, and most of these bits either are soon forgotten or are retained in ways that limit their accessibility. Similarly, skills are likely to be learned and used effectively if taught as strategies adapted to particular purposes and situations, with attention to when and how to apply them, but students may not be able to integrate and use skills that are learned only by rote and practiced only in isolation from the rest of the curriculum (Beck & McKeown, 1988; Good & Brophy, 2000; Rosenshine, 1968).

In the classroom. Whether in textbooks or in teacher-led instruction, information is easier to learn to the extent that it is coherent – the sequence of ideas or events makes sense and the relationships among them are made apparent. Content is most likely to be organized coherently when it is selected in a principled way, guided by ideas about what students should learn from studying the topic.

When making presentations, providing explanations, or giving demonstrations, effective teachers project enthusiasm for the content and organize and sequence it so as to maximize its clarity and "learner friendliness." The teacher presents new information with reference to what students already know about the topic; proceeds in small steps sequenced in ways that are easy to follow; uses pacing, gestures, and other oral communication skills to support comprehension; avoids vague or ambiguous language and digressions that disrupt continuity; elicits students' responses regularly to stimulate active learning and ensure that each step is mastered before moving to the next; finishes with a review of main points, stressing general integrative concepts; and follows up with questions or assignments that require students to encode the material in their own words and apply or extend it to new contexts.

Other ways to help students establish and maintain productive learning sets include using outlines or graphic organizers that illustrate the structure of the content, study guides that call attention to key ideas, or task organizers that help students keep track of the steps involved and the strategies they use to complete these steps.

In combination, the principles calling for curricular alignment and for coherent content imply that, to enable students to construct meaningful knowledge that they can access and use in their lives outside of school, teachers need to: (1) retreat from breadth of coverage in order to allow time to develop the most important content in greater depth; (2) represent this important content as networks of connected information structured around powerful ideas; (3) develop the content with a focus on explaining these important ideas and the connections among them; and (4) follow up with learning activities and assessment measures that feature authentic tasks that provide students with opportunities to develop and display learning that reflects the intended outcomes of the instruction.

6. Thoughtful Discourse: Questions are Planned to Engage Students in Sustained Discourse Structured Around Powerful Ideas

Research findings. Besides presenting information and modeling application of skills, effective teachers structure a great deal of content-based discourse. They use questions to stimulate students to process and reflect on the content, recognize relationships among and implications of its key ideas, think critically about it, and use it in problem solving, decision making, or other higher-order applications. Such discourse is not limited to factual review or recitation featuring rapid pacing and short answers to miscellaneous questions, but instead features sustained and thoughtful development of key ideas. Through participation in this discourse, students construct and communicate content-related ideas. In the process, they abandon naive ideas or misconceptions and adopt the more sophisticated and valid ideas embedded in the instructional goals (Good & Brophy, 2000; Newmann, 1990; Rowe, 1986).

In the classroom. In the early stages of units when new content is introduced and developed, more time is spent in interactive lessons featuring teacher-student discourse than in independent work on assignments. The teacher plans sequences of questions designed to develop the content systematically and help students to construct understandings of it by relating it to their prior knowledge and collaborating in dialogue about it.

The forms and cognitive levels of these questions need to be suited to the instructional goals. Some primarily closed-ended and factual questions might be appropriate when teachers are assessing prior knowledge or reviewing new learning, but accomplishing the most significant instructional goals requires open-ended questions that call for students to apply, analyze, synthesize, or evaluate what they are learning. Some questions will admit to a range of possible correct answers, and some will invite discussion or debate (e.g. concerning the relative merits of alternative suggestions for solving problems).

Because questions are intended to engage students in cognitive processing and construction of knowledge, they ordinarily should be addressed to the class as a whole. This encourages all students, not just the one eventually called on, to listen carefully and respond thoughtfully to each question. After posing a question, the teacher needs to pause to allow students enough time to process it and at least begin to formulate responses, especially if the question is complicated or requires students to engage in higher order thinking.

Thoughtful discourse features sustained examination of a small number of related topics, in which students are invited to develop explanations, make predictions, debate alternative approaches to problems, or otherwise consider the content's implications or applications. The teacher presses students to clarify or justify their assertions, rather than accepting them indiscriminately. In addition to providing feedback, the teacher encourages students to explain or elaborate on their answers or to comment on classmates' answers. Frequently, discourse that begins in a question-and-answer format evolves into an exchange of views in which students respond to one another as well as to the teacher and respond to statements as well as to questions.

7. Practice and Application Activities: Students Need Sufficient Opportunities to Practice and Apply what they are Learning, and to Receive Improvement-Oriented Feedback

Research findings. There are three main ways that teachers help their students to learn. First, they present information, explain concepts, and model skills. Second, they lead their students in review, recitation, discussion, and other forms of discourse surrounding the content. Third, they engage students in activities or assignments that provide them with opportunities to practice or apply what they are learning. Research indicates that skills practiced to a peak of smoothness and automaticity tend to be retained indefinitely, whereas skills that are mastered only partially tend to deteriorate. Most skills included in school curricula are learned best when practice is distributed across time and embedded within a variety of tasks. Thus, it is important to follow up thorough initial teaching with occasional review activities and with opportunities for students to use what they are learning in a variety of application contexts (Brophy & Alleman, 1991; Cooper, 1994; Dempster, 1991; Knapp, 1995).

In the classroom. Practice is one of the most important yet least appreciated aspects of learning in classrooms. Little or no practice may be needed for simple behaviors like pronouncing words, but practice becomes more important as learning becomes complex. Successful practice involves polishing skills that already are established at rudimentary levels to make them smoother, more efficient, and more automatic, not trying to establish such skills through trial and error.

Fill-in-the-blank worksheets, pages of mathematical computation problems, and related tasks that engage students in memorizing facts or practicing subskills in isolation from the rest of the curriculum should be minimized. Instead, most practice should be embedded within application contexts that feature conceptual understanding of knowledge and self-regulated application of skills. Thus, most practice of reading skills is embedded within lessons involving reading and interpreting extended text, most practice of writing skills is embedded within activities calling for authentic writing, and most practice of mathematics skills is embedded within problem-solving applications.

Opportunity to learn in school can be extended through homework assignments that are realistic in length and difficulty given the students' abilities to work independently. To ensure that students know what to do, the teacher can go over the instructions and get them started in class, then have them finish the work at home. An accountability system should be in place to ensure that students complete their homework assignments, and the work should be reviewed in class the next day.

To be useful, practice must involve opportunities not only to apply skills but to receive timely feedback. Feedback should be informative rather than evaluative, helping students to assess their progress with respect to major goals and to understand and correct errors or misconceptions. At times when teachers are unable to circulate to monitor progress and provide feedback to individuals, pairs, or groups working on assignments, they should arrange for students to get feedback by consulting posted study guides or answer sheets or by asking peers designated to act as tutors or resource persons.

8. Scaffolding Students' Task Engagement: The Teacher Provides Whatever Assistance Students Need to Enable Them to Engage in Learning Activities Productively

Research findings. Research on learning tasks suggests that activities and assignments should be sufficiently varied and interesting to motivate student engagement, sufficiently new or challenging to constitute meaningful learning experiences rather than needless repetition, and yet sufficiently easy to allow students to achieve high rates of success if they invest reasonable time and effort. The effectiveness of assignments is enhanced when teachers first explain the work and go over practice examples with students before releasing them to work independently, then circulate to monitor progress and provide help when needed. The principle of teaching within the students' zones of proximal development implies that students will need explanation, modeling, coaching, and other forms of assistance from their teachers, but also that this teacher structuring and scaffolding of students' task engagement will be faded as the students' expertise develops. Eventually, students should become able to autonomously use what they are learning and regulate their own productive task engagement (Brophy & Alleman, 1991; Rosenshine & Meister, 1992; Shuell, 1996; Tharp & Gallimore, 1988).

In the classroom. Besides being well chosen, activities need to be effectively presented, monitored, and followed up if they are to have their full impact. This means preparing students for an activity in advance, providing guidance and feedback during the activity, and leading the class in post-activity reflection afterwards. In introducing activities, teachers should stress their purposes in ways that will help students to engage in them with clear ideas about the goals to be accomplished. Then they might call students' attention to relevant background knowledge, model strategies for responding to the task, or scaffold by providing information concerning how to go about completing task requirements. If textbook reading is involved, for example, teachers might summarize the main ideas, remind students about strategies for developing and monitoring their comprehension as they read (paraphrasing, summarizing, taking notes, asking themselves questions to check understanding), distribute study guides that call attention to key ideas and structural elements, or provide task organizers that help students to keep track of the steps involved and the strategies that they are using.

Once students begin working on activities or assignments, teachers should circulate to monitor their progress and provide assistance if necessary. Assuming that students have a general understanding of what to do and how to do it, these interventions can be kept brief and confined to minimal and indirect forms of help. If teacher assistance is too direct or extensive, teachers will end up doing tasks for students instead of helping them learn to do the tasks themselves.

Teachers also need to assess performance for completion and accuracy. When performance is poor, they will need to provide reteaching and follow-up assignments designed to ensure that content is understood and skills are mastered.

Most tasks will not have their full effects unless they are followed by reflection or debriefing activities in which the teacher reviews the task with the students, provides general feedback about performance, and reinforces main ideas as they relate to overall goals. Reflection activities should also include opportunities for students to ask follow-up questions, share task-related observations or experiences, compare opinions, or in other ways deepen their appreciation of what they have learned and how it relates to their lives outside of school.

9. Strategy Teaching: The Teacher Models and Instructs Students in Learning and Self-regulation Strategies

Research findings. General learning and study skills as well as domain-specific skills (such as constructing meaning from text, solving mathematical problems, or reasoning scientifically) are most likely to be learned thoroughly and become accessible for application if they are taught as strategies to be brought to bear purposefully and implemented with metacognitive awareness and self-regulation. This requires comprehensive instruction that includes attention to propositional knowledge (what to do), procedural knowledge (how to do it), and conditional knowledge (when and why to do it). Strategy teaching is especially important for less able students who otherwise might not come to understand the value of consciously monitoring, self-regulating, and reflecting upon their learning processes (Meichenbaum & Biemiller, 1998; Pressley & Beard El-Dinary, 1993; Weinstein & Mayer, 1986).

In the classroom. Many students do not develop effective learning and problem-solving strategies on their own but can acquire them through modeling and explicit instruction from their teachers. Poor readers, for example, can be taught reading comprehension strategies such as keeping the purpose of an assignment in mind when reading, activating relevant background knowledge, identifying major points in attending to the outline and flow of content, monitoring understanding by generating and trying to answer questions about the content, or drawing and testing inferences by making interpretations, predictions, and conclusions. Instruction should include not only demonstrations of and opportunities to apply the skill itself but also explanations of the purpose of the skill (what it does for the learner) and the occasions in which it would be used.

Strategy teaching is likely to be most effective when it includes cognitive modeling: The teacher thinks out loud while modeling use of the strategy. This makes overt for learners the otherwise covert thought processes that guide use of the strategy in a variety of contexts. Cognitive modeling provides learners with first-person language ("self talk") that they can adapt directly when using the strategy themselves. This eliminates the need for translation that is created when instruction is presented in the impersonal third-person language of explanation or even the second-person language of coaching.

In addition to strategies for use in particular domains or types of assignments, teachers can model and instruct their students in general study skills and learning strategies such as rehearsal (repeating material to remember it more effectively), elaboration (putting material into one's own words and relating it to prior knowledge), organization (outlining material to highlight its structure and remember it), comprehension monitoring (keeping track of the strategies used and the degree of success achieved with them, and adjusting strategies accordingly), and affect monitoring (maintaining concentration and task focus, minimizing performance anxiety and fear of failure).

When providing feedback as students work on assignments and when leading subsequent reflection activities, teachers can ask questions or make comments that help students to monitor and reflect on their learning. Such monitoring and reflection should focus not only on the content being learned, but also on the strategies that the students are using to process the content and solve problems. This will help the students to refine their strategies and regulate their learning more systematically.

10. Cooperative Learning: Students Often Benefit from Working in Pairs or Small Groups to Construct Understandings or Help One Another Master Skills

Research findings. Research indicates that there is often much to be gained by arranging for students to collaborate in pairs or small groups as they work on activities and assignments. Cooperative learning promotes affective and social benefits such as increased student interest in and valuing of subject matter and increases in positive attitudes and social interactions among students who differ in gender, race, ethnicity, achievement levels, and other characteristics.

Cooperative learning also creates the potential for cognitive and metacognitive benefits by engaging students in discourse that requires them to make their task-related information-processing and problem-solving strategies explicit (and thus available for discussion and reflection). Students are likely to show improved achievement outcomes when they engage in certain forms of cooperative learning as an alternative to completing assignments on their own (Bennett & Dunne, 1992; Johnson & Johnson, 1994; Slavin, 1990).

In the classroom. Traditional approaches to instruction feature whole-class lessons followed by independent seatwork time during which students work alone (and usually silently) on assignments. Cooperative learning approaches retain the whole-class lessons but replace part of the individual seatwork time with opportunities for students to work together in pairs or small groups on follow-up practice and application activities. Cooperative learning can be used with activities ranging from drill and practice to learning facts and concepts, discussion, and problem solving. It is perhaps most valuable as a way to engage students in meaningful learning with authentic tasks in a social setting. Students have more chances to talk in pairs or small groups than in whole-class activities, and shy students are more likely to feel comfortable expressing ideas in these more intimate settings.

Some forms of cooperative learning call for students to help one another accomplish individual learning goals, such as by discussing how to respond to assignments, checking work, or providing feedback or tutorial assistance. Other forms of cooperative learning call for students to work together to accomplish a group goal by pooling their resources and sharing the work. For example, the group might conduct an experiment, assemble a collage, or prepare a research report to be presented to the rest of the class. Cooperative learning models that call for students to work together to produce a group product often feature a division of labor among group participants (e.g. to prepare a biographical report, one group member will assume responsibility for studying the person's early life, another for the person's major accomplishments, another for the effects of these on society, and so on).

Cooperative learning methods are most likely to enhance learning outcomes if they combine group goals with individual accountability. That is, each group member has clear objectives for which he or she will be held accountable (students know that any member of the group may be called on to answer any one of the group's questions or that they all will be tested individually on what they are learning).

Activities used in cooperative learning formats should be well suited to those formats. Some activities are most naturally done by individuals working alone, others by students working in pairs, and still others by small groups of three to six students.

Students should receive whatever instruction and scaffolding they may need to prepare them for productive engagement in cooperative learning activities. For example, teachers may need to show their students how to share, listen, integrate the ideas of others, and handle disagreements constructively. During times when students are working in pairs or small groups, the teacher should circulate to monitor progress, make sure that groups are working productively on the assigned tasks, and provide any needed assistance.

11. Goal-Oriented Assessment: The Teacher Uses a Variety of Formal and Informal Assessment Methods to Monitor Progress Toward Learning Goals

Research findings. Well-developed curricula include strong and functional assessment components. These assessment components are aligned with the curriculum's major purposes and goals, so they are integrated with the curriculum's content, instructional methods, and learning activities, and designed to evaluate progress toward major intended outcomes.

Comprehensive assessment does not just document students' ability to supply acceptable answers to questions or problems; it also examines the students' reasoning and problem-solving processes. Effective teachers routinely monitor their students' progress in this fashion, using both formal tests or performance evaluations and informal assessment of students' contributions to lessons and work on assignments (Dempster, 1991; Stiggins, 1997; Wiggins, 1993).

In the classroom. Effective teachers use assessment for evaluating students' progress in learning and for planning curriculum improvements, not just for generating grades. Good assessment includes data from many sources besides paper-and-pencil tests, and it addresses the full range of goals or intended outcomes (not only knowledge but higher order thinking skills and content-related values and dispositions). Standardized, norm-referenced tests might comprise part of the assessment program (these tests are useful to the extent that what they measure is congruent with the intended outcomes of the curriculum and attention is paid to students' performance on each individual item, not just total scores). However, standardized tests ordinarily should be supplemented with publisher-supplied curriculum-embedded tests (when these appear useful) and with teacher-made tests that focus on learning goals emphasized in instruction but not in external testing sources.

In addition, learning activities and sources of data other than tests should be used for assessment purposes. Everyday lessons and activities provide opportunities to monitor the progress of the class as a whole and of individual students, and tests can be augmented with performance evaluations using tools such as laboratory tasks and observation checklists, portfolios of student papers or projects, and essays or other assignments that call for higher order thinking and application. A broad view of assessment helps to ensure that the assessment component includes authentic activities that provide students with opportunities to synthesize and reflect on what they are learning, think critically and creatively about it, and apply it in problem solving and decision-making contexts.

In general, assessment should be treated as an ongoing and integral part of each instructional unit. Results should be scrutinized to detect weaknesses in the assessment practices themselves; to identify learner needs, misunderstandings, or misconceptions that may need attention; and to suggest potential adjustments in curriculum goals, instructional materials, or teaching plans.

12. Achievement Expectations: The Teacher Establishes and Follows Through on Appropriate Expectations for Learning Outcomes

Research findings. Research indicates that effective schools feature strong academic leadership that produces consensus on goal priorities and commitment to instructional excellence, as well as positive teacher attitudes toward students and expectations regarding their abilities to master the curriculum. Teacher effects research indicates that teachers who elicit strong achievement gains accept responsibility for doing so. They believe that their students are capable of learning and that they (the teachers) are capable of and responsible for teaching them successfully. If students do not learn something the first time, they teach it again, and if the regular curriculum materials do not do the job, they find or develop others that will (Brophy, 1998; Creemers & Scheerens, 1989; Good & Brophy, 2000; Shuell, 1996; Teddlie & Stringfield, 1993).

In the classroom. Teachers' expectations concerning what their students are capable of accomplishing (with teacher help) tend to shape both what teachers attempt to elicit from their students and what the students come to expect from themselves. Thus, teachers should form and project expectations that are as positive as they can be while still remaining realistic. Such expectations should represent genuine beliefs about what can be achieved and therefore should be taken seriously as goals toward which to work in instructing students.

It is helpful if teachers set goals for the class and for individuals in terms of floors (minimally acceptable standards), not ceilings. Then they can let group progress rates, rather than limits adopted arbitrarily in advance, determine how far the class can go within the time available. They can keep their expectations for individual students current by monitoring their progress closely and by stressing current performance over past history.

At minimum, teachers should expect all of their students to progress sufficiently to enable them to perform satisfactorily at the next level. This implies holding students accountable for participating in lessons and learning activities and turning in careful and completed work on assignments. It also implies that struggling students will receive the time, instruction, and encouragement needed to enable them to meet expectations.

When individualizing instruction and giving students feedback, teachers should emphasize the students' continuous progress relative to previous levels of mastery rather than how they compare with other students or with standardized test norms. Instead of merely evaluating relative levels of success, teachers can diagnose learning difficulties and provide students with whatever feedback or additional instruction they need to enable them to meet the goals. If students have not understood an explanation or demonstration, the teacher can follow through by reteaching (if necessary, in a different way rather than by merely repeating the original instruction).

In general, teachers are likely to be most successful when they think in terms of stretching students' minds by stimulating them and encouraging them to achieve as much as they can, not in terms of "protecting" them from failure or embarrassment.

FROM THE GENERIC TO THE SUBJECT-SPECIFIC

Given broad agreement that school subjects should be taught for understanding, appreciation, and life application, I assumed that the authors of the subject-specific chapters would view the generic guidelines just presented as applying to their subjects. Where this assumption was valid, the respective authors could simply say so and then move immediately to subject-specific issues. However, if they felt that something said here about generic features of good teaching needed to be qualified or elaborated with respect to its application to their subject, they could explain this in their chapters. In this manner, the material on generic features of good teaching included in this introduction would be assessed for its applicability to fourteen different school subjects, and at the same time, the subject-specific chapters could proceed from a common context of shared assumptions and principles.

These fourteen subject-specific chapters follow. Then in the final discussion chapter, I attempt to make sense of some of the subject-specific variation by identifying ways in which teaching methods and learning activities used in different subjects can be seen as parallel in certain respects when analyzed along common dimensions. My hope is that the volume as a whole, and its concluding chapter in particular, will help us to recognize some of the pedagogical parallels (along with the differences) that exist across subjects. This should help us to cope with the complexities identified by Sosniak (1999) as we continue to develop both generic and subject-specific theory and research about best practices in teaching.

REFERENCES

Ausubel, D. (1968). *Educational psychology: a cognitive view.* New York: Holt, Rinehart, & Winston.

Beck, I., & McKeown, M. (1988). Toward meaningful accounts in history texts for young learners. *Educational Researcher, 17*(6), 31–39.

Bennett, N., Dunne, E. (1992). *Managing small groups.* New York: Simon & Schuster.

Brophy, J. (1983). Classroom organization and management. *Elementary School Journal, 83*, 265–285.

Brophy, J. (1998). *Motivating students to learn.* Boston: McGraw-Hill.

Brophy, J. (1999). *Teaching* (Educational Practices Series No. 1). Geneva: International Bureau of Education.

Brophy, J., & Alleman, J. (1991). Activities as instructional tools: A framework for analysis and evaluation. *Educational Researcher, 20*, 9–23.

Brophy, J., & Good, T. (1986). Teacher behavior and student achievement. In: M. Wittrock (Ed.), *Handbook of Research on Teaching* (3rd ed., pp. 328–375). New York: Macmillan.

Clark, C., & Peterson, P. (1986). Teachers' thought processes. In: M. C. Wittrock (Ed.), *Handbook of Research on Teaching* (3rd ed., pp. 225–296). New York: Macmillan.

Cooper, H. (1989). *Homework.* White Plains, NY: Longman.

Cooper, H. (1994). *The battle over homework: An administrator's guide to setting sound and effective policies.* Thousand Oaks, CA: Corwin.

Creemers, B. & Scheerens, J. (Guest Eds) (1989). Developments in school effectiveness research. *International Journal of Educational Research, 13*, 685–825.

Dempster, F. (1991). Synthesis of research on reviews and tests. *Educational Leadership, 48*, 71–76.

Denham, C., & Lieberman, A. (Eds) (1980). *Time to learn.* Washington, DC: National Institute of Education.

Doyle, W. (1986). Classroom organization and management. In: M. C. Wittrock (Ed.), *Handbook of Research on Teaching* (3rd ed., pp. 392–431). New York: Macmillan.

Good, T., & Brophy, J. (1986). School effects. In: M. Wittrock (Ed.), *Handbook of Research on Teaching* (3rd ed., pp 570–602). New York: Macmillan.

Good, T., & Brophy, J. (1995). *Contemporary educational psychology* (5th ed.). New York: Longman.

Good, T., & Brophy, J. (2000). *Looking in classrooms* (8th ed.). New York: Addison Wesley Longman.

Johnson, D., & Johnson, R. (1994). *Learning together and alone: Cooperative, competitive, and individualistic learning* (4th ed.). Boston: Allyn & Bacon.

Knapp, N. (1995). *Teaching for meaning in high-poverty classrooms.* New York: Teachers College Press.

Meichenbaum, D., & Biemiller, A. (1998). *Nurturing independent learners: Helping students take charge of their learning.* Cambridge, MA: Brookline.

Newmann, F. (1990). Qualities of thoughtful social studies classes: An empirical profile. *Journal of Curriculum Studies, 22,* 253–275.

Pressley, M., & Beard El-Dinary, P. (Guest Eds) (1993). Special issue on strategies instruction. *Elementary School Journal, 94,* 105–284.

Rosenshine, B. (1968). To explain: A review of research. *Educational Leadership, 26,* 275–280.

Rosenshine, B., & Meister, C. (1992). The use of scaffolds for teaching higher-level cognitive strategies. *Educational Leadership, 49,* 26–33.

Rowe, M. (1986). Wait time: Slowing down may be a way of speeding up! *Journal of Teacher Education, 37,* 43–50.

Sergiovanni, T. (1994). Building community in schools. San Francisco: Jossey-Bass.

Shulman, L. (1986). Those who understand: Knowledge growth in teaching. *Educational Researcher, 15*(2), 4–14.

Shuell, T. (1996). Teaching and learning in a classroom context. In: D. Berliner & R. Calfree (Eds), *Handbook of Educational Psychology* (pp. 726–764). New York: Macmillan.

Slavin, R. (1990). *Cooperative learning: theory, research, and practice.* Englewood Cliffs, NJ: Prentice-Hall.

Sosniak, L. (1999). Professional and subject matter knowledge for teacher education. In: G. Griffen (Ed.), *The Education of Teachers* (98th Yearbook of the National Society for the Study of Education, Part I, pp. 185–204). Chicago: University of Chicago Press.

Stiggins, R. (1997). *Student-centered classroom assessment* (2nd ed.). Upper Saddle River, NJ: Prentice-Hall.

Teddlie, C., & Stringfield, S. (1993). *Schools make a difference: Lessons learned from a 10-year study of school effects.* New York: Teachers College Press.

Tharp, R., & Gallimore, R. (1988). *Rousing minds to life: Teaching, learning and schooling in social context.* Cambridge: Cambridge University Press.

Wang, M., Haertel, G., & Walberg, H. (1993). Toward a knowledge base for school learning. *Review of Educational Research, 63,* 249–294.

Weinstein, C., & Mayer, R. (1986). The teaching of learning strategies. In: M. C. Wittrock (Ed.), *Handbook of Research on Teaching* (3rd ed., pp. 315–327). New York: Macmillan.

Wiggins, G. (1993). *Assessing student performance: Exploring the purpose and limits of testing.* San Francisco: Jossey-Bass.

BEGINNING READING INSTRUCTION: MOVING BEYOND THE DEBATE OVER METHODS INTO THE STUDY OF PRINCIPLED TEACHING PRACTICES

James V. Hoffman and Gerald G. Duffy

INTRODUCTION

Beginning reading instruction in the twentieth century was riddled with controversy. Jeanne Chall's (1967) classic work, *Learning to Read: The Great the Debate*, captured the intensity of the positions. Chall described competing methodological approaches using both contemporary and historical lenses. Her analysis of the content of commercial materials and the research findings on competing approaches to instruction not only revealed the nature of the controversy but provided a frame for much of the research over the next two decades. Scholarly debates centered on the relative virtues of Method A vs. Method B (e.g. whether the basal text approach is better than the language experience approach, or whether either of those two approaches is better than a literature-based approach). The "First Grade Studies" (Bond & Dykstra, 1967), one of the first major programs of large-scale reading research supported by the Federal government, examined this classic Method A vs. Method B structure but failed to reveal any 'best method'. Instead, results suggested that the 'quality' of the teaching and teacher seemed to make the difference in beginning reading instruction.

Undeterred by the findings from the First Grade Studies, academic and public debates over reading methods and effectiveness continued unabated.

Subject-Specific Instructional Methods and Activities, Volume 8, pages 25–49.
2001 by Elsevier Science Ltd.
ISBN: 0-7623-0615-7

Now, as we enter the new millennium, we can only speculate whether the debate will continue along similar lines or evolve into a more meaningful focus. In this chapter, we attempt to do our part to move the discussion beyond 'methods' to consider the central role of the teacher as the key to effectiveness. We argue that improving beginning reading instruction requires that our models of teacher education move from a 'training' perspective with methods as the focus to an 'educative' perspective with teacher insights, understandings, decision-making and flexibility as the focus (Duffy, 1997, 1998; Hoffman, 1998; Hoffman & Pearson, 2000).

BACKGROUND

While teaching reading is a clear priority for primary grade teachers, it does not fit neatly into a subject matter niche. Yes, most teachers can point to a time of day when reading is taught, there are 'textbook' resources that are used, and there are learning objectives that can be taught to and assessed. But unlike history or biology or geography, reading does not have its own domain of world knowledge in the sense that an academic field conveys world knowledge. From a traditional perspective, reading is a "content-free" process rooted in skills. Hence, it is frequently described as a "basic skill" used with subject matter. Reading's unique position is revealed when one tries to apply Shulman's (1986) concept of "pedagogical content knowledge", which has been highly effective in conceptualizing content area instruction, but is of relatively less help in reading. One does not get very far by thinking of reading instruction only in terms of ". . . the most powerful analogies, illustrations, examples, explanations and demonstrations . . ." (pp. 5).

The traditional notion of reading as 'content free' was challenged over the last decade of the twentieth century with the literature-based movement in reading (Wepner & Feeley, 1993). Here it was argued that engaging texts, drawn from a wide range of literary genres, must be at the core of good instruction from the very earliest stages. The study of quality literature became the 'subject matter' to accompany the process goals of reading. The critical reading of texts was viewed not as an outcome of 'learning the basic reading skills' but as an authentic context for instruction and learning. The literature-based movement took hold in California in the late 1980s and then spread rapidly from there (Chrispeels, 1997). Its impact on the methods and materials for beginning reading instruction was profound. The carefully controlled and leveled "Sally, Dick, and Jane" texts of the 1960s and the skill-based basals of the 1970s and 80s were abandoned in favor of 'authentic' literature with little or no attention to vocabulary selection and repetition or decoding instruction

(Hoffman et al., 1994, 1995). Shared literature (Holdaway, 1979) and "whole language" (Goodman, 1989) assumed the prominent pedagogical and philosophical frames for effective teaching of reading.

Far from bringing consensus, the literature-based movement further dichotomized the debate Chall described. Method and materials remained at the center of the controversy. Those who argued for more systematic, sequential teaching of basic skills blamed a "decline" of test scores in California on literature-based approaches and the whole language philosophy (Carlos & Kirst, 1997). Research conducted under the auspices of the National Institute of Child Health and Development was used to discredit literature-based approaches and to promote systematic development of basic decoding skills (Lyon, 1995).

As we write this chapter, the debate has moved beyond academic journals and the popular press into the public policy arena. At both state and national levels, new laws require that beginning reading be taught using particular methodologies and "approved" programs (McGill-Franzen, 2000). Here again, the notion of 'method' and "materials" dominates. The debates have been heated (see, for instance, Shannon, 1990), and the apparent lack of consensus has created a vacuum that politicians filled in their search for a "quick fix" (Allington & Walmsley, 1998).

In an attempt to quiet the debate, a substantial number of scholars have recently embraced a 'balanced' perspective emphasizing both content (literature) and process (the development of basic decoding and comprehension skills) (Gambrell, Morrow, Neuman & Pressley, 1999; Pressley, 1998). The move to "balance" has been supported by recent studies of exemplary teachers (see, for instance, Pressley, Rankin & Yokoi, 1996; Pressley, Wharton-McDonald, Allington et al., 1998; Wharton-McDonald, Pressley & Hampston, 1998; Taylor, Pearson et al., 1999). Here, scholars study exemplary teachers and schools, and describe what they observe as "best practices." While these "best practices" typically cut across ideological and methodological lines, the notion of balance as "a little bit of this" and "a little bit of that" still conveys the notion that it is the method and materials that make the difference.

It is within these contexts that we embark on the task set forth for this volume. Our analysis is based on our years of studying effective reading instruction and of developing innovative teacher education programs for teachers of reading. Our experience has led us to a position that goes beyond what the media typically discusses. Our position is not tied to any particular ideology, views method as a necessary but not sufficient ingredient of effective beginning reading instruction, and finds the concept of "best practices" to be limiting because the practices themselves do not inform us about how teachers come to employ them. Therefore, we describe in this chapter "mid-range principles, methods and activities" that appear to be effective, but we do so as a means for making a case: (1) for teachers' analytical thinking as the root and fuel for "best practices", and (2) for improved teacher education, rather than improved method, as the key to teaching excellence in beginning reading.

Conceptualizing Beginning Reading

Effective "mid-range principles, methods and activities" are determined in large part by the goal being sought . That is, the method used depends on what are we trying to develop. The popular public notion is that the goal of beginning reading instruction is "decoding"; method, therefore, is often thought to be limited to teaching the alphabet and letter sounds. In reality, the goal of beginning reading instruction is much more complex and multi-faceted.

We conceptualize the goal as "engaged reading" (Guthrie & Alvermann, 1999). In this view, reading is not seen as the sum of a set of skills or competencies. Rather, engaged reading involves motivations, strategies, conceptual understandings and social interactions (Guthrie & Anderson, 1999). In concert with the core principles articulated throughout this volume – an understanding of big ideas, appreciation of the value of what is learned, and ability to apply what is learned to life outside of school – engaged reading connotes a form of literacy that goes well beyond the traditional perceptual emphasis associated with decoding. Following Hoffman and McCarthey (2000), the "engaged reader" can be described in terms of three goals: seeking meaning, applying skills and strategies, and self-monitoring. As will be seen, it is these goals that drive teachers' selection and adaptation of the "principles, methods and activities" that are the focus of this chapter.

Engaged Readers Seek Meaning

First and foremost, engaged readers seek meaning. That is, engaged readers understand that reading and writing are message-receiving and message-sending processes, executed for purposeful reasons. They understand that reading involves the seeking of a pleasurable response, or the seeking of guidance or direction, or the seeking of knowledge and greater understanding. It is this seeking that makes reading worth the effort, and that is the source of the motivational elements that sustain students as they struggle with the difficulties of learning to read. In the absence of this fundamental conceptual understanding, reading is just an empty exercise.

The basic idea of seeking meaning carries with it other important conceptual and attitudinal understandings associated with engaged readers. For instance, meaning-getting is not a passive activity. Engaged readers understand that one must be proactive; that meaning-getting is an active, constructive process of *making* meaning. The reader does not sit back and wait for meaning to come; the reader *creates* meaning.

Similarly, part of a reader's proactive stance is the understanding that engaged readers are adaptive. Because meaning comes in various forms and involves different kinds of texts, readers adjust accordingly. At times reading is a matter of interacting with narrative text for aesthetic purposes that result in emotional responses. At other times, it is a matter of interacting with expository text to extend conceptual knowledge of a subject matter such as social studies or science, or to deepen understandings, or to make critical judgments, or to be guided about how to do a task. While beginning reading instruction has traditionally emphasized narrative text, on the assumption that a shift will be made to expository forms of text at a later time, it is now understood that the flexible nature of reading requires that both narrative and expository text be emphasized from the beginning.

Engaged readers, in sum, possess certain fundamental understandings about reading. They understand the purpose of reading, why we learn to read, the function of various kinds of text, and the rewards of reading. Such conceptual foundations support and give meaning to all the other aspects of learning to read. This cannot be over-emphasized. While the lay public typically assumes reading to be a matter of learning skills and strategies, reading teachers understand that effective skill and strategy use rests upon accurate conceptual understandings.

Engaged Readers Apply Skills and Strategies to Get Meaning

As students develop conceptual understandings about the nature and function of reading, they are also learning the nitty-gritty "how to" skills and strategies required to be fluent and successful seekers of meaning. Traditionally, these skills and strategies are divided into two major categories representing what readers must do: (1) they must identify the words on the page, and (2) they must comprehend the textual meaning.

Word identification. Word identification involves two kinds of tasks (Duffy & Roehler, 1993). The first is sight word recognition, in which readers know words instantly without figuring them out. Ultimately, good readers know at sight virtually all the words on the page. Consequently, a major instructional task in word identification is the learning of a large stock of common words at sight. This involves visually discriminating among like visual forms, and storing accurate images of letters and words in one's memory. Visual discrimination, visual memory and sight words are taught as skills. That is, the intention is that students will automatize their ability to visually discriminate among and remember letters and words.

In the current national focus on phonics, sight word recognition and the prerequisite skills of visual discrimination and visual memory are often overlooked. This is a mistake. While, as we will see, phonics is a crucial part of the reading process, engaged readers do *not* sound out every word. In fact, only five percent of the words on a page should require analysis, and ninety-five percent should be recognized at sight. Consequently, good readers know most words instantly, and sight word recognition remains crucial to reading success.

However, the reality is that beginning readers recognize few words. Consequently, they must also learn the second kind of task associated with word identification: how to analyze and figure out unknown words. The major analysis tool is phonics, in which letter sounds are used to figure out the unknown word. The earliest phonics learnings are automatized. For instance, the crucial prerequisite skill of phonological awareness, in which students learn to discriminate one letter sound from another, is taught as a skill. We want students to be automatic and quick in distinguishing one sound from another. Similarly, we want students to be automatic and quick in associating letter sounds with the appropriate letter, so we teach this as a skill as well.

However, once readers encounter unknown words in text, the task becomes strategic. That is, students must problem solve and reason to figure out unknown words. For instance, readers reason with phonics-based strategies. The most heavily cited one is "decoding by analogy", or what is sometimes called "onset rimes" (Cunningham, 1995). In this strategy, readers use known words and their rime to figure out an unknown word having a different onset. For instance, if the unknown word is "chat", and the child knows the word "sat", he can substitute the onset ch- into the known rime -at to figure out the new word "chat".

But as important as phonics is, it is not the only strategic reasoning good readers use to figure out unknown words. They also use semantic and syntactic clues surrounding the unknown word (i.e. context clues) to figure out the word. Or, if the unknown word has known structural or morphemic units, such as a prefix or suffix, those can be used to figure out the unknown word. Or, most commonly, the reader will reason with combined clues. For instance, a reader may use the initial letter sound in combination with semantic or syntactic clues.

Comprehension. The second major category – comprehension, or the seeking of meaning – is always the focus of reading. Comprehension is a strategic process (Dole, Duffy, Roehler & Pearson, 1991; Keene & Zimmerman, 1997). The most dominant comprehension strategy is monitoring. That is, the reader expects that the text message will make sense, monitors the meaning-getting as he goes along to make sure that it makes sense, stops and repairs blockages when meaning breaks down and, finally, reflects about what was read. This monitoring activity goes on before reading, during reading, and after reading. For instance, good readers think about their particular purposes for reading and the clues in the text, and then make predictions about the meaning as they begin. During reading, they re-adjust their predictions as the content unfolds. After reading they reflect on what they read and decide what they learned, what is particularly useful, what can be discarded, and so on. As such, comprehension requires reasoning, particularly during beginning reading when the process is new and cannot become automatic.

This all sounds complex. And, in a sense, it is. But only a limited number of individual strategies come into play (see, for instance, Keene & Zimmerman, 1997). And even those few are really just multiple variations on one big strategy – that of accessing your prior knowledge and making connections between that and the text situation in order to make a prediction, or to create an image, or to make an inference, or to draw a conclusion, or whatever (Duffy, 1993).

Obviously, good readers are facile users of the skills and strategies associated with word identification and comprehension. However, it is important to re-emphasize that what undergirds effective use of skills and strategies is accurate conceptual understandings. For instance, it is not unusual to find third graders who have the misconception that one should sound out virtually every word on the page because they have not developed the concept of *fluent* reading where phonics is used occasionally as an emergency technique; similarly, it is not unusual to find third graders who have the misconception that comprehension is an inflexible process because they have not developed the understanding that comprehension is a process of flexibly adapting to both the purpose of the reading and to the requirements of the text.

Engaged Readers Self-Monitor

A third characteristic associated with engaged readers is their awareness of themselves as readers. This not only means that readers monitor their ability to identify words and comprehend text messages. It also means they monitor their own reading progress and mold literacy to their own needs as individuals having values, choices and perspectives (Hoffman & McCarthey, 2000). This awareness is often referred to as "metacognition", or "thinking about one's thinking" (Baker & Brown, 1984).

Self-monitoring takes on many forms. One is the reader's conscious awareness of the aforementioned conceptual understandings about the nature of reading and how it works. The reader's self-monitoring during comprehension is another example. Readers ask themselves constantly, "Does this make sense? Does this make sense? Does this make sense?" When the answer is "No", readers modify their predictions, test them, and move on. Similarly, readers achieve better when they are metacognitively aware of what strategies to use, when to use them and how they work (Duffy, Roehler, Sivan, et.al, 1987). This is not unlike "retrospective miscue analysis" in which readers and teacher become more conscious of strategy use by comparing miscues from various reading samples and discussing their significance (Goodman & Marek, 1996).

Ultimately, however, engaged readers are metacognitive about their own status as literate persons, and self-monitor their progress toward becoming literate. This means that readers must become a part of the assessment process by being made aware of what is involved in learning to be a good readers, and by being party to the process of assessing individual progress. When readers are aware of what is involved and what is at stake, they can self-monitor their literacy progress.

Summary

In sum, the methods and activities teachers choose to use in beginning reading are driven by the goal of developing "engaged readers" – independent readers who seek meaning, apply skills and strategies in the pursuit of meaning, and assume control of both the process of reading and their own development as literate members of society.

Beginning Reading: A Developmental Perspective

Engaged reading is a developmental and social process that evolves over a lifetime of experiences with text. Beginning reading is point along the path in that process. As with curricular goals, teachers' choices of instructional methods and activities are influenced by where the student is along the path. Typically, we think of beginning readers as kindergarten or first grade children. However, beginning readers are not always five and six years old. Older students who come to our schools from other countries and cultures are often beginning readers. Similarly, adults who did not learn to read as children and who enroll in adult literacy classes are also beginning readers. And third, fourth, fifth graders and even high school students whom, for whatever reason, did not learn to read in earlier grades, are beginning readers.

Learning to read unfolds in a number phases or stages of development. These phases are not lock-stepped or fixed. In fact, it is not unusual to find a student demonstrating behavior consistent with one phase while simultaneously demonstrating behavior consistent with another phase. Consequently, phases are guideposts for understanding the developmental process of learning to read, not rigid markers for labeling students. While different theoreticians have conceptualized these stages in different terms (e.g. Chall, 1996; Clay, 1991; Ehri, 1991; Juel, 1988), there are many similarities across these models. Hoffman and McCarthey (2000) have conceptualized this developmental process into four phases or stages leading to independence in reading.

The first phase is the "Emergent" phase, so-called because children are constantly emerging as readers, starting with their first contact with print. The emergent reader encounters print in the home environment, in the books read to him, in the labels at the grocery store and in well-known logos. Additionally, the emergent reader encounters language in its oral forms. The richer and more varied the early encounters with oral and written forms of language, the more easily a beginner learns to read. Hence, children tend to learn to read more easily when they come from homes where children are involved in oral discussion, are read to, are praised for doing scribble writing, and otherwise have rich language experiences. These experiences help develop sophisticated oral language and listening skills and many of the conceptual understandings about seeking meaning, as well as some of the prerequisite word identification skills, such as visually discriminating among like letter forms and the segmenting and manipulating letter sounds. Children from homes where such experiences are rare, in contrast, often have lots of ground to make up before they read. Consequently, teachers' methods and activities with emergent readers often look much like experiences advantaged children receive at home in the natural course of events.

Typically, children are considered to be at the emergent phase until they have a reading level (i.e. until they can read words in a decontextualized setting and can read a preprimer or a book having only a few words). Once they can read a simple book, they move into a phase in which they learn more sophisticated skills and strategies. Common words are recognized (*the, come, when,* etc.), and they begin to discriminate among like looking letters (such as *m* and *n* and *d* and *b*) and like looking words (*then* and *them*, for instance). In short, they begin building a sight word vocabulary. Simultaneously, they learn that English is based in an alphabetic code, and they begin to learn how to use the alphabetic principle to analyze and figure out words they have never seen before. They learn that semantics and syntax are also viable cues to be used in combination with phonics. And, of course, seeking meaning is always an emphasis, with students learning to monitor meaning-getting in both their oral listening activities and in their rudimentary reading. All this is enhanced by frequent experiences with writing. Because reading and writing are reciprocal processes, the more writing students engage in, the better readers they become (Tierney & Shanahan, 1991).

Eventually, beginning readers move to a third phase exemplified by "fluency". That is, students develop a large stock of sight words, are confident and reasonably quick in using analysis strategies such as onset rimes, are consciously applying comprehension strategies before, during and after reading, and are "reading like you talk", with appropriate inflection and intonation (Allington, 1980; Hoffman & McCarthey, 2000). The Fluency phase involves the consolidation, orchestration, and integration of skills and strategies. Fluent oral reading develops into fluent silent reading as attention to the decoding aspects of reading become less and less demanding.

At this point, students are moving out of beginning reading and into the fourth phase of "Flexibility". Here, students are able to use reading as a problem solving tool, to adapt reading to the needs of the situation, to deal effectively with a variety of text types and to use reading to solve problems and meets personal needs. They become skilled at adjusting reading strategies to purposes and to the structure of the text. They are no longer beginning readers.

In sum, beginning reading is the foundation for moving students into fluency, flexibility and independence as readers. Teachers' assessments of students' phases of development influences what methods and activities will be employed, much like the goal being sought influences what methods a teacher will choose to use.

What Mid-Range Principles and Practices are Associated with Effective Beginning Reading Instruction?

As noted above, beginning reading instruction goes well beyond "basic skills." In providing instruction, teachers select and adapt various methods and activities to develop engaged readers who seek meaning, apply skills and strategies, and self-monitor.

Methods That Support Seeking Meaning

To seek meaning one must first understand that meaning is the goal. That is, readers must develop the conceptual understanding that reading is a meaning-getting activity with the capacity to enlighten, to enliven, to enhance and to inform. Three categories of practices are illustrative.

Classroom context. Developing conceptual understandings about reading requires a supportive classroom environment. "Classroom environment" goes beyond the generic "caring and cohesive learning communities" Brophy cites as his first principle in Chapter One. Beyond being caring and cohesive, the environment must also provide students with experiences in seeking meaning. At all phases of beginning reading, there should be abundant and varied kinds of text available, and students should be engaged in using these texts in ways that help them build a schema for the fact that reading does indeed enlighten, enliven, enhance and inform. At the emergent phase, this will often take the form of teacher modeling of reading and writing, sharing reading, lap reading, reading material to students so that they can use what they hear, and so on. As students move beyond the emergent phase, teachers continue to place them in situations in which both the functional and the joyful aspects of reading are experienced.

The crucial idea here is that the most powerful way to build conceptual understandings is through experience. *Talking* about the fact that reading is enlightening, enlivening, enhancing and informing is one thing; *experiencing* it is an entirely different thing. For students to develop rich, lasting conceptions, they must *feel* what it is like to be enlightened, enlivened, enhanced and informed. That means they must experience it, not just talk about it.

One of the most powerful ways to develop such conceptions is by swinging instruction around problem-centered projects or activities. This is often referred to as "situated learning" (Putnam & Borko, 2000) or "inquiry-based instruction" (Hoffman & McCarthey, 2000) or "thematic instruction" (Valencia & Lipson, 1998). The trick is to involve students in compelling life-like problems or projects. Then, within the context of the problem, students read text that helps solve the problem or complete the project, and learn skills and strategies needed in order to read that text. Hall (1998) provides a particularly good example of such a project at the emergent phase; Duffy (1997) provides a variety of examples that span grade levels; and Guthrie's concept-oriented reading instruction provides examples that can be adapted to beginning reading (Guthrie, Van Meter et al., 1996).

A productive alternative practice is to situate beginning reading instruction in community practices. For instance, students are encouraged to write and read about family interactions and life histories of community members (Ada, 1993) and to identify community resources and use them in classrooms (Moll & Gonzalez, 1994). Such projects dramatize the connection between school learning and application to life and, in so doing, give students tangible experiences that help them conceptualize the importance of being literate.

Lots of easy reading. Another important practice for building conceptual understandings about seeking meaning is extensive engaged reading time. Surprisingly, studies have repeatedly shown that one of the least emphasized activities during reading instruction is the actual reading of connected text (as opposed to the reading of practice exercises and other short-answer activities). Some reports indicate that beginning readers read for only seven minutes a week (Hoffman & McCarthey, 2000).

Obviously, more than seven minutes of reading a week is required to develop an understanding of what it means to be fluent. Most successful beginning reading programs engage students in easy reading (that is, reading that is at the student's independent reading level, in which 99% of the words are recognized automatically and comprehension is at least at a 90% level) for at least *forty-five to sixty minutes a day.* It is only through multiple, almost habitual, experiences with easy reading that readers understand what it feels like to be fluent, and it is only through fluent reading that students understand what it means for reading to be enlightening, enlivening, enhancing and informing.

This principle is important at the emergent level, where students cannot actually read, as well as at levels where students have a reading level. For instance, emergent readers can "pretend read" familiar books that have been read to them previously, wordless books, and a variety of illustrated books. All such experiences help build desired concepts about why we learn to read.

On the negative side, students should *never* read material in which they are less than 95% fluent and in which their comprehension is less than 75%. Such material is too hard to be enjoyable. Being forced to struggle with a text that is too difficult results only in learning to hate reading.

In sum, Uninterrupted Sustained Silent Reading (U.S.S.R) and other similar periods of sustained reading of materials of choice should be a regular, featured and valued part of the classroom day. As Allington (1977) pointed out years ago, "If they don't read much, how're they ever gonna get good?"

Social occasions for reading. Reading is, among other things, social in nature. Students develop sound conceptual understandings about seeking meaning when they have experience reading with teachers and friends.

One of the best social occasions for reading is quality storytime (Roser, Hoffman & Farest, 1990). Here the teacher reads good stories orally, and teacher and students discuss them. This practice offers students opportunities to engage in extended narratives without regard for decoding ability. In doing so, students experience responses generated by good literature and, as a result, develop sound conceptual understandings.

Shared reading is another important and useful social reading occasion (Holdaway, 1979). In this situation, a favorite book is often read over and over again, with the teacher and students discussing individual responses and how language forms and conventions enhance those responses.

Another effective social occasion for reading is literature circles and book clubs (Raphael & McMahon, 1994). These can be teacher-led or student-led, and often involve students in responding to narratives, writing about narratives, and engaging in open-ended discussion about the text and their respective responses. While typically associated with the upper grades, they can be adapted for use in beginning reading.

Reader's Workshop is another effective social occasion for reading (Atwell, 1987). Like literature circles, they tend to be associated with upper grades, but can be adapted to beginning reading. This model emphasizes free choice reading by students, followed by talk about individual responses to literature in conferences with teacher and/or friends. This activity gives students experience in behaving like independent adult readers do, and helps them build a concept for independent reading.

Guided reading activities are another appropriate social occasion for reading. Two popular structures for guided reading are Stauffer's (1969) Directed Reading-Thinking Activity and Reading Recovery's version of guided reading (Fountas & Pinnell, 1996). These, as well as other forms of guided reading, involve teacher and students in public discussion of text they have all read. Teachers use these occasions to build conceptual understandings about seeking meaning, as well as to develop skills and strategies

The above examples all tend to be associated with narrative text and aesthetic responses. However, experiences with social reading occasions should not be limited to narrative text. Expository text can also be used, and teachers should look for opportunities to do so.

Methods That Support Applying Skills and Strategies

Methods for teaching skills and strategies in beginning reading are frequently debated. The debate typically pits direct teaching techniques against indirect teaching techniques. Rigid positions are sometimes taken. For instance, Fountas and Pinnell (1996) state unequivocally that strategies should *not* be directly taught (pp. 149).

Rigid positions rarely work in the reality of classrooms, however. What actually happens is that some students learn quite well when instruction is indirect, but others require more explicit instruction. Consequently, what is appropriate or inappropriate often changes depending on the situation, and cannot be rigidly prescribed ahead of time Four principles are illustrative.

Instruction must be systematic. Because different students have different needs, a major principle of effective beginning reading instruction is that instruction is planful and targeted (Duffy & Roehler,1993; Hoffman & McCarthey, 2000). That is, a system must be in place in order for the teacher to know who needs which skills and strategies, and whether direct or less direct methods are appropriate.

Systems vary from teacher to teacher. But all systems share four characteristics. First, the system makes clear what must be learned and the developmental sequence in which the learning usually occurs. As was noted above, teachers' methods and activities are chosen by reference to what beginning readers need to learn and, at least in a general sense, the sequence in which they must be learned.

Second, there has to be a way to determine who has learned these things and who has not. This means that there needs to be a diagnostic component. Informal, on-line diagnostic techniques are best for doing this because, unlike standardized measures, they provide teachers with a "window into the mind" of the reader.

Third, the system must include record-keeping. With twenty-five students operating at a variety of levels and possessing differential skill and strategy needs, it is impossible to keep all the information in one's head. Some kind of system is needed for keeping track of who is where.

Finally, a system should provide for flexible grouping so the teacher can gather together those students who need similar instruction. Traditionally, beginning reading instruction has been organized into three homogeneous groups. Recently, however, the negative consequences of such grouping for struggling readers, who often end up permanently assigned to the low group, has resulted in a move to flexible patterns of grouping. The idea is to maintain the efficiency of grouping together those students who need specific kinds of instruction while also breaking the negative expectancy associated with permanent homogeneous groups by also frequently putting students in interest groups, cooperative groups and other kinds of heterogeneous groups.

Instruction must be explicit. A large domain of research findings establish that reading instruction is most effective when the teacher provides explicit information (see, for instance, Baumann, 1984; Dewitz, Carr & Patberg, 1987; Dole, Brown & Trathen, 1996; Duffy, Roehler, Sivan et al., 1987; Pressley, El-Dinary et al., 1992; Winne, Graham & Prock, 1993). The degree of explicitness is relative to the situation. Struggling readers, for the most part, profit from very explicit information; average and above average readers who come to instruction with richer backgrounds often learn well with less explicit (or indirect) instruction.

One of the major occasions for explicit teaching is when struggling readers are grouped together to learn a specific skill or strategy. Three forms of explicitness are appropriate during such instruction.

First, struggling readers learn best when they know what they are accountable for learning. Consequently, teachers help students by making explicit statements at the beginning of lesson about what the student will be able to do by the end of the lesson.

Second, struggling readers learn best when they are provided with explicit explanations. Because reading involves "in the head" (Clay, 1991) mental activity, the processing one does to read is invisible. Therefore, explanations are most effective when presented as "think-alouds" or as "mental modeling" (Duffy, Roeher & Herrmann, 1988) that provides a window into the teacher's mind so that students can "see" how good readers think their way through text.

Third, struggling readers learn best when they have support as they try out newly-explained skills and strategies. This support is often called "scaffolding" or "coaching", but has also been called "guided practice" or "fading". Regardless of its label, the principle is that beginning readers need multiple opportunities to practice newly-learned skills and strategies in situations where they receive lots of help initially but gradually move to doing the skill or strategy without assistance.

Instruction must be public. The best beginning reading instruction makes instructional information public. In doing so, students are provided with explicit information about what is important, and are not left to figure it out for themselves.

For instance, teachers look for opportunities to make public conceptual understandings such as their aesthetic responses to literature, insights derived from sharing a story character's experiences, and the sheer fun of reading for enjoyment. They look for similar opportunities regarding skill and strategy use. For instance, they model fluency for students, making explicit the importance of automatic word identification and appropriate intonation and inflection. Sometimes they model for students how they are thinking before, during and after their reading or point out how they are using "fix-it" strategies.

Similarly, teachers make public skill and strategy information through charts, and other visuals displayed in the classroom. For instance, some teachers put up charts that display by category the major curricular learnings associated with becoming a reader (i.e. the conceptual understandings, the word identification skills, the comprehension strategies, etc.) and, as each skill or strategy is learned, it is added to the chart in the appropriate category. Similarly, teachers create displays that make public the "before-during-after" thinking progression associated with comprehension or the "95%/5%" relationship between words known at sight and words to be analyzed. In publicly displaying such instructional information, teachers help students build appropriate understandings.

Other displays provide students with direct assistance as they read. For instance, many teachers have "word walls" that display major phonogram patterns that students can refer to when decoding by analogy (Cunningham & Allington, 1999). Word banks and displays of words that everyone is expected to spell correctly are similar examples.

Instruction is pervasive. Skill and strategy instruction is not limited to the "reading period" or to a specific instructional group or time slot. To the contrary, skills and strategies are applied throughout the day. Talk about what good readers do and about the code and how it works occurs during opening exercises, during social studies, and during science. When questions are raised about writing and spelling, responses are tied to skills and strategies learned in reading. In short, the teacher looks for opportunities across the entire school day to model language connections, reading outcomes, and application of word identification and comprehension strategies. Similarly, it is not unusual to see students engaged in activities that promote fluency, such as repeated readings of familiar material, shared reading, or partner reading, throughout the school day.

Of particular importance, writing is heavily emphasized. Whereas reading and writing used to be considered separate tasks, we now understand that they are highly related (Tierney & Shanahan, 1991). The more writing beginning readers do, the better they read; the more reading they do, the better they write. Consequently, in the best beginning reading instruction, students engage in extended writing of connected text. Even when the writing is "scribble writing", students are learning to be more sensitive to the nature of the code system, to how skills and strategies apply, and to the idea that reading and writing are message-sending and message-receiving systems.

Summary. A basic fact of beginning reading instruction is that different students have different skill and strategy needs. Consequently, the key to successful instruction is the teacher's skill at orchestrating a variety of direct and indirect instructional techniques according to individual students' needs.

Methods That Support Self-Monitoring

Independent reading involves taking control of the reading process. Obviously, beginning readers are not instantly in control of their reading. Nonetheless, one of the hallmarks of good beginning reading instruction is that, even at the earliest, emergent phases, teachers involve students in activities that promote conscious awareness of how reading works and of their own progress as readers. This metacognitive element is important because it transfers responsibility from the teacher to the student.

Monitoring reading itself. From the earliest phases of instruction, teachers help beginning readers become conscious of the fact that reading is not a mysterious process. Rather, it is a system comprehensible to all. Consequently, teachers give students tangible information about the system of reading and how it works as a first step toward putting them in control of their reading. Several practices are designed to achieve this end.

For instance, the aforementioned practice of making public the basic undergirding conceptual understandings about the nature of reading and how it works is an attempt to help students become consciously aware of the process as a step toward putting them in control. Similarly, the idea of making visible the invisible mental processes involved in strategy use helps remove the mystery of how strategies work, and is a step toward putting students in control of their reading.

At another level, developing the concept of critical literacy and critical thinking helps move students to a place where they are in control of reading. Even at the earliest phases of beginning reading, teachers encourage students to use their personal background, beliefs and values as a means for making critical analyses of texts. For instance, teachers' questions about "Could this have really happened?" and "Is this story real or make believe?" are first steps toward exploring author's intentions and toward learning to make critical judgments about what is being read. The idea is to help students develop the concept that engaged readers are not passive recipients of knowledge but, instead, are in personal control of how text content is used.

Another major way to start beginning readers toward control of their own reading is to involve them in social interaction and dialogue about reading. Hence, oral sharing time and other forms of social interaction play an increasingly important role in development of self-monitoring and, ultimately, students' increased control over the process of reading.

Monitoring one's progress as a reader. It is sometimes assumed that students should be shielded from specific knowledge about their progress. However, it has recently become increasingly clear that students become engaged readers when they are intimately involved in their own progress. That is, because they know what their strengths and weaknesses are, what the next steps toward progress are, and are conscious participants in the instructional effort, they are better able to move forward (Afflerbach, 2000; Hoffman, Worthy et al., 1996).

What is required is an assessment plan that is comprehensible to students. This means two things. First, assessment should be informal and on-going rather than based in standardized tests. Second, the student must be an equal participant in the assessment process. That is, students must be privy to the same information the teacher has – what it is that must be learned, the student's particular performance relative to those criteria, and what the necessary next step is to move closer to being in control of the process.

There are a variety of ways teachers involve students in monitoring their own progress as readers. One of the best ways, however, is through use of portfolios (Valencia, 1998; Valencia & Au, 1996). Portfolios are containers; what each contains is a student's work as a reader and writer. As such, portfolios represent a student's literacy "history" because the portfolio contains literacy artifacts from one phase of development to another.

Teacher and student jointly develop portfolios, based on the diagnostic data the teacher has collected, and teacher and student negotiate together the short-term and long-term goals to be attained based on those data. As these goals are pursued, various illustrative artifacts are inserted into the portfolio as evidence.

Portfolios cannot be the entire assessment program in beginning reading. Districts and teachers also need standardized tests and other more formal measures. But, for students, portfolios are powerful devices. They require students to reflect on themselves as literate persons, and to engage in planning designed to improve their literacy. Consequently, they help put students in control of their own destinies as readers and writers.

Summary

Beginning reading instruction is a complex endeavor with multi-faceted goals (i.e. seeking meaning, applying skills and strategies and self-monitoring). Many methods and practices can be effective in developing these goals. Those we cite here reflect current professional thought, but are not exhaustive. Further, methods are inter-related, and good teachers combine and re-combine them in a variety of ways to meet situational demands. Consequently, the key to effective instruction is the teacher's analysis of what a situation demands, and subsequent orchestration of various methods and techniques to fit that situation.

What are the Implications for Improved Practice?

There are effective methods, principles and practices in beginning reading instruction. One example is the principle of providing students with experiences designed to build strong concepts about the nature and use of reading; another is the practice of having students do lots of easy reading as a means for building fluency; another is the need to be explicit, especially with struggling readers. There is not, however, a single effective *method*. That is, there is no evidence to suggest that certain combinations of principles and practices, organized into a sequential, systematic plan (i.e. method), is universally effective in teaching beginning reading (Duffy & Hoffman, 1999). The individual differences of students and the complex social interactions that characterize classrooms preclude reducing beginning reading instruction to a single method or program.

The point is not that methods are unimportant. Rather, the point is that no one method works for all students all the time. Consequently, effective teachers make choices about how to combine and re-combine different practices in different ways in different situations in order to meet the variable needs of the student clientele. Therefore, the key to effective instructions is teacher thought in making such decisions, not a particular method or set of methods.

There is a growing understanding of this phenomenon in the field. Some scholars call it "methodological eclecticism" (Shanahan & Neuman (1997), some call it "principled eclecticism" (Stahl, 1997), and some call it "conceptually adaptive teaching" (Duffy, 1991). Hoffman and McCarthey (2000) call it "principled" teaching, saying: ". . . what makes a difference for student in learning to read is clear. It is not the method or the materials. It is the teacher . . ." (pp. 11). The bottom line is that no single method or combination of methods is effective in every situation. Consequently, teachers *invent* methods by combining and re-combining various elements and techniques to fit the instructional situation.

This puts a different perspective on teaching. Power is not attributed to a method, with the teacher being a loyal follower or consumer of a certain set of "best practices". Rather, power is attributed to the teacher because what makes instruction effective is a teacher's selection and adaptation of ideas from a variety of ideologies, programs, methods and materials in a manner that meets the particular needs of students or the particular demands of a classroom situation. Hence, good teachers are *orchestraters* – that is, they create harmony by blending together many seemingly different instructional ideas. Such teachers are characterized in four ways.

First, the orchestrater of instruction is clear about whose vision counts in the classroom (Duffy, 1998). Various approaches, methods, and programs try to impose on teachers their particular visions about what is really important about beginning reading instruction. The orchestrater, however, has his/her own vision for beginning reading instruction. It is this vision, not a vision imposed from outside, that dominates the teacher's mind and guides decisions about how to combine various instructional practices and principles together. In short, power resides in the teacher's mind; methods, programs and materials are merely tools in the teacher's professional kit.

Second, the orchestrater of instruction analyzes each instructional situation to make decisions about what each student needs. Analysis is based in informal, on-line data collection. The teacher decides which practices and principles to use with which students and in which situation on the basis of data. Hence, instruction is not driven by tenets of an ideology, or by principles and practices associated with a method, or by procedures associated with a program. Rather, instruction is driven by the teacher's mind – that is, by the teacher's data-based judgments about what each student needs and how best to meet that need.

Third, the orchestrater of instruction is mentally strong. Being an orchestrater means taking risks and making judgments about what to do with students. Beginning reading, like all teaching, is confounded by numerous complexities. There are no guarantees. Further, pressure from legislators, district personnel, colleagues, parents and even students tend to put teachers on the defensive. These pressures are particularly difficult for an orchestrater, who does not have the luxury of blaming a mandated method or program when things go wrong. The decisions are the teacher's decisions, and it takes courage and strength to persist in the face of uncertainty and outside pressures.

Finally, teachers who orchestrate instruction are curious and resilient. They know there are no "sure-fire" solutions, and they constantly seek better ways to teach. When instruction fails, they re-examine the data, re-think the problem, and try something else. If a problem persists, rather than casting about for a "quick fix" solution, they initiate professional inquiry. Sometimes it involves professional reading, sometimes it involves collaboration with colleagues, and sometimes it involves semi-formal study of one's own practice. In any case, orchestraters see teaching as a series of problem-solving events, and they thrive on the ambiguity associated with the uncertainty inherent in instructional practice.

Because instructional effectiveness resides in the minds of teachers, rather than in a method or a set of "best practices", our focus must be on developing teachers who possess the psychological mind-set to be thoughtfully adaptive. In short, the pressing question about beginning reading instruction is not "What are the best methods and practices?" but, rather, "What are the best ways to develop teachers who orchestrate together all available practices and principles rather than following a particular program?"

CONCLUSION

We identify in the above sections certain principles, methods, and practices associated with effective beginning reading instruction. However, we do so with reservation. When we (or other "authorities") endorse particular methods, practices or principles, we endow them with the instructional equivalent of the Good Housekeeping Seal of Approval and imply that other methods, practices and principles are not "approved." This is a dangerous proposition because it causes people to thing that method is what counts. In actuality, however, it is the teacher – and how the teacher orchestrates the use of various methods – that is the key to successful beginning reading instruction, not the method. Because methods must be adapted to the instructional situation, and not followed prescriptively, they are only as good as the teachers using them. To imply that success lies with following certain favored methods empowers method and disempowers teachers, just the opposite of what we should be doing.

Consequently, it is time to stop searching for the mythical "perfect method" (or combination of methods, practices and principles) as if we could identify these, put them in teachers' heads ahead of time, and then sit back and watch teachers be universally effective. What effective beginning reading instruction requires is teachers who possess the personal and intellectual skills to assess an instructional situation and to then use their knowledge of various methods and practices to create instruction that fits the situation. Research on teaching, therefore, needs to stop looking for universally effective methods and study instead what empowers effective teachers with the spirit and thought to be orchestraters and inventors of instruction.

REFERENCES

Ada, A. (1993). A critical pedagogy approach to fostering the home-school connection. ERIC: ED358716.

Afflerbach, P. (2000). Our plans and our future. In: J. Hoffman, J. Baumann & P. Afflerbach (Eds), *Balancing Principles for Teaching Elementary Reading* (pp. 75-102). Mahwah, NJ: Erlbaum.

Allington, R. (1977). If they don't read much, how are ever goonna get good? *Journal of Reading*, *21*, 57–61.

Allington, R. (1980). Fluency: The neglected reading goal. *Reading Teacher*, *36*, 556–561.

Allington, R., & Walmsley, S. A. (1998). *No quick fix: rethinking literacy programs in America's elementary schools*. New York: Teachers College Press, c1995.

Anders, P., Hoffman, J. V., & Duffy, G. (2000). Teaching teachers to teach reading: Paradigm shifts, persistent problems and challenges. In: M. Kamil, R. Barr, P. Mosenthal & P. D.Pearson (Eds), *Handbook of Reading Research* (Vol. III, pp. 719–742). Mahwah, NJ: Erlbaum.

Atwell, N. (1987). *In the middle: Writing, reading and learning with adolescents*. Portsmouth, NH: Heinemann.

Baker, L., & Brown, A. (1984). Cognitive skills and reading. In: P. D. Pearson (Ed.), *Handbook of Reading Research* (pp. 352–394). NY: Longman.

Baumann, J. (1984). The effectiveness of direct instruction paradigm for teaching main idea comprehension. *Reading Research Quarterly*, *20*, 93–108.

Bond, G., & Dykstra, R. (1967). The cooperative research program in first-grade reading instruction. *Reading Research Quarterly*, *2*, 1–142.

Carlos, L., & Kirst, M. (1997). *California curriculum policy in the 1990's: "We don't have to be in front to lead"*. San Francisco: WestEd/PACE.

Chall, J. (1967). *Learning to Read: The Great Debate*. New York: McGraw-Hill

Chall, J. (1996). *Learning to Read: The Great Debate* (3rd ed.). Orlando, FL.: Harcourt Brace.

Clay, M. (1991). *Becoming literate: the construction of inner control*. Portsmouth, NH: Heinemann.

Chrispeels, J. H. (1997). Educational policy implementation in a shifting political climate: The California experience. *American Educational Research Journal*, *34*, 453–481.

Cunningham, P. (1995). *Phonics They Use: Words for Reading and Writing*. NY: HarperCollins.

Cunningham, P., & Allington, R. (1999). *Classrooms that work: They can all read and write* (2nd ed.). NY: HarperCollins.

Dewitz, P., Carr, E., & Patberg, J. (1987). Effects of inference training on comprehension and comprehension monitoring. *Reading Research Quarterly*, *22*, 99–121.

Dole, J., Brown, K., & Trathen, W. (1996). The effects of strategy instruction on the comprehension performance of at-risk students. *Reading Research Quarterly*, *31*, 62–89.

Dole, J., Duffy, G., Roehler, L., & Pearson, P. D. (1991). Moving from the old to the new: Research on reading comprehension instruction. *Review of Educational Research*, *61*, 239–264.

Duffy, G. (1991). What counts in teacher education? Dilemmas in empowering teachers. In: J. Zutell & C. McCormick (Eds), *Learning Factors/Teacher Factors: Literacy Research and Instruction* (pp. 1–18). 40th Yearbook of the National Reading Conference. Chicago: National Reading Conference.

Duffy, G. (1993). Re-thinking strategy instruction: Teacher development and low achievers' understandings. *Elementary School Journal*, *93*, 231–247.

Duffy, G. (1997). Powerful models or powerful teachers? An argument for teacher-as-entrepreneur. In: S. Stahl & D. Hayes (Eds), *Instructional Models in Reading* (pp. 351–356). Erlbaum.

Duffy, G. (1998). Teaching and the balancing of round stones. *Phi Delta Kappan*, *79*, 777–780.

Duffy, G., & Hoffman, J. (1999). In pursuit of an illusion: The flawed search for a perfect method. *Reading Teacher*, *53*, 10–37.

Duffy, G., & Roehler, L. (1993). *Improving Classroom Reading Instruction: A Decision-Making Approach* (3rd ed.). NY: McGraw-Hill.

Duffy, G., Roehler, L., & Herrmann, B. (1988). Modeling mental processes helps poor readers become strategic readers. *Reading Teacher, 41*, 762–767.

Duffy, G., Roehler, L., Sivan, E., Rackliffe, G., Book, C., Meloth, M., Vavrus, L., Wesselman, R., Putnam, J., & Bassiri, D. (1987). Effects of explaining the reasoning associated with using reading strategies. *Reading Research Quarterly, 22*, 347–368.

Ehri, L. (1991). Learning to read and spell words. In: L. Riedben & C. A. Perfetti (Eds), *Learning to Read: Basic Research and its Implications* (pp. 57–73). Hillsdale, NJ: Lawrence Erlbaum Associates.

Fountas, I., & Pinnell, G. (1996). *Guided Reading: Good First Teaching For All Children*. Portsmouth, NH: Heinemann.

Gambrell, L. B., Morrow, L. M., Neuman, S. B., & Pressley, M. (1999). *Best practices in literacy instruction*. New York: Guilford Press.

Goodman, K. S. (1989). Whole-Language research: Foundations and development. *Elementary School Journal, 90*(2), 207–222.

Goodman, Y., & Marek, A. (1996). *Retrospective Miscue Analysis*. NY: R. C. Owen.

Guthrie, J., & Anderson, E. (1999). Engagment in reading: Processes of motivated, strategic, knowledgeable, social readers. In: J. Guthrie & D. Alvermann (Eds), *Engaged Reading* (pp. 17–45). NY: Teachers College Press.

Guthrie, J., & Alvermann, D. (1999). *Engaged Reading*. NY: Teachers College Press.

Guthrie, J., Van Meter, P., McCann, A., Wigfield, A., Bennett, L., Poundstone, C., Rice, M., Faibisch, F., Hunt, B., & Mitchell, A. (1996). Growth of literacy engagement: Changes in motivations and strategies during concept-oriented reading instruction. *Reading Research Quarterly, 31*, 306–332.

Hall, N. (1998). Real literacy in a school setting: Five-year-olds take on the world. *Reading Teacher, 52*, 8–17.

Hoffman, J. V. (1998). When bad things happen to good ideas in literacy education:Professional dilemmas, personal decisions, and political traps. *Reading Teacher, 52*(2), 102–113.

Hoffman, J. V., McCarthey, S. J., Abbott, J., Christian, C., Corman, L., Dressman, M, Elliot, B., Matherne, D., & Stahle, D. (1994). So What's New in the New Basals? A Focus in First Grade. *Journal of Reading Behavior, 26*(1), 47–73.

Hoffman, J. V., McCarthey, S., Elliott, B., Bayles, D., Price, D., Ferree, A., & Abbott, J. (1995). The Literature-based basals in first grade classrooms: Savior, Satan, or same-old, same-old? *Reading Research Quarterly*, 168–197.

Hoffman J. V., & McCarthey, S. (2000). *Our principles and our practices. Balancing principles for teaching elementary reading*. Mahwah, NJ: Erlbaum.

Hoffman, J. V., & Pearson, P. D. (2000). Reading teacher education in the next millennium: What your grandmother's teacher didn't know that your granddaughter's teacher should. *Reading Research Quarterly, 35*, 28–45.

Hoffman, J. V., Worthy, J., Roser, N., McKool, S., Rutherford, W., & Stecker, S. (1996). Performance assessment in first grade classrooms: The PALM model. Yearbook of the National Reading Conference. Chicago: National Reading Conference.

Holdaway, C. (1979). *Foundations of Literacy*. Portsmouth, NH: Heinemann.

Juel, C. (1988). Learning to read and write: A longitudinal study of 54 children from first through fourth grades. *Journal of Educational Psychology, 80*(4), 437–447.

Keene, E., & Zimmerman, S. (1997). *Mosaic of Thought: Teaching Comprehension in a Reader's Workshop*. Portsmouth, NH: Heinemann.

Lyon, R. (1995). Research initiatives in learning disabilities: Contributions from scientists supported by the National Institute of Child Health and Development. *Journal of Child Neurology, 10*, 120–126.

McGill-Franzen, A. (2000). Policy and instruction: What is the relationship? In: M. Kamil, P. Mosenthal, P. D. Pearson & R. Barr (Eds), *Handbook of Reading Research* (Vol. III, pp. 889–908), Mahwah, NJ: Erlbaum.

Moll, L., & Gonzalez, N. (1994). Lessons from research with language-minority children. *Journal of Reading Behavior, 26*, 99439–99456.

Pressley, M. (1998). *Effective Reading Instruction: The Case for Balanced Teaching*. NY: Guilford.

Pressley, M., El Dinary, P., Gaskins, I., Schuder, T., Bergman, J., Almasi, L., & Brown, R. (1992). Beyond direct explanation: Transactional instruction of reading comprehension strategies. *Elementary School Journal, 92*, 511–554.

Pressley, M., Rankin, J., & Yokoi, L. (1996). A survey of instructional practices of primary grade teachers nominated as effective in promoting literacy. *Elementary School Journal, 96*, 363–384.

Pressley, M., Wharton-McDonald, R., Allington, R., Block, C., Morrow, L., Tracey, D., Baker, K., Brooks, G., Nelson, kE., & Woo, D. (1998). *The nature of effective first-grade literacy instruction*. Report 1007, National Research Center on English Learning and Achievement, University at Albany, Albany, NY.

Putnam, R., & Borko, H. (2000). What do new views of knowledge and thinking have to say about research on teacher learning? *Educational Researcher, 29*, 4–15.

Raphael, T., & McMahon, S. (1994). Book club: An alternative framework for reading instruction. *Reading Teacher, 48*, 102–116.

Roser, N., Hoffman, J., & Farest, C. (1990). Language, literature, and at-risk children. *Reading Teacher, 43*, 554–559.

Shanahan, T., & Neuman, S. (1997). Conversations: Literacy research that makes a difference. *Reading Research Quarterly, 32*, 202–211.

Shannon, P. (1990). Joining the debate: Researchers and reading education curriculum. In: J. Zutell & S. McCormick (Eds), *Literacy Theory and Research: Analyses from Multiple Paradigms*. Yearbook of the National Reading Conference. Chicago: National Reading Conference.

Shulman, L. (1986). Those who understand: Knowledge growth in teaching. *Educational Researcher, 15*, 4–14.

Stahl, S. (1997). Instructional models in reading: An introduction. In: S. Stahl & D. Hayes (Eds), *Instructional Models in Reading* (pp. 1–30). Mahwah, NJ: Erlbaum.

Stauffer, R. (1969). *Directing Reading Maturity as a Cognitive Process*. NY: Harper & Row.

Taylor, B., Pearson, P. D., Clark, K., & Walpole, S. (1999). *Beating the odds in teaching all children to read*. Ciera Report #2–006, Center for the Improvement of Early Reading Achievement. Ann Arbor: University of Michigan.

Tierney, R., & Shanahan, T. (1991). Reading-writing relationships: Processes, transactions, outcomes. In: P. D. Pearson, R. Barr, M. Kamil & P. Mosenthal (Eds), *Handbook for Reading Research* (Vol. II, pp. 246–280). NY: Longman.

Wharton-McDonald, R., Pressley, M., & Hampston, J. (1998). Literacy instruction in nine first-grade classrooms: Teacher characteristics and student achievement. *Elementary School Journal, 99*, 101–128.

Valencia, S. (Ed.) (1998). *Literacy Portfolios in Action*. Fort Worth: Harcourt Brace.

Valencia, S., & Au, K. (1996). *Portfolios across educational contexts: Issues of evaluation, teacher development and system validity.* (Research Report No. 73). Athens, GA: Universities of Georgia and Maryland, National Reading Research Center.

Valencia, S., & Lipson, M. (1998). Thematic instruction: A quest for challenging ideas and meaningful learning. In: T. Raphael & K. Au (Eds), *Literature-Based Instruction: Reshaping the Curriculum* (pp. 95–123). Norwood: Christopher-Gordon.

Wepner, S. B., & Feeley, J. T. (1993). *Moving forward with literature*. New York: Macmillan.

Winne, P., Graham, L, & Prock, L. (1993). A model of poor readers' text-based inferencing: Effects of explanatory feedback. *Reading Research Quarterly, 28*, 52–69.

CONTENT AREA READING AND LITERATURE STUDIES

Donna E. Alvermann and George G. Hruby

INTRODUCTION

Instructional methods and learning activities within content area reading and literature studies are distinct from those found in the area of beginning reading. In fact, textbook authors, curriculum planners, college professors, and any number of other interested parties build entire careers around this distinction. Although an oft repeated reason for the difference is that students learn to read before they read to learn, we take exception to this outdated notion that would somehow separate the act of reading from one of its functions – reading to learn *something*. Developmentally, beginning readers are different from skilled readers, but the difference lies more with the content or subject matter materials the two read than with their purpose for reading.

In this chapter, we focus on the reading methods and activities that teachers and students use in the middle grades and high school to foster learning in four subject areas (English language arts, mathematics, science, and social studies). In synthesizing what the research and wisdom of practice suggest about these methods and activities, we have divided the chapter into three parts. In the first part, we trace the differences in the historical development of content area reading and literature studies. This is an important first step because reading at the secondary level is heavily discipline-based, and the methods and activities associated with teaching literature vary considerably from those used to teach social studies, mathematics, and science. In part two, we connect the research and practice in content area reading and literature studies to the principles of

---

Subject-Specific Instructional Methods and Activities, Volume 8, pages 51–81.

ISBN: 0-7623-0615-7

good teaching outlined in the book's introductory chapter. Where necessary, we qualify and elaborate on those principles that do not fit neatly into the overall scheme of instruction in content area reading and literature studies. Finally, in part three of the chapter, we draw implications from our synthesis of the literature for teacher education. At the same time, we comment on what we see as gaps in the research literature that will need to be addressed if the field of reading education is to grapple effectively with the new literacy challenges that living in the 21st century afford.

HISTORICAL DEVELOPMENT OF CONTENT READING AND LITERATURE STUDIES

What is content area reading and how does it differ from literature studies? As a specialty area within literacy education, content area reading traces its roots to the early part of the 20th century and more specifically to the work of William S. Gray, a psychologist at the University of Chicago (Moore, Readence & Rickelman, 1983). Beginning with Gray's (1919) work on the relation of studying to reading, a long line of reading educators (e.g. Bond & Bond, 1941; Strang, 1937; Yoakam, 1928) established through their research and writing the reading profession's awareness of the need to teach students at the secondary and college levels how to comprehend and retain the content of their subject matter textbooks. Such instruction frequently involved pull-out programs designed specifically to remediate students deemed "deficient" in their use of basic reading and writing skills.

In 1970, Harold Herber published the first textbook devoted exclusively to integrated content area reading instruction at the high school level. In this text he introduced a method for teaching reading, *not* separately by a reading specialist in a pull-out program, but rather as a process that classroom teachers could use in guiding their students' comprehension of the subject matter textbooks they regularly assigned. Since the 1970s, numerous literacy teacher educators have written textbooks on integrating reading and writing with subject matter instruction at the secondary level. Following Herber's lead, a maxim underlying many of those books is this familiar phrase: "content determines process." Briefly, it refers to the fact that implicit within the content of the subject matter texts that teachers expect students to read lie the reading skills students need to comprehend the material. By giving content the upper hand over skills, Herber made it clear that he did not equate content area reading instruction with teaching reading as a separate subject.

More recently, the term content area reading has given way to adolescent literacy. Although there is still a focus on the instructional methods and strategies used in subject matter classrooms to help students grapple with their assigned readings, the field seems to be redefining itself (Alvermann, Hinchman, Moore, Phelps & Waff, 1998; Moore, Bean, Birdyshaw & Rycik, 1999). Adolescent literacy encompasses both the psychological perspectives on reading introduced during the cognitive revolution of a few decades ago and the sociocognitive and sociocultural perspectives that are prevalent today. Current methods texts on adolescent literacy tend to emphasize, along with strategy instruction, issues related to technology and teaching in culturally and linguistically diverse classrooms.

Only a few address the critical literacies outlined by the New London Group (1996). These new literacies are concerned with ways of teaching and learning that go beyond simply mining a textbook for its literal or inferential meaning. No longer willing to think of reading as primarily a psychological phenomenon – one in which individuals who can decode and have the requisite background knowledge for drawing inferences are able to arrive at the "right" interpretation of a text – the New London Group has documented how the "right" interpretation of a text rarely holds for different individuals reading in different contexts. Their work and that of others who are similarly focused on students' critical literacies has become known as the New Literacy Studies (Willinsky, 1990). Despite the changing terminology, for the purposes of this chapter, we have chosen to stay with content area reading because the literature from which we draw in synthesizing the research on instructional methods and learning activities in subject matter classrooms is historically rooted in the practice of teaching reading in the content areas.

History of Literature Study in American High Schools

Two general perspectives on best practice compete within secondary literature study. Their differences on best practice are matched by different goals and differing models of learning. One holds for the direct instruction in literary classics and their cultural significance, as well as direct instruction in the elements of good writing and literature, a position predicated on a transmission model of learning. The other view, grounded in constructivist principles, claims experience-oriented classroom practices facilitate students' development of literary judgment, critical acumen, and moral sensibility. While research indicates that actual secondary English teachers draw on both schools of thought in their classrooms, a long history of theoretical debate, both in language arts education and literary criticism, fosters perhaps unreasonable partisanship in language arts teacher education. It is therefore impossible to address the question of best practice in secondary literature study without some review of the history behind these two pedagogic perspectives that have taken turns holding sway over professional wisdom.

At the turn of the last century, secondary education was primarily for those wishing preparation for college. Thus, the literary works taught were often those expected of entering freshmen by the universities, as identified by the Committee of Ten (1894). As high schools began attracting more students who were not destined for college, and as the education profession began to come into its own as a social science, friction developed between the teaching profession and higher education over the question of appropriate curriculum (Cremins, 1961). Educators stressed the need for age-appropriate, pedagogically useful, and morally uplifting texts. Academics stressed the need for a familiarity with those classic works deemed to be of longstanding literary and intellectual value. The choice had implications for teaching practice.

Organizations such as the NCTE and the Progressive Education Association (PEA) came to champion the teacher's professional judgement in the early decades of the 20th century (Hosic, 1921). The PEA's position, populist and pragmatic, was that America, as a nation of laborers, farmers, and factory workers, needed a general, not an academic, education. They also held that appropriate teaching practice could overcome a teacher's lack of content knowledge, and that students' skill development and grasp of general principles were more important than content mastery (Cremins, 1961). For literacy instruction, this Life Adjustment approach meant developing reading and writing skills suitable for the presumed social and economic roles students would play after graduation. Reading practice using technical manuals, newspapers, Farm Bureau bulletins and so on was commonly recommended (Cremins, 1961).

The academy's counter-argument was articulated in works such as *The Higher Learning in America* (Hutchins, 1936), calling for an immersion in "the great books," *How To Read a Book* (Adler, 1940), on the rudiments of close reading, and *Liberal Education* (VanDoren, 1943), a trenchant case for why all Americans should have a familiarity with the lessons of history and the common nature of the human condition.

The debate between these two camps escalated throughout the 1940s and 1950s, and shifted to the public sphere where rhetoric often got the better of reason (e.g. Bestor, 1953; Smith, 1954; Flesch, 1955). The general thrust of these politicized attacks was that America's schools were inherently anti-intellectual and unmeritocratic and thus unsuited for the historically unprecedented challenges facing post-war and cold war America. While the furor over this debate seemed at times histrionic, the hysteria was at least in part encouraged by the seeming intransigence of some education organizations, a move which proved less than strategic (Cremins, 1961).

"New Criticism" in Secondary Literature Study

By the 1960s, the PEA had disbanded, the progressivist emphasis on process had been overturned, and a heightened regard for content knowledge, both for teachers and students, brought literary theory to literature instruction. At about the same time that Herber (1970) was calling for content to lead skills in content area reading, the College Entrance Examination Board (1965) was calling for the training of high school English teachers in literary criticism and analysis. Having previously been groomed as social engineers by professional pedagogues and psychologists, language arts instructors were now to become humanities specialists. Moving educators' professional training from independent teachers' institutes to departments of education within the universities made this shift seem feasible.

Although this is often referred to as the "new criticism" influence in language arts education, it seems to be only loosely related to the literary movement of the same name that successfully challenged the elitist hegemony of New England's academic literati in the 1930s and 1940s (Brooks, 1947; Brooks & Warren, 1938; Crane, 1952; Frye, 1957; Ransom, 1941; Wellek & Warren, 1949). What was being promoted in language arts education in the 1960s and 1970s, well after New Criticism had been superceded in the academy by alternative critical perspectives, was rather different. Indeed, from current descriptions of educational "new criticism" by scholars of language arts education (Pennfield, 1994; Squire, 1991), it would seem to be an amalgam of Structuralism, Chicago School cultural criticism, basic poetics, and formalist linguistics, wrapped in the most polemical New Critical assertions (e.g. Wimsatt & Beardsley, 1954). This content-oriented approach to secondary literature study in teacher education reigned until the 1970s, when it began to give way to more student-oriented theories and methodologies for many of the same reasons that it had been abandoned earlier in the century. But by the 1980s calls for direct instruction in the "Great Books," and a common curriculum celebrating the mainstream Western cultural tradition were once again being heard (Bloom, 1987; Hirsch, 1996), echoing the thoughts of Van Doren (1943) and Hutchins (1936) and the Committee of Ten (1894) before them.

If the quick immersion approach to making New Critics out of school teachers had been a failure, the distinction between content area reading (still informed by educational psychology) and literature study (now also informed by literary theory) has lived on (Applebee, Burroughs & Stevens, 2000;

Monseau, 1999). Today's literature education academics ground their work on the literary theories of Louise Rosenblatt (1994a, 1994b), Kenneth Burke (1966), Mikhail Bakhtin (Morson & Emerson, 1990), Judith Langer (1995), Wolfgang Iser (1993), Stanley Fish (1980), Peter Rabinowitz (Rabinowitz & Smith, 1998), and others. These scholars share a focus on the reader, although, like the New Critics before them, their individual approaches vary widely – from reception theory and textual response, to transactional and interpretive response, to authorial readings, aesthetics, and prosaics. In general, however, informing literature study by these perspectives has allowed language arts educators to return from the teaching of the text to the teaching of the student.

It should be borne in mind that to be informed by literary theory is not the same as doing literary theory. The secondary teaching of literature is more influenced by the demands of the curriculum, the constraints of the classroom, the desires of the teacher, and the needs and interests of the students and their community than by arguments among literary analysts. In addition, the sociocultural turn in education research has had an impact on literature study as well (Scholes, 1998; Stock, 1995; Wolf, 1995). It is not strictly accurate, therefore, nor pedagogically helpful, to conflate and confuse the teaching of the elements of literature with New Criticism, or general discussion with Reader Response, and to pit these two against one another in futile theoretical debate. Classroom teachers draw from both (Applebee, Burroughs & Stevens, 2000).

Comparing the Psychological Perspective with the Literary Perspective

We have attempted in the preceding sections to explain the differences between content area reading and literature study by the lights of their respective histories. Content area reading has been greatly informed by research in educational psychology. Literature study has been less informed by educational psychology than it has by research in literary theory, and on this score is of two minds. Because Brophy's principles of good teaching are informed by educational psychology, they are well suited to content area reading, but there is a less snug fit between those principles and the wisdom of practice in literature study.

We would suggest that the psychological perspective gives cognition pride of place, sees cognition as the serial solving of problems, views texts as containers of information that can be mined for answers to questions, and is therefore concerned with teaching students strategies that will allow them to do this effectively. The traditional assessment for reading ability within this perspective has been, and continues to be, the standardized reading comprehension test.

The literary perspective, by contrast, centers on experience, critical imagination, and emotional valence, sees reading literary works as the having of aesthetic experiences for evaluation and appreciation, views literary texts as a means of transactionally generating such experiences, and is therefore concerned with developing students' open mindedness, tasteful discrimination, and moral sensibility (Rosenblatt, 1994b). Assessment is more often gauged by written essays, journals, portfolios, or group projects.

To use Louise Rosenblatt's transactional terminology (1994a), literature study is as concerned with a student's aesthetic response to a text as it is to the student's efferent response. The psychological research has studied how readers access the denotations of facts invoked in a text. The literary research is concerned with this also, but, in addition, it stresses how readers evaluate the connotations and associations evoked by the experience of transacting with a text. Diamondstone and Smith (1999) have asserted that the purpose of literature study ought to be "to engage students in developing an understanding of how texts work so that they can effectively assert their textual power" (p. 194). While they go on to state that the development of such an understanding begins "by recognizing that [texts] were produced by someone for someone" (p. 194), other scholars might note that the true first step is actually experiencing a text that works.

Thus, Brophy's description of appreciation as "students valu[ing] what they are learning because they understand that there are good reasons for learning it" is entirely bound up in the psychological instrumentalist idiom. Clearly, literary appreciation, which is meant to foster the development of sensitive, ethical, evaluative skills, is not about utilitarian reasons, but about quality of life. Appreciation of literature, as it applies to living, is about the qualitative incorporation of experiences, the development of connoisseurship focused by moral vision, and thus is about being ever more richly and meaningfully, not just effectively alive.

CONNECTING READING RESEARCH AND PRACTICE TO THE PRINCIPLES OF GOOD TEACHING

Our focus here is on the research and wisdom of practice in content area reading and literature studies that support the principles of good teaching outlined in the introduction to the book. In instances where a synthesis of the quantitative research on a particular method exists, we use that information to make connections between what content reading specialists espouse and what Brophy (this book) perceives as the generic aspects of effective teaching. We also cite key studies from the qualitative research on content area reading and literature studies as a way of contextualizing the quantitative research findings. Finally, in instances where the general principles of good teaching seem not to square with the research or wisdom of practice in content area reading and literature studies, we make note of these differences and provide qualifying statements related to the principle(s) in question.

In all, we examine four methods of instruction thought to represent much of the research on content area reading and literature studies. These four methods are also common to most, if not all, of the textbooks that literacy teacher educators use in their classes. The methods include teaching students text comprehension strategies, preteaching content area vocabulary, using text-based discussions to engage students in reflecting on what they have read, and integrating literacy instruction across the curriculum – a method, interestingly enough, that is grounded more in the wisdom of practice than in the research literature per se.

Text Comprehension Strategies

The intended goal of teaching students to use text comprehension strategies is to enable them to become independent readers – the assumption being that a strategy learned is a strategy students will apply on their own. For obvious reasons, this assumption does not always hold up in day-to-day practice. Nonetheless, teaching students strategies for comprehending their textbook assignments is an instructional method that is emphasized in more than 25 published methods texts on content area reading and literature studies currently in circulation. The rationale for teaching these strategies is tied to the notion that "comprehension is an active process that requires an intentional and thoughtful interaction between the reader and the text" (National Reading Panel, 2000, p. 13). This rationale, coupled with the widespread advocacy in reading methods texts for teaching comprehension strategies, would seem well founded in the research literature (Alvermann & Moore, 1991; Bean, 2000; Pearson & Fielding, 1991; Tierney & Cunningham, 1984; Wade & Moje, 2000).

Charged by Congress in 1997 to assess the effectiveness of various instructional methods for teaching students to read, members of the National Reading Panel (2000) identified 205 studies on text comprehension instruction that met the stringent criteria the Panel had laid out prior to selecting the studies. In its review of these studies, the Panel concluded that the following seven categories of text comprehension instruction have a solid scientific basis for improving students' understanding of what they read in the academic areas, such as social studies, English language arts, mathematics, and the sciences:

- *Comprehension monitoring*, where readers learn how to be aware of their understanding of the material;
- *Cooperative learning*, where students learn reading strategies together;
- *Use of graphic and semantic organizers (including story maps)*, where readers make graphic representations of the material to assist comprehension;
- *Question answering*, where readers answer questions posed by the teacher and receive immediate feedback;
- *Question generation*, where readers ask themselves questions about various aspects of the story;
- *Story structure*, where students are taught to use the structure of the story as a means of helping them recall story content in order to answer questions about what they have read; and
- *Summarization*, where readers are taught to integrate ideas and generalize from the text information. (National Reading Panel, 2000, p. 15.)

It is not surprising that these same seven categories of effective text comprehension instruction overlap with five of the twelve principles of good teaching outlined in the introduction to this book: Principle No. 5 (coherent content achieved through the use of graphic and semantic organizers, including story maps); Principle No. 6 (thoughtful discourse in the form of question answering and question generation); Principle No. 8 (scaffolding students' engagement through comprehension monitoring); Principle No. 9 (strategy teaching in which teachers model and provide explicit instruction in general study skills such as summarization); and Principle No. 10 (cooperative learning with opportunities for students to work together in pairs or small groups on authentic comprehension tasks).

Although these seven types of instruction are known to accelerate students' reading comprehension, they tell us little if anything about the contexts in which such instruction occurs. Nor do they offer information as to how the various approaches fit within the larger curricular frameworks at the middle and high school levels. Studies that might have addressed these concerns are available. They are found largely in the qualitative literacy research literature (e.g. Bean, 2000; Ivey, 1999; Moje & O'Brien; 2000; Moore, 1996) – a body of literature excluded from consideration by the National Reading Panel because it did not meet the experimental and quasi-experimental design criteria that the Panel had specified as evidence of highly rigorous research. Highlights from the qualitative research literature on text comprehension instruction are included here, however, in an effort to provide a context for the National Reading Panel's findings.

Qualitative research on text comprehension instruction. This body of research relies on qualitative methodologies to flesh out some of the ways in which teachers are using text comprehension strategies to create developmentally responsive literacy instruction in the content areas. It is a contextually rich body of literature in that it focuses attention on how various social, cultural, and physical arrangements affect students' uses of text comprehension strategies, and, in turn, how this usage affects, or modifies, those arrangements. Typically, such arrangements involve teachers and students jointly interacting in ways that influence each other's classroom literacy practices (Prentiss, 1998; Santa Barbara Classroom Discourse Group, 1992) and the manner in which they respond to each other (Bloome, 1987).

Moore (1996) reviewed and commented on several qualitative studies that showed how the joint actions of teachers and learners created contexts for effective text comprehension instruction. For example, Mr. Ruhl, a high school biology teacher in a qualitative cross-case analysis by Dillon, O'Brien, Moje, and Stewart (1994), used his understanding of science and his goals as a teacher to co-construct with his students a learning context in which text comprehension played a major role. In summarizing this study, Moore (1996) concentrated on a particular comprehension strategy known in reading circles as the study guide. Below, we quote from Moore's summary in an effort to show how qualitative research can flesh out the quantitative research findings on text comprehension instruction:

> Mr. Ruhl believed there were sets of science-related information that all informed citizens needed to know. Along with grasping basic scientific knowledge, students needed to understand how science affected their lives and how informed decisions could be based on this understanding. Mr. Ruhl also believed in the personal value of each individual. He saw his role as a biology teacher to facilitate each student's acquisition of domains of knowledge. He believed that mastery learning systems provided the opportunity for every student to learn successfully.
>
> Mr. Ruhl produced study guides for students to complete. The guides contained learning objectives, key vocabulary, directions for reading and responding to print materials, diagrams that clarified difficult concepts, and page number references to designated information. These guides were meant to mediate reading difficulties and focus students on important concepts. Students worked on the guides daily in cooperative groups, reading, writing, discussing, and consulting with their teacher (Moore, 1996, p. 29).

In another qualitative study of a high school biology teacher, Thomas (1999) focused almost exclusively on how the teacher and students socially constructed a classroom learning environment in which students' voices became part of both process and content. According to Thomas,

> Student voices were heard in relation both to classroom process and scientific content. Students gave the teacher feedback about ways the classroom process could be most

> helpful to their learning. Students asked content questions to clarify confusion and find links to prior knowledge and experiences. Students used talk as a tool for modifying knowledge in light of new ideas and experiences. In addition, many teacher questions went beyond low-level factual questions and were designed to encourage students to think through their answers.
>
> The extended use of cooperative learning groups in this class promoted student-to-student talk, some involving "talking science" (Lemke, 1990) – using scientific vocabulary, assisting each other in finding scientific information or understanding procedures for scientific labs, and formulating and trying out scientific questions and hypotheses. Cooperative learning groups also allowed most students to find their own best method for learning, whether that meant working in groups, working alone, or sharing answers after completing work (Thomas, 1999, p. 326).

In addition to fleshing out the quantitative findings on comprehension monitoring, scaffolding information through study guides, cooperative learning groups, question answering, and question generation, these two qualitative studies lend further support for several of the principles of good teaching outlined in the introductory chapter to this book. They also point to the encompassing nature of text comprehension instruction, which includes aspects of vocabulary instruction, classroom discussion, and integrated literacy instruction – all of which are treated separately here for the sake of organization and emphasis but which occur naturally and in indivisible ways in actual classroom practice.

Some qualifying statements about the principles. For reasons mentioned earlier in this chapter, comprehension of literary works is not thought of in the same way that comprehension is understood elsewhere in reading. In early reading generally, and in later reading for informational content, comprehension involves surface features of the text, or the information it contains. As a result, comprehension studies focus on word or concept identification, word calling, and so forth, and are monitored by verbal protocols and miscue analysis (in a sense, the basis for question answering and generation). But in later reading, particularly of literary works, the meaning of the text is something that is said to either emerge from a textual transaction (Rosenblatt, 1994b), or is crafted by the reader (Beach, 1993). It can often be ambiguous, ironic, or otherwise beyond the grasp of explicitly taught strategies for simplification and summarization (Brooks, 1947). That summation may be best that complexifies the text most (Fish, 1980). The meaning of a literary text may be specific to the specific reader encountering it, or to a particular reading (depending on which theoretical frame you wish to bring to bear).

Moreover, what a reader comprehends of a literary work is a personal matter. It is the personal element in reading response that can make reading pleasurable and motivate further reading. Preempting a student's personal response to a text with the explicit instruction in what a poem means, or what an author is trying to say, or the significance of a particular trope or leitmotif, is poor pedagogy, according to the general wisdom of practice in literature study (Beach, 1993; Foster, 1994). Finally, the sort of knowledge that emrges from the integration of literary experiences is larger than the sum of the parts would indicate (Applebee, 1996; Scholes, 1998). Therefore, assessing the parts is not likely to be an accurate quantification of the qualitative whole.

Preteaching Content Area Vocabulary

Teachers who preteach vocabulary necessary for understanding the content of their subject matter areas do so with this intended goal in mind: that students will recognize a new term when they see it in print and will attach the appropriate meaning to it. The importance of vocabulary knowledge to subject matter reading has been recognized since the 1920s (Whipple, 1925). Along with this recognition has come the generally accepted belief in the need for both direct and indirect instruction in content area vocabulary (Herber, 1970, 1978). The rationale for such instruction is that school reading materials often present many unfamiliar words, which if left untaught will cause students to experience difficulty in comprehending the materials. Or, if familiar, the words may carry special meanings. Depending on the content area in which the word *table* is used, meanings will vary; for example, one might encounter a water table in a geography text, a periodic table in a chemistry text, or a multiplication table in a math text.

In an effort to determine the best approach for teaching vocabulary, the National Reading Panel (2000) evaluated 50 studies that met their design criteria. Within those 50, the Panel identified 21 different approaches. Due to the relatively large number of variables represented in the small number of studies evaluated, the Panel could not conduct a formal meta-analysis of the results of these studies. Therefore, the information that we present here represents what the Panel called trends across studies.

Basically, the National Reading Panel (2000) found that preteaching vocabulary in a variety of subject matter areas does lead to improved text comprehension, with computer-assisted instruction edging out traditional methods of instruction in a few cases. Although students' vocabulary knowledge can be enhanced incidentally through reading or listening to others read, the Panel's findings suggest that teachers will find it advantageous to structure tasks and learning activities so that they provide students with multiple exposures to the same word but in different contexts. In addition, there was some evidence to suggest that teachers should consider substituting easy words for more difficult words when planning learning activities that involve low-achieving students.

Similarly, in a review of the literature that focused solely on classroom-based applications of the research on vocabulary instruction, Blachowicz and Fisher (2000) found support for focusing on teaching and learning activities that support students' word growth in a variety of contexts. They concluded from their analysis of the last two decades of research that four main principles ought to guide teachers' vocabulary instruction. These principles are:

- That students should be active in developing their understanding of words and ways to learn them.
- That students should personalize word learning.
- That students should be immersed in words.
- That students should build on multiple sources of information to learn words through repeated exposures (Blachowicz & Fisher, 2000, p. 504).

Although these principles support the generic aspects of Brophy's (this volume) fourth principle of good teaching – establishing learning orientations – they do not spell out in any detail what the preferred strategies for teaching vocabulary in the content areas might be. Instead, they point vaguely to what the members of the National Reading Panel (2000) concluded: namely, that while much is known about the importance of student involvement and vocabulary knowledge in accelerating reading achievement, the research says little about the best instructional approaches or combinations of approaches teachers should use with their classes. This conclusion would seem to corroborate the literacy field's growing awareness of the futility in looking for the one best "fix" or combination of "fixes" when it comes to literacy instruction at the middle and high school levels.

Qualitative research on pretaching vocabulary. In reviewing the qualitative research literature on vocabulary instruction in the middle and high school grades, one thing becomes immediately clear. Evidence that supports such instruction comes mostly from observational studies and surveys of teachers' beliefs regarding the importance of preteaching vocabulary (Blachowicz & Fisher, 2000), or from content analyses of practitioner journals that regularly publish articles proclaiming the central role of vocabulary instruction in subject matter learning (Alvermann & Swafford, 1989; Swafford & Hague, 1987). The number of studies on vocabulary instruction published in rigorously reviewed research journals and yearbooks actually declined in the last two decades (Dunston, Headley, Schenk, Ridgeway & Gambrell, 1998), or remained stable but accounted for a relatively small proportion of the studies published (Guzzetti, Anders & Neuman, 1999).

In selecting a qualitative research study to feature in this section of the chapter, we were partial to one that addressed a gap in the National Reading Panel's (2000) analysis of effective vocabulary instruction. That is, we chose to highlight a case study (Harmon, 1998) in which four middle grades students targeted unfamiliar words during self-selected reading materials. Although this research does not qualify strictly as a preteaching vocabulary study, our rationale for including it here is that it offers some valuable insights into how adolescents incidentally learn vocabulary through independent reading – an area inadequately explored in the National Reading Panel's report.

Working from the perspective that little is known about the strategies learners use to construct meanings for unfamiliar words while reading independently from self-selected materials, Harmon (1998) collected verbal protocols (think-alouds) for a period of six months from four middle grades students enrolled in a large suburban district with a predominantly English and Spanish-speaking population. The four readers (two average, one proficient, and one struggling) represented the range of ethnicities present in the school. These students were recommended by their seventh-grade teacher, who used a literature-based approach to teach reading.

As a participant observer in the study, Harmon sat with individual students as they read silently from the books they had chosen. When students came to an unfamiliar word, they were asked to talk about how they would figure out its meaning. Harmon offered general probes, such as "What makes you say that?" and "Please tell me more." For example, in the following transcript from a think-aloud involving the word *ornate* from a passage in *Roll of Thunder, Hear My Cry* (Taylor, 1976), Harmon asked Lynette, a proficient reader, to explain why she initially thought the word meant "delicate." The passage appears first, followed by the think-aloud:

> The furniture, a mixture of Logan-crafted walnut and oak, included a walnut bed whose ornate headboard rose halfway up the wall toward the high ceiling, a grand chiffonier with a floor-length mirror, a large rolltop desk which had once been Grandpa's but now belonged to Mama, and the four oak chairs, two of them rockers, which Grandpa had made for big Ma as a wedding present (p. 36).

Lynette: The girl was explaining her parents' bedroom and . . . she's saying it "included a walnut bed whose ornate . . ." that's the word ". . . headboard." Maybe it's like a type of wood or . . . but they've already explained the wood so maybe it's like . . . OK . . . it says "the headboard rose halfway up the wall toward the high ceiling . . . ." Maybe like . . . delicate . . .

Researcher: What makes you think of delicate?

Lynette: I don't know. It's like made out of wood and it's pretty old. And like . . . maybe crafted because it's like when things are made out of wood that somebody like had to make it so like it's a craft. I don't think it has anything to do with size because then she [author] goes on to explain the size. So I don't think . . . she'd [author] do that twice.

Researcher: So right now you would say "crafted" or "delicate."

Lynette: Yeah or like . . . detailed maybe.

Researcher: What made you say detailed?

Lynette: Well, like on some headboards there's all these designs on it. Like wood headboards at furniture stores or whatever. So it's like maybe it's very detailed with . . . um . . . I don't know just the wood is like cut out to be, you know, detailed and stuff (Harmon, 1998, pp. 566–567).

In addition to collecting the think-aloud protocols, Harmon interviewed each student on several occasions to ascertain their attitudes toward vocabulary teaching and learning as well as reading in general. A final round of interviews focused on students' perceptions of what she [Harmon] was observing about their strategies for unlocking the meanings of unfamiliar words. She also interviewed the students' seventh-grade teacher to learn more about her approach to vocabulary instruction. Following common qualitative research design procedures, Harmon collected and analyzed the data concurrently as she engaged in inductive, individual and cross-case comparisons.

Among the many findings from the observational portion of this study, three in particular stand out. First, regardless of reader type, all four students used similar strategies to construct meanings of unfamiliar words. These strategies included the use of context clues, structural or word-level analysis, and outside help (e.g. asking someone or going to the dictionary) to determine a word's meaning. Second, whether or not they were able to generate accurate meanings for unfamiliar words, all students used multiple strategies during single encounters with a word and accepted the definitions they worked out for themselves, claiming that their comprehension was still intact. Third, all of the student participants indicated they believed that learning new words was critical to becoming good readers, and all but one (the struggling reader) used contextual analysis as an initial strategy for unlocking meaning.

Finally, Harmon's (1998) analysis of the interview data revealed that there was general agreement among the four focal students that "teachers help them learn new vocabulary by highlighting and talking about [the words] before a reading assignment" (pp. 588–589). They also agreed that during sustained silent reading periods, it was inappropriate to seek outside help when they did not understand a word; interestingly, this included not using a dictionary. The implications of this research for content area reading and literature studies teachers would seem considerable, if for no other reason than the insider perspective this study provides is crucial to understanding how middle school students use what they know to construct meaning from unfamiliar words.

Some qualifying statements about the principles. The strategy of pre-teaching vocabulary is a common and apparently useful one, but there are response-oriented scholars who would prefer reliance on learning vocabulary through reading, not separately beforehand. The research does not privilege either approach (Ruddell, 1994), and a more balanced view is one that embeds new vocabulary learning in prereading activities, tapping and crafting, by turns, prior knowledge of historical period, genre, theme, and so on (Beck, 1998; Foster, 1994; Lorenz, 1998). In this way, the student's constructed comprehension of the vocabulary is embedded within a richly interconnected web of signification, not merely inserted into the student's head by the teacher. For instance, the teaching of Romeo and Juliet may be preceded by a week's worth of study in Elizabethan history and language, sonnet forms, thinking activities built around themes such as star-crossed love, gang rivalries, and obedience/disobedience to parents, engaging in other media that might familiarize, or bolster the student's pertinent schemata – in a sense, "tilling" (Block, 1999) the student's mind in preparation for planting the vocabulary of a crafted comprehension.

Using Text-Based Discussions

Teachers no doubt have several goals in mind when they plan their instruction around text-based discussions. One such goal is to enrich and refine students' understandings of what they have read. Another goal is to guide students' higher order thinking abilities; without discussions that take students to increasingly higher levels of understanding of the ideas presented in texts, it is likely that many would remain at the literal level – that is, recalling and regurgitating facts. Yet another goal of text-based discussions is to provide opportunities for students who do not speak English as their first language to hear ideas explained by their peers rather than only through the language of their teachers. The rationale behind class discussion is based largely on the work of Vygotsky (1978), who believed that "mental functioning in the individual originates in social, communicative processes" (Wertsch, 1991, p. 13). In other words, a Vyygotskian perspective on discussion does not assume that students will learn independently from their textbooks, but rather, that they will benefit from having someone more knowledgeable than themselves guide their learning.

In the past two decades, the field of reading education has awakened to the facilitative effect of text-based classroom discussions on students' understanding of subject matter material. Practically speaking, this awakening was sparked by the work Bridges (1979) and Dillon (1988) had been doing in England and the United States to foster interest in the instructional possibilities of discussion. Contrasting discussion with the more traditional recitation pattern of classroom interaction in which the teacher asks a question, the student answers, and the teacher evaluates the student's response, Bridges (1979) noted that ideally discussions are give-and-take exchanges between teachers and students for the express purpose of enriching and refining students' understanding of subject matter material. Dillon (1988, 1994) wrote prolifically on the topic of classroom discussion and included practical suggestions for engaging students who had been schooled in the questioning and answering techniques of old in more informal dialogue.

To our knowledge, no large-scale analyses have been conducted to determine the effectiveness of text-based classroom discussions on older students' subject matter learning. However, there are any number of smaller studies in the sociolinguistic literature that look at the effects of classroom oral language on literacy learning (for an overview of these studies see Wilkinson & Silliman, 2000). Most of the early 1970s sociolinguistic research on classroom language was conducted for the purpose of analyzing the communicative competence of young children and the complex demands of classroom communications involving both teachers and students at the elementary level. Although this research was informative and led to speculations about the nature of classroom discussions in the upper grades, by and large the early descriptive studies were limited in what they had to say about text-based classroom discussions, especially in subject matter areas.

Nonetheless, the assumptions that came out of this early work in sociolinguistics did provide a starting point for the work with older populations of students. For example, researchers interested in content area reading and literature studies at the middle and high school levels (Alvermann & Hayes, 1989; Alvermann et al., 1996; Dillon & Moje, 1998; Finders, 1997; Guzzetti, 1996; Hinchman & Zalewski, 1996; Langer, 1995) based their studies in part on these common assumptions from the sociolinguistic literature:

- Learning is a *social activity* – interpersonal behaviors, both observed and enacted in the classroom, are the basis for new conceptual understandings in cognition and communication.
- Learning is *integrated* – strong interrelationships exist between oral and written language learning.
- Learning requires *active student engagement in classroom activities and interaction* – engaged readers are motivated for literacy learning and have the best chance of achieving full communicative competence across the broad spectrum of language and literacy skills (Wilkinson & Silliman, 2000, pp. 337–338).

The appeal of text-based classroom discussion as a method of engaging adolescents in content area reading and literature studies can be seen in the ever increasing number of professional texts on the subject. For example, educators have the option of choosing from among a wide assortment of professional texts that range in scope and content. There are first-hand accounts of using discussion to engage unmotivated adolescents (Allen, 1995; Finders, 1997), edited collections that describe multidisciplinary teaching with discussion (Dillon, 1988), as well as full-blown descriptions of ways to adapt special reading programs, such as the Book Club, to various students' needs (McMahon, Raphael, Goatley & Pardo, 1997). A whole genre of "how-to" books have sprung up around using text-based discussions in content area classrooms (Dillon, 1994; Hill, 1977) and the art of facilitating "grand conversations" in literature study groups (Peterson & Eeds, 1990).

A common aspect of all these texts is the appeal they have for teachers who value the supportive kind of classroom learning environment that Brophy (this volume) describes in his first principle of good teaching. As well, Brophy's seventh principle of good teaching – the one that he refers to as practice and application activities – would seem to be reflected in this literature on text-based discussion, for it is through teachers' use of discussion activities that students are motivated to engage in subject matter learning. It is through these same kinds of activities that dialogue develops around the content – dialogue that is rich in students' shared background experiences as well as their cultural and linguistic differences. In the qualitative research studies that are highlighted next, we attempt to show how such dialogue works to create a supportive and literate classroom environment in which students find ample opportunities to practice and apply what they are learning from their content area texts.

Qualitative research on using text-based discussions. An important and useful insight to come out of the qualitative research literature on text-based discussions at the middle and high school levels is the marked contrast in what counts as discussion within various classroom activities. As Hinchman and Zalewski's (1996) work carefully documents, there are four main types of classroom activities involving talk of one kind or another within the conventional transmission approach to teaching and learning in subject matter areas. These activities include lectures, question-and-answering sessions (sometimes called recitations), small group discussions, and assessment-related talk.

Although theoretically these categories of classroom talk would seem to indicate fairly distinctive interaction styles, in truth, they often blend in curious ways in the everyday world of teaching content area reading and literature studies. For example, in one study of text-based discussions involving five teachers in a grade 7–12 rural, comprehensive secondary school (Alvermann & Hayes, 1989), the eighth-grade general science teacher counted lecturing as a form of discussion because she "discussed" the previous night's textbook assignment as she lectured on the text's main points. For other teachers in that study and a subsequent one conducted with a similar group of teachers (Alvermann, O'Brien & Dillon, 1990), the notion of text-based discussion spanned a continuum of classroom talk, ranging from whole-class, open-ended discussions to small-group literature study discussions to question-and-answer (recitation-like) drill sessions, which purportedly prepared students for their end-of-unit and end-of-year assessments.

Unlike discussions held by teachers with a transmission approach to teaching content, classroom talk about texts was more student-centered in classes where teachers saw themselves as adhering to more participatory-like instructional approaches. For example, Wade and Moje (2000) found in their synthesis of the research on the role of text in classroom learning that what counted as discussion within participatory teachers' classes more often than not included students working cooperatively in peer-led discussion groups, reading/writing workshops, and project-based pedagogies (Garner & Gillingham, 1996; Goldman, 1997).

However appealing the participatory approach might be, it is important to consider the tradeoffs when using one or the other of the two approaches to teaching and learning. For as Wade and Moje noted, "teachers who have adopted literature-based, participatory approaches have found that they and their students struggle to negotiate a balance between a focus on skills and strategies and a focus on personal response to the literature" (p. 618). A similar type of struggle can erupt in so-called transmission classrooms when students' perceptions of worthwhile discussions do not match their teachers' perceptions. Examples of this struggle were observed in two of the multicase cross-site analyses of middle and high school students' perceptions of content reading discussions (Alvermann et al., 1996). Wade and Moje's (2000) characterization of those struggles follows:

> Studying the oral texts generated by students in peer-led discussions, Alvermann et al. (1996) found that talk was likely to be unrelated to the topic when students found the topic to be boring and meaningless or when the task was unchallenging and did not require debate or collaboration. For example, in one of the research sites – a 10th-grade college preparatory class in which the curriculum and final examination were controlled by state requirements – students divided the task up among group members to complete independently with minimal talk except to call out answers to other group members, even though students reported that they would have preferred meaningful discussions. A similar

> situation occurred in an 8th-grade language arts class in which discussion tasks consisted of questions and activities from the published resource book accompanying the literature textbook. Students in this class also rarely complied with the teacher's instructions to discuss tasks in groups because collaboration was not necessary. Both classrooms are interesting examples of how elements of participatory models, such as peer-led discussions that simply are added onto transmission approaches, are transformed by students to conform to the larger classroom context (p. 615).

What is important to keep in mind is that regardless of teachers' approaches – whether they be transmission-like, participatory, or some blend of the two – the fact remains that it is *students'* perceptions of the types of teaching and learning that go on in the name of text-based discussions that account for how supported they feel and how willing they are to participate in activities designed to engage them in practicing and applying subject matter concepts. Keeping this in mind, it is easy to see that the qualitative research literature on text-based discussions, while generally supporting Brophy's (this volume) first and seventh principles of good teaching, does not do so unqualifiedly.

Some qualifying statements about the principles. The use of text-based discussion in literature study can take many forms, from rote, question and answer exchanges straight from the teacher's edition of the anthology, to free-form small group literature discussion circles. But the whole language influence on literature study has emphasized employing all modes of language as pertinent pedagogical venues. This reliance on discussion has developed in sophistication with the sociological and sociocultural turn in educational research (Applebee, 1998; Fecho, 2000; Foster, 1994; Scholes, 1998; Stock, 1995; Wolf, 1995). Whether classroom discussions are coherently structured into larger curricular themes, in keeping with Brophy's third principle, is still another matter (Applebee, Burroughs & Stevens, 2000). Current research on discussion in literature study falls into one of two categories: observation of discourse patterns in classrooms, and the propounding of theoretical frameworks for organizing discussion. There seems to be no empirically driven comparative research on best practice.

Integrated Literacy Instruction

The intended goal of integrated literacy instruction is to teach the English language arts (reading, writing, speaking, and listening) functionally. For example, these skills are taught functionally when students learn to use them as a means of understanding the content of their subject matter texts (Herber, 1978). The rationale for this method of instruction dates back to Parker and Rubin's (1966) notion that content determines process. That is, subject matter material is central and should drive literacy instruction. In reality, of course, not all teachers follow this principle; in the lower grades, the situation is often reversed such that language processing is central and the choice of reading materials is only incidental. Where this reversal is seen most frequently at the middle and high school levels is in special pull-out programs where reading specialists work one-on-one with students who are not able to keep up with the regular content load.

More than a decade ago, Venezky (1987), in his history of reading instruction in the United States, cautioned that if we accept the fact that schools are vulnerable to external pressures, then we must also acknowledge the double vulnerability of reading instruction. In his words, "No other component of the curriculum has been subjected throughout its history to such intense controversy over both its basic methods and its content" (Venezky, 1987, p. 159). We believe that this cautionary note, reiterated by Gaffney and Anderson (2000) in their review of trends in reading research over the past three decades, has some bearing on why integrated literacy instruction is a method highly regarded by some educators in content area reading and literature studies, all but dismissed by others, and rarely studied in any systematic way by still others.

Paradoxically, if one were to examine the professional literature on integrated literacy instruction – works such as Nancie Atwell's (1998) *In the Middle*, Janet Allen's (1995) *It's Never Too Late*, and Susan Hynds's (1997) *On the Brink* – one might come away with the mistaken impression that any right-minded secondary level teacher would surely rely to some extent on this method of instruction. However, that is not the case, and apparently for reasons well grounded in the history of integrated literacy instruction.

According to Gavelek, Raphael, Biondo and Wang (2000), the history of this method can be traced as far back as 1894 to the National Education Association's Committee of Ten and its finding that high school students were underprepared in language skills needed for interdisciplinary study and entry into college. From there, the nature of integrated literacy instruction assumed several guises, although never entirely losing its interdisciplinary focus. For example, there were calls for integrated literacy instruction during the progressive education movement beginning in the early 1900s, during the Eight Year Study of the 1930s, the open school movement of the 1970s, and again during the whole language movement and new school reform efforts – e.g. Sizer's (1984) Coalition of Essential Schools – in the 1980s and 1990s. What is significant about the long and convoluted history of integrated literacy instruction is that the definition of this method is imprecise, hence making it practically impossible to attempt any systematic study of its effectiveness. As Gavelek et al. (2000) deftly pointed out,

> Although integrated approaches have a long history, those supporting them have not clearly delineated the construct. Integrated curricula are often based in life experiences, but it is not clear whether integrating experiences should be the basis for exploring curriculum content, or if the content itself should be presented as an integrated fait accompli. Across the decades, our field has confounded these two orientations (p. 589).

In an attempt to bring some clarity to the situation, Gavelek et al. (2000) overviewed the literature on integrated literacy instruction and identified three categories that composed their conceptual map of what this method entails: namely, integrated language arts, integrated curriculum, and integration of learning experiences in and out of school. The integrated language arts category focuses on instruction that organizes around either language processes (e.g. reading, listening, speaking, writing) or a literary selection. In the first case, language processes are central and the choice of reading materials tends to be incidental; in the second case, the situation is reversed – literary texts drive language instruction. The category called integrated curriculum refers to interdisciplinary instruction across content areas, and the last category, integration in and out of school, references the need to merge classroom experiences with the learning experiences students encounter in their homes and larger communities.

Research on integrated literacy instruction. After completing their review of the literature on integrated literacy instruction, Gavelek and his colleagues (2000) noted their surprise at finding so few data-driven articles on the topic, at least compared to the large and sprawling professional literature on integrated literacy instruction. So few in fact, that they were left wondering if "any push toward integration of any kind might be premature or even ill-founded" (p. 604) at this time. However, on further consideration, they concluded that albeit small, the empirical research on integrated literacy instruction did offer some interesting ways of rethinking the school curriculum. On this basis, they recommended not only further research but also a concerted effort among researchers to develop a strong, conceptual framework that will help the literacy field unpack the messiness of the construct called integrated literacy instruction. We note with interest here that if Gavelek et al. (2000) found the research on integrated literacy instruction lacking, we must admit to having found it doubly so. For of the 25 or more data-driven pieces included in their overview, only two (Guzzetti, Kowalinski & McGowan; Raphael & Brock, 1997) pertained to instruction at the middle and high school levels.

Some qualifying statements about the principles. In its simplest form, integrated literature study combines speaking, listening, and discussion skills (see above section) with reading and writing. The logic of this integration is as follows: The more students read, the more exposure they have to the tropes, literary devices, rhetorical structures, narrative organizations, and so forth, employed by effective authors. Students can then attempt these in their own writing. The more students write, the more opportunity they have to develop a facility with these elements of literature, and the greater their appreciation of these elements in meeting the linguistic challenges of effective communication when they come across them in their reading. Thus, broad reading makes for better writing, and motivated writing makes for better reading (Atwell, 1987; Foster, 1994).

Far from taking time away from the study of "great literature," such an integrated approach teaches students how to recognize "greatness" in a text, and to effectively and confidently express their informed appraisal. Integrated literature study therefore confirms Brophy's second and third principles of good teaching: opportunity to learn, the maximization of time devoted to curricular activities, and curricular alignment, a cohesive program for accomplishing instructional goals. However, because such informed responses are personal and non-objective, they cannot be adequately assessed by standardized reading comprehension tests, which stand as a violation of Brophy's eleventh principle, goal-oriented assessment.

IMPLICATIONS FOR TEACHER EDUCATION AND FURTHER RESEARCH

A pervasive and legitimate concern of middle and high school content area teachers is how to help students learn from texts. In the English language arts curriculum that concern is broadened to include an emphasis on how to help readers evaluate the connotations and associations evoked by the experience of transacting with texts. In all areas of the curriculum, the goal is to support adolescents' literacy growth by providing them with access to materials they can and want to read. This support is channeled through instructional strategies that build students' skill and desire to read as well as through teachers' modeling of these strategies across the curriculum (Moore, Bean, Birdyshaw & Rycik, 1999).

Based on the research and wisdom of practice reviewed in this chapter on content area reading and literature study, we found considerable though not total support for a number of Brophy's (this volume) principles of good teaching. In earlier sections of this chapter, when Brophy's principles and our field's research (or wisdom of practice) on text comprehension, vocabulary, discussion, and integrated literacy instruction parted company, we noted the differences and offered a few caveats in the spirit of calling attention to the different psychological and literary perspectives that inform the field. Next, however, unlike previously, we offer our own personal assessment of this body of research in an effort to highlight what we think it says (or does not say) to practitioners and researchers in the area of content area reading and literature studies.

Implications for Practitioners

A number of instructional strategies for teaching students to comprehend texts have been validated in the research literature. These include comprehension monitoring, cooperative learning, using graphic and semantic organizers, answering and generating questions, using story structure, and summarizing. Although some of these apply more to content area reading instruction than to literature study, all are known to improve students' comprehension.

What we do not know, however, is how these strategies play out when used with young adolescents from a wide range of cultures and linguistic backgrounds. Although up-to-date reviews of the literature exist on second language reading (Bernhardt, 2000; Garcia, 2000), they provide only meager information on how to teach the adolescent bilingual reader. Such information is completely missing in the report of the National Reading Panel (2000). As a result, we are left to wonder about the instructional implications for teaching reading to an ever-increasing number of second language learners in this country's schools.

The report of the National Reading Panel (2000) was equally silent on teachers' beliefs and understandings regarding the usefulness of transactional strategy instruction and the direct explanation model of instruction. This seems problematic to us. Knowing that the Panel found such instruction to be effective is only part of the story. The other part – knowing how teachers' belief systems interact with their implementation (or non-implementation) of effective instructional models – is an important but missing part of the equation.

A further limitation of the National Reading Panel's (2000) report is its restricted view of the reading process. For example, six of the seven approaches that the Panel concluded had a solid research backing in the cognitive literature are representative of the methods teachers would use if they believe reading comprehension instruction consists of teaching strategies that enable individual students to work by themselves in extracting information from printed texts. As Wade and Moje (2000) have pointed out elsewhere, this rather narrow view of the reading comprehension process risks "disenfranchising large groups of students for whom print texts are not paramount because they hold different social or cultural values" (p. 623). Moreover, as Wade and Moje noted, this view of the comprehension process "privileges the learning and textual practices of some students and devalues the practices of others" (p. 623).

In similar fashion, the research on vocabulary instruction speaks to a narrow range of teacher concerns, with most of it focusing on methods for teaching students to define or recognize specific words in a given context. This rather restricted view of vocabulary instruction has prompted some scholars in the field of reading education to question the value of generalizing from a synthesized body of knowledge that is linked to a specific objective, such as pre-teaching vocabulary words in a specific textbook chapter or passage. For example, echoing Graves and Prenn's (1986) admonition that there is no one best way to teach vocabulary, Baumann, Kame'enui and Ash (in press) have argued the necessity of aligning one's instructional objectives with the instructional means to achieving those objectives:

> [That is], if one's objective were to teach the meanings of a relatively few specific words in a content subject like science, the least costly approach might be to use a definitional method. However, if one wished to teach meanings for many words, or if the goal were to enhance passage comprehension, another method, perhaps a semantic relatedness procedure, would be preferred. If one's goal were long-term, expansive, independent vocabulary learning, regular independent reading combined with instruction in the use of contextual and morphemic analysis would be the logical approach. In short, the simplicity of Graves and Prenn's statement, "there is no one best method of teaching words," should not mask its importance (Baumann, Kame'enui & Ash, in press).

Implications for Future Research

Researchers working within both the quantitative and qualitative paradigms have much work to do if they are to address adequately the issues that literacy educators in the middle and high school grades are grappling with on a daily basis. Although large numbers of studies exist on how to teach reading comprehension, only a few select topics within this domain were included in the rigorous meta-analyses that the National Reading Panel (2000) recently conducted. Among those topics that the Panel did address, questions still remain as to the applicability of certain findings for teaching middle and high school students who speak a language other than English as their primary language. Partial or provisional answers to some of those questions might be forthcoming, however, if the findings from qualitative studies on second language reading and literacy instruction were to be analyzed through cross-case comparisons. In turn, such comparisons might inform new experimental or quasi-experimental research projects designed to address questions that arise from the more in-depth and close-up qualitative work.

Given the lack of large-scale studies of best teaching practice for literature instruction, it might be tempting to suggest such studies be done. It could be argued for instance that the results of these studies, if meta-analyzed, might point to patterns of instructional effectiveness or perhaps a single, promising approach. The downside of such an argument, of course, is that meta-analytic techniques often find such studies incomparable due to a number of reasons, including the wide-ranging nature of the variables and theoretical positions that make up their design (National Reading Panel, 2000). Moreover, it is unlikely that any one approach would work best with all teachers, all students, all curricular demands, or in all school settings. To give a ludicrously simplistic example, if 60% of teachers were found to teach literature best using method A, and 40% taught best using method B, would it make sense to claim that method A is better? And if so, should all teachers be forced to use method A (or be forced out of their position)? And how do we determine what is best in literature study when the standard quantitative measures do not even assess the experience most central to literature study, the experience of being engaged by a text?

Finally, questions concerning the degree to which the available knowledge base on text comprehension, vocabulary instruction, discussion, and integrated literacy instruction is being translated into practice remain largely unanswered. Studies are needed that both quantitatively and qualitatively investigate what characterizes a school in which teachers, administrators, and supervisory personnel actively engage in applying relevant findings from the available knowledge base to their school's curriculum, and, in particular, to literature study and teaching reading in the content areas. A major focus of any such inquiry should be on how well, if at all, the research on second-language learners' reading development and instructional needs is being implemented. To continue to ignore this gap in the literature makes absolutely no sense given the diversity of languages spoken in today's schools.

REFERENCES

Adler, M. J. (1940). *How to read a book: the art of getting a liberal education.* New York: Simon and Schuster.

Alexander, P. A. (2000). Toward a model of academic development: Schooling and the acquisition of knowledge. *Educational Researcher, 29*(2), 28–33, 44.

Allen, J. (1995). *It's never too late: Leading adolescents to lifelong literacy.* Portsmouth, NH: Heinemann.

Alvermann, D. E., & Hayes, D. A. (1989). Classroom discussion of content area reading assignments: An intervention study. *Reading Research Quarterly, 24*, 305–335.

Alvermann, D. E., Hinchman, K. A., Moore, D. W., Phelps, S. F., & Waff, D. R. (Eds) (1998). *Reconceptualizing the literacies in adolescents' lives.* Mahwah, NJ: Lawrence Erlbaum Associates.

Alvermann, D. E., & Moore, D. W. (1991). Secondary school reading. In: R. Barr, M. L. Kamil, P. Mosenthal & P. D. Pearson (Eds), *Handbook of Reading Research* (Vol. II, pp. 951–983). New York: Longman.

Alvermann, D. E., O'Brien, D. G., & Dillon, D. R. (1990). What teachers do when they say they're having discussions of content reading assignments: A qualitative analysis. *Reading Research Quarterly, 24*, 296–322,

Alvermann, D. E., & Swafford, J. (1989). Do content area strategies have a research base? *Journal of Reading, 32*, 388–394.

Alvermann, D. E., Young, J. P., Weaver, D., Hinchman, K. A., Moore, D. W., Phelps, S. F., Thrash, E. C., & Zalewski, P. (1996). Middle and high school students' perceptions of how they experience text-based discussions: A multicase study. *Reading Research Quarterly, 31*, 244–267.

Applebee, A. N. (1996). *Curriculum as conversation*. Chicago: University of Chicago Press.

Applebee, A. N., Burroughs, R., & Stevens, A. S. (2000). Creating continuity and coherence in high school literature curricula. *Research in the Teaching of English, 34*, 396–429.

Atwell, N. (1998). *In the middle: New understandings about writing, reading, and learning* (2nd ed.). Portsmouth, NH: Boynton/Cook.

Baumann, J. F., Kame'enui, E. J., & Ash, G. E. (in press). Research on vocabulary instruction: Voltaire Redux. In: J. Flood, J. M. Jensen, D. Lapp & J. R. Squire (Eds), *Handbook of Research on Teaching the English Language Arts* (2nd ed.). New York: Macmillan.

Beach, R. (1993). *A teacher's introduction to reader-response theories*. Urbana, IL: National Council of Teachers of English.

Bean, T. W. (2000). Reading in the content areas: Social constructivist dimensions. In: M. L. Kamil, P. B. Mosenthal, P. D. Pearson & R. Barr (Eds), *Handbook of Reading Research* (Vol. III, pp. 629–644). Mahwah, NJ: Lawrence Erlbaum Associates.

Beck, C. R. (1998). The poet's inner circle: Gaming strategies based on famous quotations. *English Journal, 87*(3), 37–44.

Bestor, A. E. (1953). *Educational wastelands: the retreat from learning in our public schools*. Urbana, IL: University of Illinois Press.

Blachowicz, C. L. Z., & Fisher, P. (2000). Vocabulary instruction. In: M. L. Kamil, P. B. Mosenthal, P. D. Pearson & R. Barr (Eds), *Handbook of Reading Research* (Vol. III, pp. 503–523). Mahwah, NJ: Lawrence Erlbaum Associates.

Block, C. C. (1999). Comprehension: Crafting understanding. In: L. B. Gambrell, L. M. Morrow, S. B. Neuman & M. Pressley (Eds), *Best Practices in Literacy Instruction* (pp. 98–118). New York: Guilford Press.

Bloom, A. D. (1987). *The closing of the American mind*. New York : Simon and Schuster.

Bloome, D. (1987). Reading as a social process in a middle school classroom. In: D. Bloome (Ed.), *Literacy and Schooling* (pp. 123–149). Norwood, NJ: Ablex.

Bond, G., & Bond, E. (1941). *Developmental reading in high school*. New York: Macmillan.

Bridges, D. (1979). *Education, democracy and discussion*. Windsor, U.K.: National Foundation for Educational Research in England and Wales.

Brooks, C. (1947). *The well wrought urn: studies in the structure of poetry*. New York: Reynal & Hitchcock.

Brooks, C., & Warren, R. P. (1938). *Understanding poetry*. New York: Henry Holt.

Burke, K. (1966). *Language as symbolic action: Essays on life, literature, and method*. Berkeley, CA: University of California Press.

Commission on English. (1965). *Freedom and discipline in English*. New York: College Entrance Examination Board.

Committee of Ten on Secondary School Studies (1984) *Report of the Committee of ten on secondary school studies: with the reports of the conferences arranged by the committee*. New York: National Education Association.

Crane, R. S. (Ed.) (1952). *Critics and criticism, ancient and modern*. Chicago: University of Chicago Press.

Cremin, L. A. (1961). *The transformation of the school: Progressivism in American education*. New York: Vintage Books.

Dillon, D. R., & Moje, E. B. (1998). *Listening to the talk of adolescent girls: Lessons about literacy, school, and life*. In: D. E. Alvermann, K. A. Hinchman, D. W. Moore, S. F. Phelps & D. R. Waff (Eds), *Reconceptualizing the Literacies in Adolescents' Lives* (pp. 193–223). Mahwah, NJ: Lawrence Erlbaum Associates.

Dillon, D. R., O'Brien, D. G., Moje, E. B., & Stewart, R. A. (1994). Literacy learning in secondary school science classrooms: A cross-case analysis of three qualitative studies. *Journal of Research in Science Teaching, 31*, 345–362.

Dillon, J. T. (Ed.) (1988). *Questioning and discussion: A multidisciplinary study*. Norwood, NJ: Ablex.

Dillon, J. T. (1994). *Using discussion in classrooms*. Buckingham, U.K.: Open University Press.

Diamondstone, J., & Smith, W. M. (1999). Teaching literature and composition in secondary schools. In: L. B. Gambrell, L. M. Morrow, S. B. Neuman & M. Pressley (Eds), *Best Practices in Literacy Instruction* (pp.193–209). New York: Guilford Press.

Dunston, P. J., Headley, K. N., Schenk, R. L., Ridgeway, V. G., & Gambrell, B. (1998). National Reading Conference research reflections: An analysis of 20 years of research. In: T. Shanahan & F. V. Rodriguez-Brown (Eds), *National Reading Conference Yearbook, 47* (pp. 441–450). Chicago, IL: National Reading Conference.

Eliot, T. S. (1936). *Hamlet and his problems. Selected essays 1917–1932* (pp.124–141). London: Faber and Faber.

Fecho, B. (2000). Critical inquiries into language in an urban classroom. *Research in the Teaching of English, 34*, 368–395.

Finders, M. J. (1997). *Just girls*. New York: Teachers College Press.

Fish, S. E. (1980). *Is there a text in this class?: The authority of interpretive communities*. Cambridge, MA: Harvard University Press.

Flesch, R. F. (1955). *Why Johnny can't read – and what you can do about it*. New York: Harper.

Foster, H. M. (1994). *Crossing over: Whole language for secondary English teachers*. New York: Harcourt Brace.

Frye, N. (1957). *Anatomy of criticism: four essays*. Princeton: Princeton University Press.

Gaffney, J. S., & Anderson, R. C. (2000). Trends in reading research in the United States: Changing intellectual currents over three decades. In: M. L. Kamil, P. B. Mosenthal, P. D. Pearson & R. Barr (Eds), *Handbook of Reading Research* (Vol. III, pp. 53–74). Mahwah, NJ: Lawrence Erlbaum Associates.

Garner, R., & Gillingham, M. G. (1996). *Internet communication in six classrooms: Conversations across time, space, and culture*. Mahwah, NJ: Lawrence Erlbaum Associates.

Gavelek, J. R., Raphael, T. E., Biondo, S. M., & Wang, D. (2000). Integrated literacy instruction. In: M. L. Kamil, P. B. Mosenthal, P. D. Pearson & R. Barr (Eds), *Handbook of Reading Research* (Vol. III, pp. 587–607). Mahwah, NJ: Lawrence Erlbaum Associates.

Goldenberg, C. & Patthey-Chavez, C. (1995). Discourse processes in instructional conversations: Interactions between teacher and transition readers. *Discourse Processes, 19*, 57–73.

Goldman, S. R. (1997). Learning from text: Reflections on the past and suggestions for the future. *Discourse Processes, 23*, 357–398.

Graves, M. F., & Prenn, M. C. (1986). Costs and benefits at various methods of teaching vocabulary. *Journal of Reading, 29*.

Gray, W. S. (1919). *The relation between studying and reading*. (Proceedings of the fifty-seventh annual meeting of the National Education Association, pp. 580–586). Washington, DC: National Education Association.

Guzzetti, B. (1996). Gender, text, and discussion: Examining intellectual safety in the science classroom. *Journal of Research in Science Teaching*, *33*, 5–20.

Guzzetti, B., Anders, P. L., & Neuman, S. B. (1999). Thirty years of JRB/JRL: A retrospective of reading/literacy research. *Journal of Literacy Research*, *31*, 67–92.

Guzzetti, B., Kowalinski, B., & McGowan, T. (1992). Using a literature-based approach to teaching social studies. *Journal of Reading*, *36*, 114–122.

Harmon, J. M. (1998). Constructing word meanings: Strategies and perceptions of four middle school learners. *Journal of Literacy Research*, *30*, 561–599.

Herber, H. L. (1970). *Teaching reading in content areas*. Englewood Cliffs, NJ: Prentice-Hall.

Herber, H. L. (1978). *Teaching reading in content areas* (2nd ed.). Englewood Cliffs, NJ: Prentice-Hall.

Hill, W. F. (1977). *Learning thru discussion*. Beverly Hills, CA: Sage.

Hinchman, K. A., & Zalewski, P. (1996). Reading for success in a tenth-grade global-studies class: A qualitative study. *Journal of Literacy Research*, *28*, 91–106.

Hirsch, E. D. (Ed.) (1996). *A first dictionary of cultural literacy: what our children need to know*. Boston : Houghton Mifflin.

Hosic, J. F. (1921). The National Council of Teachers of English. *English Journal*, *10*(1), 1–10.

Hutchins, R. M. (1936). *The higher learning in America*. New Haven: Yale University Press.

Hynds, S. (1997). *On the brink: Negotiating literature and life with adolescents*. New York: Teachers College Press/International Reading Association.

Iser, W. (1993). *The fictive and the imaginary: charting literary anthropology*. Baltimore: J. Hopkins University Press.

Ivey, G. (1999). A multicase study in the middle school: Complexities among young adolescent readers. *Reading Research Quarterly*, *34*, 172–192.

Langer, J. A. (1995). *Envisioning literature: Literary understanding and literature instruction*. New York: Teachers College Press.

Lemke, J. (1990). *Talking science: Language, learning and values*. Norwood, NJ: Ablex.

Lorenz S. L. (1998). Romeo and Juliet: The movie. *English Journal*, *87*(3), 50–51.

McMahon, S. I., & Raphael, T. E. (with Goatley, V. J., & Pardo, L. S.) (Eds) (1997). *The book club connection: Literacy learning and classroom talk*. New York: Teachers College Press/ International Reading Association.

Moje, E. B., & O'Brien, D. G. (Eds) (2000). *Constructions of literacy: Studies of teaching and learning in secondary classrooms and schools*. Mahwah, NJ: Lawrence Erlbaum.

Monseau, V. R. (Ed.) (1999). Our love affair with literature. *English Journal*, *89*(2), entire issue.

Morson, G. S., & Emerson, C. (1990). *Mikhail Bakhtin: creation of a prosaics*. Stanford, CA: Stanford Univ. Press.

Moore, D. W. (1996). Contexts for literacy in secondary schools. In: D. J. Leu, C. K. Kinzer & K. A. Hinchman (Eds), *Literacies for the 21st Century: Research and Practice* (Forty-fifth yearbook of the National Reading Conference, pp. 15–46). Chicago, IL: National Reading Conference.

Moore, D. W., Bean, T. W., Birdyshaw, D., & Rycik, J. R. (1999). Adolescent literacy: A position statement. *Journal of Adolescent & Adult Literacy*, *43*, 97–112.

Moore, D. W., Readence, J. E., & Rickelman, R. J. (1983). An historical exploration of content area reading instruction. *Reading Research Quarterly*, *8*, 419–438.

National Reading Panel. (2000). *Report of the National Reading Panel*. Washington, DC: National Institute of Child Health and Human Development.

New London Group. (1996). A pedagogy of multiliteracies: Designing social futures. *Harvard Educational Review, 66*, 60–92.

Parker, J. C., & Rubin, L. J. (1966). *Process as content*. Chicago: Rand McNally.

Pearson, P. D., & Fielding, L. (1991). *Comprehension instruction*. In: R. Barr, M. L. Kamil, P. Mosenthal & P. D. Pearson (Eds), *Handbook of Reading Research* (Vol. II, pp. 815–860). New York: Longman.

Pennfield, E. (1994). New criticism. In: A. C. Purves (Ed.), *Encyclopedia of English Studies and Language Arts*. New York: Scholastic.

Peterson, R., & Eeds, M. (1990). *Grand conversations: Literature groups in action*. New York: Scholastic.

Prentiss, T. M. (1998). Teachers and students mutually influencing each other's literacy practices: A focus on the student's role. In: D. E. Alvermann, K. A. Hinchman, D. W. Moore, S. F. Phelps & D. R. Waff (Eds), *Reconceptualizing the Literacies in Adolescents' Lives* (pp. 103–128). Mahwah, NJ: Lawrence Erlbaum Associates.

Rabinowitz, P. J., & Smith, M. W. (1998). *Authorizing readers: resistance and respect in the teaching of literature*. New York: Teachers College Press and NCTE.

Ransom, J. C. (1941). *The new criticism*. Norfolk, CN: New Directions.

Ransom, J. C., Davidson, D., Owsley, F. L., Fletcher, J. G., Lanier, L. H., Tate, A., Nixon, H. C., Lytle, A. N., Warren, R. P., Wade, J. D., Kline, H. B., & Young, S. (1930). *I'll take my stand; The South and the agrarian tradition, by twelve southerners*. New York: Harper.

Raphael, T. E., & Brock, C. H. (1997). Instructional research in literacy: Changing paradigms. In: C. Kinzer, D. Leu & K. Hinchman (Eds), *Inquiries in Literacy Theory and Practice* (pp. 13–36). Chicago, IL: National Reading Conference.

Richards, I. A. (1926). *Principles of literary criticism*. London: Routledge & K. Paul.

Rosenblatt, L. M. (1994a). The transactional theory of reading and writing. In: R. B. Ruddell, M. R. Ruddell & H. Singer (Eds), *Theoretical Models and Processes of Reading* (pp. 1057–1092). Newark, DE: International Reading Association.

Rosenblatt, L. M. (1994b). *The reader, the text, the poem: The transactional theory of the literary work*. Carbondale, IL: Southern Illinois University Press.

Ruddell, M. R. (1994). Vocabulary knowledge and comprehension: A comprehension-process view of complex literacy relationships. In: R. B. Ruddell, M. R. Ruddell & H. Singer (Eds), *Theoretical Models and Processes of Reading* (pp. 414–447). Newark, DE: International Reading Association.

Santa Barbara Classroom Discourse Group (1992). Do you see what we see? The referential and intertextual nature of classroom life. *Journal of Classroom Interaction*, *27*(2), 29–36.

Scholes, R. (1998). *The rise and fall of English: Reconstructing English as a discipline*. New Haven: Yale University Press.

Sizer, T. (1984). *Horace's compromise: The dilemma of the American high school*. Boston, MA: Houghton Mifflin.

Smith, M. B. (1954). *The diminished mind: a study of planned mediocrity in our public schools*. Chicago: H. Regnery Co.

Squire, J. R. (1991). The history of the profession. In: J. Flood, J. M.Jensen, D. Lapp & J. R. Squire (Eds), *Handbook of Research on Teaching the English Language Arts*. New York: Macmillan Publishing.

Stock, P. L. (1995). *The dialogic curriculum: Teaching and learning in a multicultural society*. Portsmouth, NH: Heineman.

Strang, R. (1937). The improvement of reading in high school. *Teachers College Record, 39*, 197–206.

Swafford, J., & Hague, S. (1987). *Content area reading strategies: Myth or reality?* Paper presented at the annual meeting of the College Reading Association, Baltimore, MD.

Taylor, M. (1976). *Roll of thunder, hear my cry*. Philadelphia, PA: Dial Books.

Thomas, H. K. (1999). The social construction of literacy in a high school biology class. In: T. Shanahan & F. V. Rodriguez-Brown (Eds), *National Reading Conference Yearbook, 48* (pp. 317–328). Chicago, IL: National Reading Conference.

Tierney, R. J., & Cunningham, J. W. (1984). Research on teaching reading comprehension. In: P. D.Pearson, R. Barr, M. L. Kamil & P. Mosenthal (Eds), *Handbook of Reading Research* (pp. 609–655). New York: Longman.

Van Doren, M. (1943). *Liberal Education*. New York: H. Holt and Company.

Venezky, R. L. (1987). Steps toward a modern history of American reading instruction. *Review of Researh in Education, 13*, 129–167.

Vygotsky, L. S. (1978). *Mind in society: The development of higher psychological processes*. Cambridge, MA: Harvard University Press.

Wade, S. E., & Moje, E. B. (2000). The role of text in classroom learning. In: M. L. Kamil, P. B. Mosenthal, P. D. Pearson & R. Barr (Eds), *Handbook of Reading Research* (Vol. III, pp. 609–627). Mahwah, NJ: Lawrence Erlbaum Associates.

Wellek, R., and Warren, A. (1949). *Theory of literature*. New York: Harcourt, Brace.

Wertsch, J. V. (1991). *Voices of the mind*. Cambridge, MA: Harvard University Press.

Willinsky, J. (1990). *The new literacy: Redefining reading and writing in the schools*. New York: Routledge.

Wimsatt, W. K., & Beardsley, M. (1954). The affective fallacy. In: W. K. Wimsatt (Ed.), *The verbal icon: studies in the meaning of poetry*. Lexington, KY: University of Kentucky Press.

Whipple, G. (Ed.) (1925). *The twenty-fourth yearbook of the National Society for the Study of Education: Report of the National Committee on Reading*. Bloomington, IL: Public School Publishing Company.

Wilkinson, L. C., & Silliman, E. R. (2000). Classroom language and literacy learning. In: M. L. Kamil, P. B. Mosenthal, P. D. Pearson, & R. Barr (Eds), *Handbook of reading research* (Vol. III, pp. 337–360). Mahwah, NJ: Lawrence Erlbaum Associates.

Wolf, D. P. (1995). Pacesetter revisited. *English Journal, 84*(1), 59–82.

Yoakam, G. A. (1928). *Reading and study: More effective study through better reading habits*. New York: Macmillan.

INSTRUCTIONAL METHODS AND LEARNING ACTIVITIES IN TEACHING WRITING

Sarah Warshauer Freedman and Colette Daiute

INTRODUCTION

Helping students learn to write requires activities that shift their perspectives between those of speaker and listener, writer and reader, creator and critic, skeptic and persuader, to name a few. Student writers, like experienced writers, must learn to do such role shifting because writing is at once thought and communication, cognitive and social, content and process. Effective writing teachers address these complexities by offering informative coaching on challenging writing tasks and extensive opportunities to practice multiple types of writing. They organize instruction in ways that promote language-specific processes and attitudes, as well as adhere to the basic assumptions and generic learning principles described by Brophy in Chapter 1 of this volume. After a brief review of major assumptions about writing development that influence instruction, we present three principles that underlie the creation of effective instructional methods and learning activities in writing classrooms. The principles are derived from research literature on the teaching and learning of writing as well as the work and writings of expert practitioners. As we discuss each principle, we provide examples of instructional methods and learning activities that follow from it.

Subject-Specific Instructional Methods and Activities, Volume 8, pages 83–110.
2001 by Elsevier Science Ltd.
ISBN: 0-7623-0615-7

ALTERNATE PERSPECTIVES OF WRITING DEVELOPMENT

Implicit in all writing programs are ideas about the ideal components and sequences of this fundamental literacy skill. Historically, the most effective programs have based instruction in a theory of writing development. Several different notions of writing development have dominated the field since the 1980s, notions that often co-occur as underlying assumptions of writing curricula and instruction. The prevailing views of writing development have been what we refer to as the print-based view, the maturation view, the expertise view, and most recently, the sociocultural view. Instruction differs across these views, but rarely is any single notion of development the basis for an entire writing program.

Defining student progress in terms of print means following a sequence of teaching letters, words, grammar, paragraph types, and extended genres. Since these written forms are essential to writing, the print view is always relevant, but classroom practice that moves beyond a strict adherence to this sequence shows that children write better if they have other kinds of instruction as well (Graves, 1983; NAEP,1999).

The maturation view is based generally in cognitive-developmental theory and proposes a sequence of writing instruction that follows children's evolving abilities to analyze phenomena in the world, in this case focusing on print, its attendant features and functions, and children's increasing abilities to manipulate print to meet age-appropriate goals (Clay, 1991; Ferriero & Teberosky, 1982). Skills like being able to form letters, mastering the orthographic code, writing grammatical sentences, and organizing arguments are believed to mature as the child interacts with written language. Children are encouraged to symbolize their ideas and communications in spontaneous graphic forms such as invented spelling, drawing, or dramatic performances (Clay, 1991; Graves, 1983; Dyson, 1989, 1993; Wagner, 1998). According to this view, children can express ideas before they have mastered all the mechanics of standard orthography, sentence and paragraph structure. Educators and researchers working from this view also explain that writing instruction begins in pre-school and includes generating interesting content for purposes of discovery, self expression, and communication. Descriptions for instruction up through middle school have been offered (Atwell, 1987; Calkins, 1986). Research on the effectiveness of this approach has indicated that many children have, in fact, begun to write at earlier ages than was ever believed possible, achieving adequate fluency up through the fourth grade (NAEP,

1999). These successes, however, have been relatively limited for children from minority backgrounds.

Related but more specific processes, especially for older students, have been based in the view that, once the basic orthography, sentence structure, and awareness of purposes for writing are established, at least to some extent, developing as a writer involves gaining expertise with composing processes. In this view, writing is a problem-solving process involving strategies for generating, organizing, and reflecting on ideas as they are expressed in text form (Flower & Hayes,1980). Writing researchers compared the processes of expert and novice writers and recommended instructional activities aimed at helping novice writers adopt more expert-like behaviors, such as planning by establishing goals throughout the composing process, thinking about the needs of potential readers, and revising across sentences and paragraphs – not just within words and sentences (Bereiter & Scardamalia, 1987; Flower & Hayes, 1980). Writing instruction focuses on processes that engage writers in stating and following goals, defining the relevant strategies to meet goals, and deliberately delaying editing until late in the process. Interestingly, this approach teaches students to work with aspects of writing that will never actually appear in the text, before they pay attention to the specifics of text form and to delay text-based revising and editing until relatively later in the process (Bereiter & Scardamalia, 1987; Flower & Hayes, 1980).

The sociocultural view of development is consistent with the need for student writers to learn expert practices, but conceptualizing writing development as culture implies more instructional attention to the issue that there are myriad values and forms of written language and that each is a particular cultural discourse forms (Dyson, 1989; Dyson & Freedman, 1991, in press; Freedman, Dyson, Flower & Chafe, 1987; Lee, 1993). This view implies making a place in the writing curriculum for relating oral and written language, in large part, to account for cultural differences in language, symbolization, expression, and values. It also means emphasizing the importance of making linguistic diversity and multiplicity central to writing instruction. Finally, the sociocultural view implies that instruction requires that writing be meaningful to the writer and that there be time for translation among different modes of written language, such as expressing a news item in a school news paper, in a text book, and in a song for audiences that prefer Standard English, Black English, and/or a foreign language like Spanish.

Writing activities that follow from a sociocultural view must thus involve the examination and creation of texts that are true to different aspects of the students' linguistic experiences and that attend to what is involved for students as they move between their everyday oral language and the more formal written language varieties of the schools and ultimately the workplace. These activities are built on the understanding that while some children are not prepared for the requirements of school, others come to school unfamiliar with the valued discourses because their home cultures do not share language with mainstream institutions and teachers or because their families, even if from diverse cultures, do not have educational backgrounds that prepare them for home to school language transitions (Michaels, 1991; Au, 1980; Heath, 1983; Ogbu, 1987). It follows that the purposes and practices around learning to write must be made explicit so that everyone has the chance to participate. Achieving some common language is an important goal of education in a democracy, but the unique nature of writing instruction relative to other aspects of the curriculum requires a more subject-specific interpretation of equity. Research on writing, especially from the sociocultural perspective, indicates that diversity must be addressed rather than transcended.

INSTRUCTIONAL METHODS AND LEARNING ACTIVITIES FOR THE TEACHING OF WRITING

This section shows how sociocultural notions of writing development and the generic principles Brophy outlined in Chapter 1 can guide the creation of learning activities for writing classrooms. We limit our focus to methods and activities aimed at helping students learn to produce school-based written texts. Although other kinds of writing (most commonly diaries and journals) may be used as a means to this end, our concern is only how such writing functions to help students achieve the goal of improving their writing for school. We also consider how students might use writing to acquire subject-matter knowledge across the curriculum. To supplement and make Brophy's generic teaching-learning principles more specific to writing, we build on sociocultural notions of writing development and propose three instructional principles specific to the teaching and learning of writing.

Principle 1: Productive writing activities employ varied types of classroom language that support the development of varied cognitive strategies necessary to writing development.

Principle 2: Productive writing activities take into account the different demands of different types of written language and the fact that students must learn to write in a variety of ways, for a variety of purposes, and for a variety of audiences.

Principle 3: Productive writing activities include classroom-based reflection and assessments by teachers and students to monitor students' development and achievement in writing.

Principle 1: Productive Writing Activities Employ Varied Types of Classroom Language That Support the Development of Varied Cognitive Strategies Necessary to Writing Development

All forms of communication are needed for the teacher to understand and to help support children's writing. A major advancement from writing research is that classroom talk is not just talk. Rather, there are different kinds of talk that relate to writing in different ways. So developing ways of thinking and communicating occurs through different ways of delivering instruction. Much research on writing has focused on the purposes and functions of talk as it occurs in different instructional contexts to support students' writing development (Cazden, 1988; Daiute, Campbell, Griffin, Reddy & Tivnan, 1993).

Classroom talk has become increasingly purposeful in writing classes as ideas about the close relationship between talk and writing have increasingly guided practice. Writing instructors have, for example, organized social interactions that support a specific writing role, alternating teacher recitation in front of a full class (Cazden, 1988; Nystrand, Gamoran, Kachur, Prendergast, 1997) with conferencing in small groups (Graves, 1982; Freedman et al., 1987; Nystrand et al., 1997; Sperling, 1990), and peer collaboration (Daiute, 1986; Di Pardo & Freedman, 1988; Dyson, 1986; Freedman, 1992). Such arrangements engage students in different ways as members of communities that motivate and respond to writing. They highlight, moreover, that writing is social even though authors typically must imagine their audiences as they compose individually.

In addition to supporting writing instruction by varying activity contexts in class, effective teachers structure those contexts in ways that plan for different kinds of talk. In other words, teacher/student conferences and collaborative writing are designed for teachers to engage individual students in the specific expert composing strategies they need to improve their writing. In addition, small group work is designed to increase student writers' motivation, sense of audience, and use of personally-meaningful language, all of which transfer from group to individual composing contexts.

In summary, teachers take advantage of different classroom arrangements to impart information about writing, to guide students in thinking like a writer, to respond to students' writing, or to guide students in practicing different writerly roles. These different functions of talk promote different kinds of writing abilities.

Questioning to Deepen Reflection

Nystrand and his colleagues have analyzed instructional talk among adolescents in English classes to learn about the results of different types of instructional questions on students' writing (Nystrand et al., 1997). In particular, the researchers examined whether and how teachers engaged students to deepen their thinking about topics and texts by asking "authentic questions". Authentic questions require students to respond with original thoughts, extensions between literary texts they were studying, what had already been said in class discussion, and their own evolving ideas about the topic. Students' responses to authentic questions contrast with responses to questions that probe for specific, limited answers as in fill-in-the blank work sheets (Nystrand et al., 1997). The following brief example illustrates a relatively open-ended question that requires the student to probe his/her own understanding and the teacher to respond in spontaneous ways as well. The class is discussing a poem about a sibling relationship during the 1920s depression in the U.S.

> Mr. Kramer: Now what about the final stanza? . . . Look at the final stanza.
>
> Student 1 (interrupts): She's showing that she's not as sure of everything as she says she is.
>
> Mr. Kramer (jumping in to read a segment of the poem): "He sees her . . .".
>
> Student 2: Does that mean that she is nervous and trying to hide her feelings? She doesn't want him to think she's nervous.
>
> Mr. Kramer: Yeah, she's successfully hid her feelings, until like the neon sign that continually beats.

Such exploration of ideas is a process that writers must use when working on their own to create texts that are original, transformative, and interesting to read. Authentic questioning should be a recurring, sustained activity in writing classes so that students can internalize such a process for solo use.

Teacher/Student Interaction to Develop Expert Composing Strategies

Classroom interaction with younger students has emphasized the use of talk to model expert composing strategies, like planning. In the following example from a third grade open classroom, the teacher wrote collaboratively with each student (when the aide was in the room to guide small group work with the other students) to let them in on the ways that she as an experienced writer, makes decisions about the organization, information, and phrasing of texts. In short, this teacher thinks aloud while writing with her students and guides them as apprentices in the composing process (Daiute et al., 1993).

In the following excerpts from a collaborative composing session, the teacher guided her student Gary in how to compose text with attention to opening sentences and rich content. The first excerpt illustrates the strategy of creating an opener with the reader in mind.

> Teacher: Got it! What a good idea you have! Now we need a catcher sentence at the beginning, that's going to get everybody to want to read this story in the newspaper!
> Gary: Hmm (Daiute & Griffin, 1993, p. 116).

The next excerpt illustrates how the teacher emphasized including important information related to the content and the process of selecting information based on its significance.

> Teacher: What was significant? What was the most important thing about the printing press? Why was that an important invention?
> Gary: It made it easy, it made it easier to copy things, like umm you wouldn't have to write every page or something. Or if you wanted to have copies, you could just put it on the printing press, and it would go easier.
> Teacher: Right. And then people could, and then more ideas could speak, because people could write down their ideas and pass it along.
> Gary: Mm-hmm (Daiute & Griffin, 1993, p. 116).

The teacher then shifts to focus Gary's attention on the form of the narrative.

> Teacher: And then we can talk about what happened, how Guttenberg wanted to make books beautiful and how this other person wanted to just mass-produce books.
> Gary: Yeah.

When done two or three times in the course of five months, such teacher-student talk has proved effective for helping students write more explicitly organized texts. Knowing whether students use the composing strategies modeled by the teacher is not easy, but Gary's writing as well as the writing of other students in the class showed this teacher's influence. Further interactions between peers indicates that children, including Gary and his peer partner, when working alone together, also use some of the language of expert composing processes introduced by the teacher. Interestingly, student pairs transformed the teacher's strategies into their own unique approaches.

Peer Collaboration Transforms Ideal Language into Children's Voices and Real Audiences

Teachers can promote yet another aspect of interaction to support writing. Carefully designed and monitored peer collaborative writing groups bring youth genres into the writing classroom (Daiute, 1993). Children doing the same types of writing activities for the same class newspaper as Gary and his teacher talked about different topics than the teacher and Gary. While the teacher guided students to plan, revise, and use detailed subject-matter content, the children playfully and interdependently reflected on how they felt about topics and different ways of expressing these topics. Interestingly, such peer interactions led to improvements in children's writing although these differed from the characteristic improvements after children collaborated with the teacher (Daiute et al., 1993).

Young people express their concerns and goals spontaneously, through playful and seemingly aimless talk, in contrast to the deliberate nature of different types of teacher talk. Nevertheless, the consistency in patterns of play raises issues that challenge commongly accepted notions of how writing and knowledge develop. Student composing dyads vary greatly in topic and form as they compose, yet they persist at finishing the task. Most importantly, they express deep, albeit idiosyncratic, motivation for writing in the context of their social interaction, although they are not able to explain their major motivations as the teachers do.

The difference between teacher-student collaboration, focusing on composing strategies, and student peer collaboration, focusing on socially- and emotinally-based composing (Daiute, et al., 1993) is one that played out in different ways in a cross-cultural study of middle schools students in the U.S. and the U.K. (Freedman, 1994). Teachers in the U.K. built writing instruction out of a range of interests and needs expressed by students as they progressed in their own individual ways. This type of negotiated curriculum evolved through talk which the British teachers used to shape instructional goals into an evolving curriculum for a particular group of students, making adjustments for individuals as needed. They did not adhere to pre-planned and teacher developed curriculum, which was the main way teachers in the U.S. proceeded. In some ways, the interpersonal quality of the British "negotiated curriculum" was like what the young U.S. peers in the study described above did, which may offer insights about why the peers' interactions led to improvements in their writing.

Engaging Students in Metalinguistic Talk as the Basis of Interpretive Writing

Metalinguistic talk – talk about language – occurs in integrative writing activities that combine talking, reading, writing, and other forms of communication. Maintaining integration between talk and text in high school literacy instruction is atypical yet a forward-looking example of a sociocultural approach. Although writing instruction in high school and middle school tends to be dis-integrated into traditional disciplines of English, literary study, composition, and specialized subject matters, the need for integration has become increasingly salient, especially because it is in the isolated disciplines that children from non-mainstream language backgrounds begin to fall behind (NAEP, 1999).

Because writing is social, any specific piece of writing is the expression of a particular culture, its values, forms, and purposes of expression. For this reason, learning to write involves understanding how different instances of writing relate to other forms of communication and thought, including oral language, non-verbal communication, and the languages of specific disciplines. Integrating across diverse forms of expression also makes more likely that multiple perspectives are invited into the classroom. Integrative writing activities must be well-planned to focus around the creation and interpretation of written texts. Such activities tend to evolve over extended periods of time and require checkpoints along the way for reflection on how the different modes support the creation of text. The most innovative practices in writing instruction integrate across oral and written language in ways that support students from diverse backgrounds. Toward this end, writing can be related to reading, for example, with the close examination of literary and academic texts representing different cultures and the ideas, symbolic forms, and language forms in these texts. Thus, reading provides models and motivation for writing.

Analyzing multicultural literature has become the focal point for integrated writing activities in many K through 12 classrooms. An example from a high school English class in a school serving predominantly African American students illustrates how analysis of everyday language can be the basis for complex exposition. In this class students came in reading and writing significantly below grade level and many had never even read an entire book. They were taught to write about and analyze complex pieces of literature by building on their everyday African American English Vernacular (AAVE) language practices The instructional unit shows these students how AAVE possesses the metaphorical quality and critical orientation that characterizes literary analysis and effective expository writing (Lee, 1993).

The teacher focused on signifying – an AAVE practice involving ritual insults through figurative, playful language. The activity began and ended with students writing about literature. Instructions for the final writing sample asked for analysis ("identify", "discuss specifically what they meant", "in your own words"):

> On page 90, what does Pauline mean when she makes the following statement? "All them colors was in me". Be sure to identify the colors she describes and to discuss specifically what they meant to her in your own words (Lee, p. ?).

Given the beginning level of these students, the following written response was rated as proficient.

> *When Pauline said that all them colors was in me she means that when she met Chloe she thought about when she was little and the berries mashed in her dress and when her mother made lemonade and Cholly made her feel like a little girl again and when he whistle it brought shivels down her skin. I believe the colors stood for beauty and it also made her feel good inside the colors just made her happy* (Lee, 1993, p. 178).

This students' analysis is marked clearly with phrases like "she means". The student also offers details, a metaphor, and some lively albeit colloquial language – "shivels".

In whole class discussions, before students completed this writing, the teacher explicitly guided them in analyzing signifying as a critical, poetic form and then in extending this "speakerly" practice to the interpretation of literature. The following excerpt from a discussion about *Their Eyes Were Watching God* by Zora Neale Hurston shows how the teacher guided students in such a metalinguistic analysis:

> Teacher: Can you find any examples of anything that these characters have said where their words are like pictures?
> Pat: When he was talking about that mule. They said he was scrubbing by his [unintelligible] bone [laughing]. If you picture that, it was just too funny.
> Teacher: While they were signifying about that mule, if you heard somebody say that, if you closed your eyes, could you see that mule? Could you see it just like it was a little movie or cartoon, just from the way they described it?
> Students: [groups of students respond] Can I read? Can I read?
> Teacher: No. Let's read it out loud together to get a good feel for the language.
> [Students read conversation on page 49 out loud. Students laugh at section about mule and women.]
> Mary: He said they was using his ribs for a washboard. Can you imagine someone doing that to your mule or dog or something?
> Teacher: What's the saying?
> Charles: He feed him out of a . . .
> Mary: A tea cup . . .
> Teacher: What do you think the writer Zora Neale Hurston means when she describes the way these people talk as thought pictures and crayon enlargements of life?
> Mary: They talk about a person so bad that you could picture it.
> Teacher: That you could picture it from their words. If they are crayon enlargements of life – crayons are what?
> Harry: Then you could actually see it.
> Teacher: It's almost like you could see the what of it?
> Charles: The entire thing, like what he say, the color, everything.
> Teacher: Every aspect of what this picture might look like. And if what they are saying is enlargements of life, how is their talk enlargements of life?
> Pat: Things that happen are for real, but it's just a little exaggerated.

The students "laughed, paraphrased the text, and provided examples from the text to support their claims" (Lee, 1993, p 113). Meeting the students on their linguistic turf, the teacher worked in school language and its implications with phrases like "it means", "how is . . . like . . . ?", and the practice of imagining the words on the page, which the student who wrote the paragraph about Cholly seems to have internalized.

This class discussion is integrative in several ways. The teacher and students talk about the figurative language of signifying ("He said they was using his ribs for a washboard"), linking this to visual images in the text, and relating these symbols to the author's greater meaning in the text. Such a discussion moves across language familiar to several cultures in the U.S. as well as across oral and written language. In addition to being literary talk, such metalinguistic reflective language became the basis for analytic writing that is organized around practices like comparison and contrast, making an argument, and supporting generalizations with examples. This session created the kind of analytic language that students needed for their expository writing assignment.

Principle 2: Productive Writing Activities Take Into Account the Different Demands of Different Types of Written Language and the Fact That Students Must Learn to Write in a Variety of Ways, for a Variety of Purposes, and for a Variety of Audiences

Learning to write by imitating model texts was a common practice for centuries, but as writing instruction has evolved to address the needs of diverse students who are required to learn to write at younger ages than ever believed possible, the concept of "genre" or writing type has come to be defined by the dynamic interaction of text in context (Cope & Kalantzis, 1993). While model texts are still invaluable cultural exemplars, teaching student writers how to create letters, essays, stories, and reports involves teaching not just the elements but also the processes for creating such texts. Writing teachers have moved away from focusing exclusively on structural aspects of paragraphs, for example, like topic sentences and details, to thinking about contextual factors like the purpose of the paragraph in the larger text, the situation in which the paragraph will be read, and the readers who eventually will be informed, entertained, moved, or otherwise engaged by it. Teaching written genres now means making the less visible but utterly formative rhetorical demands of purpose, audience, and context explicit as the motivation and processes for specific text forms. Assessments indicating that student writers often master the mechanics of writing but fail to convey compelling meanings underscore the impetus for making such contextual factors central in writing instruction.

Focusing on the formal characteristics of written genres, teachers might organize curricula around classically-defined rhetorical purposes of written language (Britton et al., 1975), around text forms like essays, stories, and letters; around more basic elements like paragraphs (descriptive paragraphs, cause and effect paragraphs, etc.); or around theses, also referred to as the central idea (Sommers, 1980). Increasingly, however, such forms are taught as moments in socially-situated activities, like identifying and solving a social problem via writing letters to influential people in the community, summarizing existing knowledge about the problem, and writing reports including facts and proposals crafted to express a specific point of view and to persuade readers of the effectiveness of a specific solution. In this way, writing is embedded in broader social issues. In such contexts, focus on text forms augmented by analysis of specific nested goals, the motivating interests served by the project, and the diverse readers to be informed or persuaded.

Creation of the specific text forms also requires close attention since students are prepared differently by the ways of thinking and talking in their homes and communities. For this reason, effective writing instruction devotes attention to specific text forms via cycles of practice writing and reading writing aloud. Text-based instruction has persisted, for example, with a focus on writing narratives, essays, and reports. Teaching such text forms as ends in themselves is not the best practice. Rather these forms can best be taught when framed as modes used in educational materials, newspapers, scientific/scholarly publications, literature, and civic documents in Western written rhetorical traditions.

Teaching Purposeful Narrative Writing

The primacy and function of narrative (or story) as a mode of thought and communication has been intensely discussed in recent years. Teachers working in kindergarten through high school devote considerable attention to written narrative. While teachers differ greatly in whether and how they explain the rationales for narrative writing, they have put narrative to work for their students' development as self-aware individuals, as citizens, and as effective test-takers.

Teachers engage students in narrative writing as a pre-writing technique since writing about events from memory is one way to generate content. Some scholars and educators have argued that narrative is the primary organizer of memory (Bruner, 1986; Graves, 1983; Nelson, 1993), and thus it is a "natural" mode for at least the beginning stages of generating material for reporting and reflecting on topics that may later be examined in essay or some other form. Narrative writing, like speaking, has been found to be a mode of sense-making since the reporting of sequences of events in time order engages even the youngest writers in making choices about which details to include, how to sequence them, how to use language to mark their significance, and how to use grammar and punctuation in the service of these functions.

Using Personal Experience Narratives to Generate Content

In the popular process approach to teaching writing, personal narrative writing is the core of the curriculum in the elementary grades (Calkins, 1986; Graves, 1983). Teachers set aside time every day for students to write about their lives in journals where unfettered expression in the young writer's authentic voice is a resource for subsequent crafted pieces. Since journal writing is designed to capture spontaneous thoughts and feelings, teachers encourage students to invent spellings based on how the words sound to them (rather than to fuss about correct spellings) and to write whatever comes to mind, uncensored by concerns for structure or coherence beyond individual words (Elbow, 1981). Such narrative writing is a process of researching one's life which makes it foundational in the composing process. More mature writers often use phases of journal writing throughout the development of stories, essays, or other professional genres.

Researchers have described the daily process of writing personal narratives in elementary school classrooms. Journal writing typically begins with a specific focus each day, like "write about something that surprised you" and progresses to one-on-one conferences in which the teacher guides a student to re-read the journal for events, insights, or other ideas that can form the core of personally-meaningful story or an expository report to share with classmates or readers beyond the classroom. While research indicates that such wide spread practice of writing personal experience narratives has succeeded in helping huge numbers of children gain basic written literacy, especially in narrative form, by the fourth grade, research has also identified limits of the use of narrative writing as pre-writing. Assumptions about the naturalness of time-ordered narrative sequences as the basic mode of expression, thought, and communication have been questioned. Studies of children's early spontanous writing have shown that many begin in an expository mode, albeit cryptically, writing lists and letters, rather than event sequences (Newkirk, 1987). Moreover, children of diverse cultures and genders have been socialized to narrative in different ways, which raises questions about any notion that narrative or any particular form of narrative is a basic skill (Heath, 1983; Newkirk, 1987). While valuable as one method for generating content, narrative writing should be taught in other contexts, especially as children mature.

Narrative Writing to Reflect on Important Social Issues

The view that narrative writing is a specific cultural tool for examining self in society is the basis for activities that teach narrative writing in the upper elementary grades. In this context, instruction in narrative writing teaches the canonical form required in mainstream U.S. education, news reporting, scholarly reporting, and much literary discourse for the purpose of reflection and social development (Daiute, 2000). A series of narrative writing activities guided teachers and children in the examination of social conflicts based on racial/ethnic discrimination, issues that face all children in the diverse U.S. society and which, unfortunately, become salient in elementary school. In the curriculum, narrative writing activities were the focus of a violence prevention program designed to equip children with strategies for dealing with conflicts that they might encounter with peers. In this program, children learned to write standard narratives with the broader purpose of using narrative to describe conflicts with peers and to role play different ways of dealing with conflicts. Third and fifth graders wrote a series of fictional and reality-based narratives to be compiled in a book to help the next year's third and fifth graders gain insights about dealing with peer conflicts.

The curriculum used literary narratives and children's own stories as bridges between their classrooms, their lives outside of school, and their imagined worlds. The context was one that introduced issues of racial and ethnic discrimination as a theme, with high-quality children's literature as the point of departure for class discussions, peer group activities, and children's writing around social conflict (Daiute, 2000; Walker, 1998). For example, in the fall, the fifth grade classes each spent several weeks reading, discussing, and extending the novel *Felita* by Nicholasa Mohr (1979) about a Puerto Rican girl in the 1960s in New York City. In this novel, Felita and her family suffer many explicit and implicit discriminatory assaults by neighbors, family, friends, and community representatives, and the characters deal with these issues in diverse ways. Interspersed with sessions devoted to reading and discussing this and other novels, the young people participated in a range of collaborative and individual written narrative experiences designed to extend the social conflict themes in literature to their own real and imagined lives. About half way through each book, students were asked to work with a partner on a "Literary Re-construction" task, which explicitly foregrounded issues of discrimination in a quotation from the novel and asked the author teams to write their own endings after discussing different options. This reality-based "Personal Conflict Story" was designed to extend the work classes had been doing around social conflict literature to events in their own lives.

When asked to write about a time when he or someone he knew had a disagreement or argument with someone his age, John, an African-American boy wrote the following story in the fall of his fifth grade year.

> *My conflict is when me and my best friend Robert and I was fight because I didn't want to be on his team so and we soloved it later that day no body got herti. and that is when we became best friends.*

This narrative is a brief rendition of a fight focused on elements highlighted in the task prompt. "Tell about what happened, how the people involved felt, and how it all turned out," which used narrative elements of time, character motivation, and resolution to build on central elements of the conflict resolution strategy curriculum. Later in the year, John crafted a narrative in terms of central aspects of conflict resolution.

> *One day I was walk in the croner store and manni [manager/clerk] said get out for no resana at all so I walked out of the store and manni was walking right behid me and he tryed to hit me so he hit me and I truned a round and I punched him in the face and he stared to cry. He was feeling bad and I was not so happy me self becaues I now that I could have talked it out and not punched him. He walked away crying.*

In the spring narrative John offered deeper reflection about "what happened," as an event with a reason (rather than just a fight), and the participants' complex mixture of suspicion, intentions, hurt, confusion, regret, and empathy, rather than as polarized positions as in the fall story. Interestingly, John's spring narrative did not provide a classic resolution to the fight but used the resolution to express the character's increased awareness. When reading their stories aloud in large and small groups, teachers supported children's narrative writing as a social tool and a literacy skill by guiding them to examine the description of events and evaluations of events closely as they indicated the characters' approaches to social conflict. Discussions of improving narrative writing, thus hinged on social awareness and responsibility as well as on verb tense, character description, and different types of narrative resolution (Daiute, 2000).

Principle 3: Productive Writing Activities Include Classroom-Based Reflection and Assessments by Teachers and Students to Monitor Students' Development and Achievement in Writing. Writing Portfolios Provide Important Structures That Can Help Teachers and Their Diverse Students Find Ways to Achieve This Goal

Students learn through classroom talk and through practice writing in varied ways and for varied purposes, but they also need to learn to assess their progress for themselves. In this way, they come to know what progress they have made and what further progress they need to make. Such academic self-knowledge is a building block for learning across the curriculum.

In the field of writing, an elaborate set of classroom activities has evolved specifically for the purpose of helping students develop self-knowledge about their writing abilities and about their progress. At the core of these activities is the writing portfolio.

A writing portfolio, simply put, is a folder or binder that each student keeps and that contains either all of the writing the student produces, both in class and outside of class, sometimes even including all drafts, or more commonly, a selection of varied pieces that showcase the students' accomplishments. Portfolios also commonly include the students' reflections on his or her work – which pieces the student likes best and why, what aspects of writing the student wants to continue to work on, the process the student followed to compose particular pieces, what changes the student sees in his or her writing across time. Once the student's writing and reflections on that writing have been collected, the work is available for further teacher and student analysis and reflection.

Instructional activities surrounding portfolios are particularly effective when they are designed to help both teachers and students better understand the nature of student growth in writing and the kinds of teaching and learning that best lead students to progress. Such activities provide the teachers with an opportunity to monitor student achievement and to make expectations clear, but they additionally provide students with practice learning to monitor their own achievement and internalizing those expectations. As Brown (1978) explains, the development of metacognitive skills, coming to know what one knows, is essential to the learning process.

Teachers who introduce portfolios can learn a great deal from case studies of successful portfolio use inside classrooms. Many of these case studies show portfolios as part of large-scale assessment systems and school reform efforts that are attempting to influence classroom teaching in positive ways (e.g., Calfee & Perfumo, 1996; Lucas, 1988a; Lucas, 1988b; Murphy & Underwood, 2000; Wolf, 1988). These case studies show substantial variation in how portfolios are used (e.g., Belanoff & Dickson, 1991; Cooper & Odell, 1999; Murphy & Underwood, 2000; Tchudi, 1997; Tierney, Carter & Desai, 1991; Underwood, 1999; Yancey, 1992). Whether portfolios are used in a single classroom or across a school or district, those who implement portfolios have to make decisions about: (a) what student writing to include in the portfolio; (b) what reflective pieces to include; and (c) how to evaluate and to whom to communicate the evaluation of the portfolio. In the rest of this section we concentrate on how educators who use portfolios to help students learn to write make sound instructional decisions along these dimensions.

The Kinds of Writing

Students learn best from portfolios when they have choice about the kinds of writing to include. At Mt. Diablo High School, where the English department has been using portfolios for over ten years, the teachers who designed the portfolio program, adhering to relatively standard notions of formally-defined written genres, first asked students to include specific types of writing, such as "a personal memory" and "a piece of descriptive writing." Bergamini, the department chair, writes that the Mt. Diablo staff discovered early on that "categories, types, genres, and labels often stultify students' efforts" (1993, quoted in Murphy & Underwood, 2000, p. 28). Showing the deficiencies of genre categories when conceptualized independent of their rhetorical purposes (see Principle 2), she explains that many students felt that their best writings did not fit into the categories. Mt. Diablo teachers abandoned these categories in favor of more open and more purposeful choices, asking students to include:

a "personal best"
a "most imaginative"
a paper from another discipline
a paper that shows process and revision
a piece that shows potential for further work
a paper that states and supports an opinion
a reflective letter that focuses on one's self as a writer" (Murphy & Underwood, p. 28).

Similarly, the Arts PROPEL project in Pittsburgh, which promoted project-based instruction and portfolio assessment in fine arts and imaginative writing, developed relatively open ways of asking for varied pieces of work. Project participants asked students to include:

a writing inventory based on questions about the students past experiences with writing
a piece of writing selected by the student as "important," with reflections about why and about the experience of writing the piece
two pieces, one the student considers "satisfying" and the other "unsatisfying," with reflections on the qualities of each
a piece that illustrates the students' processes and strategies, with a description of the creation process
a "free pick," with a description of reasons for the selection

a final reflection on the portfolio, including changes the student sees across the year (Murphy & Underwood, p. 76).

The Kinds of Student Reflection

A major goal of most portfolio programs is to help students learn to reflect on their writing and their learning. Consistent with Brown's (1978) general theory of the key role of metacognition in learning, Camp (1992) points out that "In writing as in other performances, we learn in part by looking back on what we have done. In this sense, looking back – reflecting – on the experience of writing a piece or on the written piece itself is an integral part of our becoming more accomplished writers" (p. 61). Yancey (1996) argues that reflection "is most insightful, most generative, when it draws on what Vygotsky (1962) called *spontaneous* knowledge and belief, and when it then juxtaposes these with formal, usually school-based knowledge; and that such juxtaposition is required for the problem-solving that contextualizes and enables learning" (p. 84).

To reflect on their writing in their portfolios, students must learn how to see and evaluate their work. They must be taught how to develop standards for their writing and how to meet those standards. Most portfolio systems encourage formal written reflective pieces as a way to help students develop and articulate their understandings of their progress. Students commonly write one or more pieces in which they reflect on the quality of their writing, their writing process, and/or their growth across time. But simply being asked to write reflective pieces is not enough. The quality of the reflections students produce vary as well, depending on how well teachers teach students to reflect.

The portfolio project with the most fully developed ways of teaching reflection is Arts PROPEL (Camp, 1992; Wolf, 1988). Reflection is deeply embedded in almost every task students do. In the list of tasks in the previous section, students reflect on: (a) past experiences with writing, (b) the quality of varied pieces, (c) their writing process, (d) their reasons for selecting particular pieces, and (e) their changes as a writer across a year's time. Camp (1992) writes about the teaching required to help students reflect in these ways. She reports that it took four years to perfect this system for teaching reflection.

Arts PROPEL teachers begin by modeling reflection and asking students to reflect orally before asking them to reflect in writing even on individual pieces. In the first months students collect all of their writing in folders, including all notes and drafts. They spend time reading their writing silently to themselves and aloud to one another, and listening and responding to others. In whole class discussions, teachers model questions students might answer to help them reflect: " 'What did you like best about the piece?' or 'What in the piece would you like to know more about?' or 'What did you most want your reader to get from this piece?' " (p. 65). Students then carry the lessons they learned in class into small group meetings with their teacher and several peers or discussions with a partner. Teachers, meanwhile, reinforce the students' responses with written comments focused on two points: "one thing that is done well in the writing, and one thing to focus on in future writing" (p. 66). The idea is to help students see the "strengths that can be built upon, and to think about next steps that are likely to yield a return for effort expended" (p. 66).

By October or November students are usually ready to move to the next level, the written reflection. At this point teachers ask them to answer two questions in writing, before the teachers respond:

> What do you like best about this piece of writing?
> Which of your writing skills or ideas are you least satisfied with in this piece? Why? (p. 67)

Teachers prepare their students to answer these questions by using class time to read samples of the writing of past students and by discussing those students' responses to these questions. Teachers illustrate the importance of providing specific answers but also show that one can take many perspectives in answering such questions and that no single perspective is right or best. Once the students answer these questions for their own writing, the teacher engages in a written dialogue with each student, taking that student's answers as the starting point. This process is followed for every major piece of writing. From time to time teachers collect the responses of all the students in the class and share them so that students can see a range of ways to think about their writing and can increase their reflective vocabularies. Although not conceptually difficult to answer, these questions require students to be candid. Camp shows something of the range of student responses to these questions. The answers to the second question, especially the "why" portion, are especially interesting because they provide insight into the students' "criteria for writing and into the ways that they apply them to their own writing" (p. 67):

> I think the way the piece flows is the skill I'm least satisfied with. I think it could have been more smoothly written.
> The arrangement of ideas and what I put down wasn't exactly what I thought in my head. It takes away from the writing and what I'm trying to prove, because I assume people know what I'm talking about.
> I really didn't have a way to conclude this piece because I couldn't find the right words that would make sense (p. 69)

Once students have collected a body of writing and have developed some sense of how to reflect on individual pieces, they begin to reflect in writing on their past experiences with writing. Their teacher might illustrate how she would respond to such questions about her own writing as:

> What kinds of writing have you done in the past?
> What do you like to do most in writing?
> What do you like to do least?
> Where do you get your ideas for writing?
> What do you think is important to know about you as a writer? (p. 69)

Students then answer these questions for themselves.

As time goes on, students develop the skills to look back across a number of samples of their writing and to select pieces for their formal portfolios. They use the criteria they have been developing for judging their work first to select a piece of writing that is important to them and to explain their reasons for their selection. The following questions help them expand their perceptions about the piece:

> Why did you select this particular piece of writing?
> What do you see as the special strengths of this paper?
> What was especially important when you were writing this piece?
> What have you learned about writing from your work on this piece?
> If you could go on working on this piece, what would you do?
> What kind of writing would you like to do in the future? (p. 72)

Around March or April, students must apply more challenging reflective criteria to select a second set of writings for their portfolio, a piece they found satisfying and a piece they found unsatisfying. According to Camp, the contrast helps them "move beyond the view that one is either a good writer or a poor writer" (p. 73) and requires them to have a clear set of standards for their writing.

Toward the end of the year, students look over all the pieces they have written to select an additional piece for their portfolio, the free pick and to specify why they made the choice they did. Finally, they look back over their portfolios and respond to the following questions, or some variation, designed to help them evaluate their growth across the year:

> What do you notice about your earlier work?
> How do you think your writing has changed?
> What do you know now that you did not know before?
> At what points did you discover something new about writing?
> How do the changes you see in your writing affect the way you see yourself as a writer?

> Are there pieces you have changed your mind about – that you liked before, but don't like now, or didn't like before but do like now? If so, which ones? What made you change your mind?
> In what ways do you think your reading has influenced your writing? (p. 76)

Klimenkov and LaPick (1996) provide another well-elaborated case study of how students learn to assess themselves, this time with young children. Childen in LaPick's K-1 class develop goals based on their identification of what they can do now and what they can't do yet. Children in Klimenkov's sixth grade class set goals and learn to use scoring rubrics to assess how well they have met their goals. The younger students in LaPick's class have an older student buddy from Klimenkov's class. When they meet together, the older students show the younger students their best work and explain why it is their best. They also learn how to help the younger students choose showcase pieces and articulate what is good about the showcase work. Finally, they help their little buddies identify what they do well in school. Ultimately, these reflective meetings are designed to help the older and younger students get ready for a major conference in which, together with their teacher, they will share their portfolios and their progress with their parents.

The kinds of reflective activities embedded in the Arts PROPEL curriculum and those described by Klimenkov and LaPick help students develop standards for their writing as they come to see what they do well and what they still need to learn. These activities also promote communication between teachers and students as well as among students. Student reflections yield important information for teachers about how students understand what they are learning and provide a base for further teaching. Teachers can use the information found in students' reflective writings to better meet the needs of their varied students. Finally, the understandings that come from reflective self-evaluation encourage students to take ownership of their writing and see it as something more than an assignment for school or a test of their competence. This kind of atmosphere helps writers build on the opinions and expertise of others.

Most students need a great deal of support to learn what is involved in reaching the high levels of reflection obtained by students in the Arts PROPEL project and in Klimenkov and LaPick's classrooms; one cannot simply tell students to reflect and feel confident that high quality reflection will be the result. Unless students are taught to reflect and have numerous opportunities to do so, they will continue to rely on their teachers to tell them how they are performing and what they need to do to perform better.

Evaluating the Portfolio and Broadening the Audience

Classroom teachers who think about evaluating student portfolios need to answer two main questions: What kinds of useful information can portfolios yield? And what is the best way to get and communicate that information?

Some of the most productive portfolio evaluation schemes involve preparing students to assess themselves. If grades are given, they generally are negotiated between the teacher and student as part of teaching-learning dialogues (Klimenkov & LaPick, 1996; Murphy & Camp, 1996). Teachers can play a collaborative role, helping the student succeed rather than just judging the student.

Educators in Great Britain developed a large-scale formative assessment system specifically to provide a structure for gathering systematic information about the growth and development of individual students and for keeping records of that growth: the Primary Language Record (PLR). The PLR covers all areas of language – writing, reading, listening, and speaking. The main goal is to produce "information which would be directly useful informing decisions about teaching and learning" (Barrs, 1990, p. 244). The educators who created the PLR developed a detailed handbook for teachers (Barrs, Ellis, Hester & Thomas, 1987). Drawing from techniques used in studies of child development and in studies inside classrooms, the PLR Handbook authors provide a matrix to help teachers keep organized notes on individual students' uses of language, including the child's behavior over time, patterns of learning, and the adequacy of the contexts for learning. Teachers also take regularly scheduled samples of each child's language and analyze those samples in some depth (Barrs et al., 1987). For writing, students create portfolios, which become part of their cumulative record and which teachers study from one year to the next. Teachers routinely meet with each student for a 20 minute conference, during which the child reflects on him or herself as a language user and learner. According to Barrs, these conferences are useful not only for helping students learn to reflect on their learning but especially for understanding the particular backgrounds and needs of bilingual students. Teachers can discover the "bilingual pupil's linguistic range and their literacy in their first language." Barrs claims, that this information "will, in turn, illuminate their development in English" (p. 247). Besides the classroom-based observations and the input from the child, the PLR includes teachers' and parents' perspectives on the child's development. These records of student progress provide ways of managing and organizing observations and can be shared with others in ways that help teachers establish common standards and increase their professional knowledge. The PLR "is multidimensional in that it facilitates observation

- on a variety of occasions, and over time
- in a variety of contexts
- using a number of different techniques of recording and assessment" (pp. 251–52).

Koelsch and Trumbull (1996), who introduced portfolio assessment to schools in the Chinle, Arizona School District on the Navajo Nation, also argue that "Portfolio assessment is especially appealing for use with ethnolinguistically nondominant students because of its ability to contextualize student performance and because of its flexibility to include a range of types of students performance" (p. 263). In the Chinle district, the "long term goal for schooling is that Navajo culture and the Navajo way of being walk side by side with non-Native culture . . . and ways of being throughout students' K-12 education" (p. 265). Thus, portfolios can be evaluated for how well students exhibit both non-Native and Native communication skills.

When portfolios have been institutionalized as part of school, district or state accountability systems much attention is focused on creating valid and reliable scoring systems, with teachers commonly involved in scoring portfolios. In this way, many teachers have gained valuable experience in setting standards and learning to respond specifically to student portfolio writing. The scoring rubrics are usually keyed to standards which then are keyed to common features of writing, such as purpose; audience; organization; development of ideas and support for points; voice or tone; sentence variety and construction; language, usage, mechanics, and grammar. Sometimes teachers borrow from the scoring rubrics created for large-scale evaluations and teach their students to apply them, in this way making standards explicit.

Mt. Diablo teachers introduced an interesting evaluation innovation in conjunction with school-level scoring of their students' portfolios. Students get a letter at the end of each year in which a teacher or other adult comments on their portfolios. These letters are not written by their classroom teacher, so students gain a new perspective and a fresh response to their portfolios. Murphy and Underwood explain that the letters serve the goal of teaching in that they function "to give focused feedback on student writing, to remind students about what they had learned, and to make suggestions about what they might work on next" (p. 37). Some students even write back to the letter writers. Although the letter writing proved time consuming for the writers, who were mostly teachers, even after the formal state and district funding for the portfolio project came to an end, the Mt. Diablo staff considered the portfolios and the letters so worthwhile that they found creative ways to continue these activities.

CONCLUSION

Research and practice have shown that writing development is far from a linear process involving the acquisition of a set of discrete skills. Consequently, writing instruction must take into account the contexts for student writing – the context of oral language and culture, the context of purpose and audience, the context of personal meaning. Students must learn that being a writer means being a responsible part of a community, and this takes place in concrete ways in the best writing classes – when teachers relate written language to oral language, when they connect the teaching of varied genres to personal and social issues, and when they teach students how to be responsible for ideas and expression in their writing across time.

Among the numerous challenges of teaching writing in context is that context must be related to composing processes and written forms. In addition, the central contextual elements discussed here – of culture, language, purpose, audience, and content – occur in school contexts where issues of educational accountability and physical resources, to name a few, exert their own influences on how students and teachers can work together to achieve academic goals. Writing assessments, for example, do not always acknowledge issues of culture, process, or purpose. When such tests carry high stakes, they can dominate classroom life in ways that diminish instructional opportunities (Freedman, 1993; Loofbourrow, 1994), for example by forcing teachers to spend inordinate amounts of time drilling students on grammatical rules out of context, something that has been found not to support writing development (Elley, 1979).

Although we have written about a range of functions for writing in this chapter, writing is part of the teaching and learning of most school subjects, since it can function to help students remember, synthesize, and process complex information of all kinds. For this reason writing can play an important role in learning activities in most of the disciplines represented by chapters in this volume. We conclude by suggesting that the type of meaningful writing instruction we advocate, based on our readings of research and practice, can be strengthened by connections across the subject areas in school – an effort this volume could foster.

ACKNOWLEDGMENTS

Freedman worked on this chapter while she was a Fellow at the Center for Advanced Study in the Behavioral Sciences. She gratefully acknowledges the support she received from grants to the Center from The Spencer Foundation and the William and Flora Hewlett Foundation. Daiute gratefully acknowledges The William T. Grant Foundation and The Spencer Foundation for their support of her research related to this chapter.

REFERENCES

Atwell, N. (1987). *In the middle: Writing, reading, and learning with adolescents*. Portsmouth, NH: Heinemann.

Au, K. (1980). Participation structures in a reading lesson with Hawaiian children. *Anthropology and Education Quarterly, 11*, 91–115.

Barrs, M. (1990). *The primary language record*: Reflection of issues in evaluation. *Language Arts, 67*(3), 244–253.

Barrs, M., Ellis, S., Hester, H., & Thomas, A. (1987). *The Primary Language Record handbook for teachers*. Portsmouth, NH: Heinemann.

Belanoff, P., & Dickson, M. (1991). *Portfolios: Process and Product*. Portsmouth, NH: Boynton/Cook Publishers.

Bereiter, C., & Scardamalia, M. (1987). *The psychology of written composition*. Hillsdale, NJ: Lawrence Erlbaum Associates, Publishers.

Britton, J., Burgess, T., Martin, N., McLeod, A., & Rosen, H. (1975). *The development of writing abilities: 11–18*. London: Macmillan Education Ltd.

Bruner, J. (1986). *Actual minds, possible worlds*. Cambridge, MA: Harvard University Press.

Brown, A. L. (1978). Knowing when, where, and how to remember: A problem of metacognition. In: R. Glaser (Ed.), *Advances in Instructional Psychology*. New York: Halsted Press.

Calkins, L. M. (1986). *The art of teaching writing*. Portsmouth, NH: Heinemann.

Cazden, C. B. (1988). *Classroom discourse: The language of teaching and learning*. Portsmouth, NH: Heinemann.

Clay, M. (1991). *Becoming literate: The construction of inner control*. Portsmouth, NH: Heinemann.

Calfee, R., & Perfumo, P. (Eds) (1996). *Writing portfolios in the classroom: Policy and practice, promise and peril*. Mahwah, NJ: Lawrence Erlbaum Associates.

Camp, R. (1992). Portfolio reflections in middle and secondary school classrooms. In: K. B. Yancey (Ed.), *Portfolios in the Writing Classroom: An Introduction*. Urbana, IL: National Council of Teachers of English.

Cooper, C., & Odell, L. (Eds) (1999). *Evaluating writing: The role of teachers' knowledge about text, learning, and culture*. Urbana, IL: National Council of Teachers of English.

Cope, B., & Kalantzis, M. (Eds) (1993). *The powers of literacy: A genre approach to teaching writing*. Pittsburgh, PA: University of Pittsburgh Press.

Daiute, C. (2000). Narrative sites for youths' construction of social consciousness. In: M. Fine & L. Weis (Eds), *Construction Sites: Excavating Class, Race, Gender, and Sexuality Among Urban Youth*. New York. Teachers College Press.

Daiute, C. & Griffin, T. M. (1993). The social construction of written narrative. In: C. Daiute (Ed.), *The Development of Literacy Through Social Interaction*. San Francisco, CA: Jossey Bass.

Daiute, C., Campbell, C., Cooper, C., Griffin, T., Reddy, M., Tivnan, T. (1993). Young authors' interactions with peers and a teacher: Toward a developmentally sensitive sociocultural literacy theory. In: C. Daiute (Ed.), *The Development of Literacy Through Social Interaction*. San Francisco, CA: Jossey Bass, Publishers.

Davis, B., Scriven, M., & Thomas, S. (1987). *The evaluation of composition instruction* (2nd ed.). New York: Teachers College Press.

Diederich, P. (1974). *Measuring growth in English*. Urbana, IL: National Council of Teachers of English.

DiPardo, A., & Freedman, S. (1988). Peer response groups in the writing classroom: Theoretic foundations and new directions. *Review of Educational Research, 58*(2), 119–149.

Dyson, A. H. (1989). *Multiple worlds of child writers: Friends learning to write*. New York, NY: Teachers College Press.

Dyson, A. H. (1993). *Social worlds of children learning to write in an urban primary school*. New York, NY: Teachers College Press.

Dyson, A., & Freedman, S. W. (1991). Writing. In: J. Flood, J. Jensen, D. Lapp & J. Squire (Eds), *Handbook of Research on Teaching the English Language Arts* (pp. 754–774). New York: Macmillan Publishing Company.

Dyson, A., & Freedman, S. W. (in press). Writing. In: J. Flood, J. Jensen, D. Lapp & J. Squire (Eds), *Handbook of Research on Teaching the English Language Arts* (2nd ed.). Mahwah, NJ: Lawrence Erlbaum Associates.

Elbow, P. (1981). *Writing with power*. London: Oxford University Press.

Elley, W. B., Barham, I. H., Lamb, H., & Wyllie, M. (1979). *The Role of Grammar in a Secondary School Curriculum*. Wellington: New Zealand Council for Educational Research.

Ferriero, E., & Teberosky, A. (1982). *Literacy before schooling*. Portsmouth, NH: Heinemann.

Flower, L. S., & Hayes, J. R. (1980). The dynamics of composing: Making plans and juggling constraints. In: L. W. Gregg & E. R. Steinberg (Eds), *Cognitive Processes in Writing* (pp. 31–50). Hillsdale, NJ: Lawrence Erlbaum Associates, Publishers.

Freedman, S. W. (1992). Outside-In and Inside Out: Peer Response Groups in Two Ninth-Grade Classes. *Research in the Teaching of English, 26*(1), 71–107.

Freedman, S. W. (1993). Linking large-scale testing and classroom portfolio assessments of student writing. *Educational Assessment, 1*(1), 27–52.

Freedman, S. W. (1994). *Exchanging writing, exchanging cultures: Lessons in school reform from the United States and Great Britain*. Cambridge, MA and Urbana, IL: Harvard University Press and National Council of Teachers of English.

Freedman, S. W., Dyson, A. H., Flower, L., & Chafe, W. (1987). *Research in writing: Past, present, and future* (Technical Report No.1), Berkeley, CA: Center for the study of writing.

Graves, D. (1983). *Writing: Teachers and children at work*. Portsmouth, NH: Heinemann.

Guitierrez, K., Badquedano-Lopez, P., Tejeda, C. (1999). Rethinking diversity: Hybridity and hybrid language practices in the third space. *Mind, Culture, and Activity, 6*, 286–303.

Heath, S. B. (1983). *Ways with words: Language, life, and work in communities and classrooms*. New York, NY: Cambridge University Press.

Klimenkov, M., & LaPick, N. (1996). Promoting student self-assessment through portfolios, student-facilitated conferences, and cross-age interaction. In: R. Calfee & P. Perfumo (Eds), *Writing Portfolios in the Classroom: Policy and Practice, Promise and Peril* (pp. 239–259). Mahwah, NJ: Lawrence Erlbaum Associates.

Koelsch, N., & Trumbull, E. (1996). Portfolios: Bridging cultural and linguistic worlds. In: R. Calfee & P. Perfumo (Eds), *Writing Portfolios in the Classroom: Policy and Practice, Promise and Peril*. Mahwah, NJ: Lawrence Erlbaum Associates.

Lee, C. D. (1993). *Signifying as a scaffold for literary interpretation: The pedagogical implications of an African American discourse genre*. Urbana, IL: National Council of Teachers of English.

Loofbourrow, P. (1994). Composition in the context of CAP: A case study of theinterplay between composition assessment and classrooms. *Educational Assessment, 2*(1), 7–49.

Lucas, C. K. (1988a). Recontextualizing literacy assessment. *The Quarterly of the National Writing Project and the Center for the Study of Writing, 10*(2), 4–10.

Lucas, C. K. (1988b). Toward ecological evaluation. *The Quarterly of the National Writing Project and the Center for the Study of Writing, 10*(1), 1–3, 12–17.

Matsuhashi, A. (1982). Explorations in the real-time production of written discourse.

Michaels, S. (1991). The dismantling of narrative. In: A. McCabe & C. Peterson (Eds), *Developing Narrative Structure*. Hillsdale, NJ: Lawrence Erlbaum Associates, Publishers.

Murphy, S., & Camp, R. (1996). Moving toward systemic coherence: A discussion of conflicting perspectives on portfolio assessment. In: R. Calfee & P. Perfumo (Eds), *Writing Portfolios in the Classroom: Policy and Practice, Promise and Peril* (pp. 103–148). Mahwah, NJ: Lawrence Erlbaum Associates.

Murphy, S., & Underwood, T. (2000). *Portfolio practices: Lessons from schools, districts and states*. Norwood, MA: Christopher-Gordon Publishers.

NAEP 1998 writing report card for the nation and the states. (1999). Compliled by Greenwald, E., Persky, H., Campbell, J., & Mazzeo, J. Washington, DC: U.S. Department of Education.

Nelson, K. (1986). Event knowledge: Structure and function in development. Hillsdale, NJ: Lawrence Erlbaum Associates.

Newkirk, T. (1987). The non-narrative writing of young children. *Research in the Teaching of English, 21*, 121–145.

Nystrand, M. (Ed.). *What writers know: The language, proces, and structure of written discourse* (pp. 269–290). New York: Academic Press.

Nystrand, M. with Gamoran, A., Kachur, R., & Prendergast, C. (1997). *Opening dialogue: Understanding the dynamics of language and learning in the English classroom*. New York, NY: Teachers College Press.

Ogbu, J. (1987). Variability in minority school performance: A problem in search of an explanation. *Anthropology and Education Quarterly, 18*, 312–341.

Sommers, N. I. (1980). Revision strategies of student writers and experienced writers. *College Composition and Communication, 31*, 378–387.

Sperling, M. (1990). I want to talk to each of you: Collaboration and the teacher-student writing conference. *Research in the Teaching of English, 24*(3), 279–321.

Simmons, W., & Resnick, L. (1993). Assessment as the catalyst of school reform. *Educational Leadership, 50*(5), 11–15.

Tchudi, S. (Ed.) (1997). *Alternatives to grading student writing*. Urbana, IL: National Council of Teachers of English.

Tierney, R. J., Carter, M., & Desai, L. (1991). *Portfolio assessment in the reading-writing classroom*. Norwood, MA: Christopher-Gordon Publishers.

Underwood, T. (1999). *The portfolio project: A study of assessment, instruction, and middle school reform*. Urbana, IL: National Council of Teachers of English.

Wagner, B. J. (1998). *Educational drama and language arts: What research shows*. Portsmouth, NH: Heinemann.

Wolf, D. P. (1988). Opening up assessment. *Educational Leadership, 45*(4), 24–29.

Yancey, K. (Ed.) (1992). *Portfolios in the writing classroom*. Urbana, Il: National Council of Teachers of English.

Yancey, K. (1996). Dialogue, interplay, and discovery: Mapping the role and the rhetoric of reflection in portfolio assessment. In: R. Calfee & P. Perfumo (Eds), *Writing Portfolios in the Classroom: Policy and Practice, Promise and Peril* (pp. 83–102). Mahwah, NJ: Lawrence Erlbaum Associates.

TEACHING AND LEARNING MATHEMATICS: HOW INSTRUCTION CAN FOSTER THE KNOWING AND UNDERSTANDING OF NUMBER

Mary Kay Stein

INTRODUCTION

My charge in writing this chapter was to provide a systematic review of the classic teaching methods and learning activities known or thought to be effective for the portion of mathematics broadly referred to as "number." According to the recently released *Principles and Standards for School Mathematics (PSSM)*, all of mathematics, from algebra to geometry to data analysis, is strongly grounded in number (National Council of Teachers of Mathematics, 2000, p. 32). I will confine my review to grades K-8, however, because the pedagogical emphasis on number tends to be considerably greater during these years (National Council of Teachers of Mathematics, 2000). For the purposes of this chapter, I include the following in my treatment of the teaching and learning of number: number sense (reasoning with and about numbers); properties of and operations with integers and rationale numbers; ratio and proportion; and representing number in concrete, graphic, and symbolic forms.

My focus is further refined by the charge to overview what is known or thought to be effective in teaching mathematics for "the development of student understanding of its big ideas, appreciation of its value, and capability and

Subject-Specific Instructional Methods and Activities, Volume 8, pages 111–144.

ISBN: 0-7623-0615-7

disposition to apply it throughout their lives" (page 7, introductory chapter, this volume). I view this perspective on knowing and understanding mathematics as essentially congruent with the goals of the National Council of Teachers of Mathematics (NCTM) which recommend that students "learn important mathematical concepts and processes *with understanding*" (NCTM, 2000, p. ix). More specifically, the *PSSM* stresses that powerful performances in mathematics draw upon a combination of conceptual understanding, factual knowledge, and procedural proficiency.

Also stressed in NCTM reform documents, as well as in other consensus documents (National Research Council, 1989; Mathematical Association of America, 1991), is the need to develop students' appreciation for mathematics and the predisposition to apply and use it. In short, mathematical expertise is seen as extending beyond mastery of the structures, concepts, procedures and facts of the discipline to include the "habits of mind" that underlie robust problem-solving in communities of mathematical practice: framing problems, seeking solutions, searching for patterns, formulating conjectures, and appealing to mathematical logic and reasoning as evidence for one's reasoning. This view of mathematical expertise stresses becoming an expert user and maker of mathematics through the gradual appropriation of the discourse style, values, and norms of the mathematical community.

This chapter treats students' development of mathematical understanding and their socialization into the practice of making and using mathematics as equally important and complementary goals of mathematics instruction. However, the underlying paradigms that guide how educators and learning theorists think about these two goals are quite distinct. Typically, cognitive psychological models are summoned to help us think about mathematical understanding, while sociocultural models are used to aid our thinking with regards to becoming a member of a mathematical community of practice. Because both constitute an important basis for understanding the rationale for various teaching methods to be discussed in this chapter, I will begin with a brief review of the primary learning paradigms that underlie what it means to know and understand mathematics. I will then turn to the main part of this chapter, in which I review what is known about effective teaching methods and learning activities in the teaching of number. Although discussed under the headings of tasks, discourse, and classroom norms, I will from time to time refer back to the cognitive psychological and sociocultural paradigms discussed earlier to provide rationales for these recommended teaching practices. I will conclude with a discussion of what makes effective teaching in mathematics distinct from generic teaching methods.

HOW IS MATHEMATICS LEARNED?

Learning Mathematics with Understanding

The most widely accepted view of mathematical understanding defines it in terms of the way information is represented and organized in the minds of individuals. Simply stated, "a mathematical idea or procedure or fact is understood if it is part of an internal network" (Hiebert & Carpenter, 1992, p. 67). Networks that contain multiple and strong connections among mathematical ideas, among different representations of the same idea, and between mathematical concepts and procedures are considered to represent more sophisticated mathematical understanding than are sparsely delineated networks – those with few, non-reinforced linkages and linkages that are not multiply connected to more than one piece of information.

In this view of mathematical understanding, new knowledge builds on prior knowledge. Hence, individuals with well-structured, rich, and coherent networks are able to more readily build bridges to newly encountered information, thereby allowing them to integrate and understand the new information more easily. According to Hiebert and Carpenter (1992), well-organized networks are efficacious because they promote remembering, reduce the amount that must be remembered, enhance transfer and influence beliefs that mathematics is a well-ordered, meaningful discipline (pp. 74–77).

Cognitive psychological approaches to learning contend that individuals develop more sophisticated and elaborated understandings in two ways: by adding to and by reorganizing their internal networks of facts, procedures, and concepts. Students can simply add new pieces of information when it fits neatly into an already organized scheme, such as when second graders encounter the hundreds place for the first time and incorporate it into their understanding of place value as an interconnected system of multiples of ten. They reorganize their existing network, on the other hand, when incoming information clashes with their existing network. For example, students who have conceptualized multiplication as always making a number larger will need to reorganize their understandings of the concept of multiplication when they encounter rationale numbers. In general, reorganizations are much more difficult, lengthily, and anxiety provoking than are simple additions to one's network; they also produce new insights into the structure of mathematics that make the difficulty worthwhile, however.

Becoming an Expert Practitioner in a Mathematical Community of Practice

Thinking about mathematical expertise as a practice entails shifting from the view of learning as something that happens within the minds of individual students to a view of learning as something that happens in the "fields of interaction" (Bredo & McDermott, 1992, p. 35) among students and between students and teacher. Knowing mathematics means *valuing* mathematics, appropriately *using* and *making* mathematics, and *developing an identity* for oneself as an expert member of a community whose practices embody mathematical values and modes of thinking (e.g. modeling; using symbols; engaging in inference, logical analysis and mathematical argumentation).

As with cognitive psychological models of mathematical understanding, this view of expertise also recognizes a distinct role for disciplinary knowledge, but places a greater emphasis on the "socially developed and patterned ways" (Scribner & Cole, 1981, p. 236) of using the discipline of mathematics (e.g. the genres of mathematical discourse, the norms of mathematical knowledge production) than on the content and structure of mathematical knowledge. Mathematics is seen as a way of viewing and making sense of the world, not as a finished record of knowledge to be acquired. As such, the mathematically astute are those who understand the rules by which mathematics is made and applied and who display the predispositions to talk about and use mathematics in appropriate ways.

Sociocultural views of learning contend that individuals learn through participation in the social practice of mathematics. In the process of discussion and debate, students develop the intellectual tools needed to think and reason mathematically. Following the Vygotskian tradition (Vygotsky, 1978; 1986; Wertsch, 1985), sociocultural theorists believe that thinking and reasoning first appear in social interactions that take place between actors on the *inter*mental plane and gradually become appropriated by the individual on the *intra*mental plane. Classroom communication that is patterned, structured, and supported by jointly constructed norms is a crucial – if not the crucial – element of classrooms that support students' learning in this way.

The present reform movement incorporates elements of both of the above ways of thinking about mathematical expertise. Students are expected to develop powerfully organized knowledge bases *and* identities as knowers, valuers, and users of mathematics. Writing from a research perspective, Sfard (1998) contends that neither paradigm alone is sufficient for examining the teaching and learning of mathematics. She notes that, while the "acquisition" metaphor suffers from the "learning paradox" (how to acquire knowledge of something not yet known) and potentially unhealthy norms surrounding knowledge as property, the "participation" metaphor suffers from an inability to construe a reasonable explanation of (or substitute for) transfer and from the gradual disappearance of subject matter knowledge when the focus is held for too long and too intensely on actions and context. The best judgment seems to be that the field can benefit from the energetic development of both paradigms (Greeno, 1997). This includes finding synergistic ways to coordinate them and/or choosing which to draw upon based upon one's goals at the time (Cobb, 1994; Sfard, 1998).

EFFECTIVE INSTRUCTION IN NUMBER

Over the past several decades, evidence has mounted that American students are not learning mathematics with understanding. Nor are they acquiring a sense of the discipline as a useful sensemaking pursuit. Reporting on the findings from the Sixth National Assessment of Educational Progress (NAEP), Silver (1997) notes the continuation of a historical trend: improvements in students' basic knowledge and skills, but little progress on students' development of more advanced, higher-level knowledge and skills. Most recently, results from the Third International Mathematics and Science Study (TIMSS) reveal that our students are being outperformed by their international counterparts (Schmidt, McKnight, Cogan, Jakwerth & Houang, 1999). Although, the U.S. fourth graders placed somewhat above the international mean, the United States scored below the international mean in every area of mathematical competence for eighth graders.

Many trace the poor performance of American students to an instructional style that does little to encourage understanding or develop active minds. Over the past 15 years, findings from a number of different classroom studies have converged to produce a consistent portrait of the nature of instruction typically found in mathematics classrooms. From these studies, it is clear that conventional mathematics instruction places a heavier emphasis on memorization and imitation than on understanding, thinking, reasoning, and explaining. For example, data on instructional practices gathered as part of the last two NAEP examinations indicate that mathematics teachers are more concerned with students' rote use of procedures than with their understanding of concepts and development of higher-order thinking skills (Dossey, Mullis, Lindquist & Chambers, 1988; Grouws & Smith, in press). This finding is consistent with the observational research of Stodolosky (1988), who found that mathematics instruction in fifth-grade classrooms was almost entirely devoted to learning facts and skills, and the survey-based research of Epstein and MacIver (1989) who identified a "conservative emphasis on basic skills" (p. 31). Moreover, studies that have focused on economically disadvantaged students have suggested that mathematics instruction in these settings tends to place even more emphasis on rote computation and less on concepts and applications (Porter, Floden, Freeman, Schmidt & Schwille, 1988).

When concepts are the focus of instruction, they tend to be directly taught rather than developed in American classrooms (Stigler & Hiebert, 1997). Typically, teachers explicitly tell their students the important ideas and mathematical relationships that they want them to learn, with little or no emphasis on how those concepts were derived. Even less frequent are opportunities for students to derive mathematical concepts and procedures through their own problem solving efforts. One of the most striking findings of the videotape portion of TIMSS was the sharp difference between Japanese and German lessons vs. U.S. lessons in this regard. Concepts were significantly more likely to be stated (simply provided by the teacher or students but not explained or derived) in the United States (78% stated vs. 22% developed). In Japan and Germany, on the other hand, concepts were more often developed (83% developed vs. 17% stated in Japan and 77% developed vs. 23% stated in Germany) (Stigler & Hiebert, 1997).

With the development and broad dissemination of the NCTM Standards (1989; 1991; 1995, 2000),[1] the country's attention has turned to transforming the way in which our students learn mathematics. The NCTM documents have set into motion an unprecedented degree of professional development, curriculum development and classroom-based reform efforts.

The overall image of instruction recommended by NCTM is one of using problem contexts for developing concepts and procedures. Alternatively referred to as problem-centered teaching, investigations, inquiry-based teaching, or project-based pedagogy, this form of instruction places heavy emphasis on highly designed problem situations and students' use of models and oral language to describe, represent, and make meaning from those situations. Students are asked to solve problems and, in the course of doing so, to observe patterns and relationships, conjecture, test, and discuss what they are observing and hypothesizing. Only after they've developed some understanding of the situation (often qualitatively), are they encouraged to connect their work with mathematical symbols and conventional procedures. The goal is to develop and build on the personal and collective sensemaking of the class rather than to encourage the rote memorization of procedures disconnected to underlying meaning.

We are beginning to amass evidence that this approach to mathematics instruction is associated with increases in student performances on assessment tasks that measure students' capacity to think, reason, and communicate. (Carpenter, Fennema, Peterson, Chiang & Loef, 1989; Cobb, Wood, Yackel, Nicholls, Wheatley, Trigatti & Perlwitz, 1991; Hiebert & Wearne, 1993; Griffin, Case & Stigler, 1994). Although most of the evidence has been obtained with elementary school students on fairly simple mathematical topics (e.g. addition and subtraction of whole numbers, place value), evidence that instruction focusing on higher level cognitive demands leads to improved student performance has also been shown for the middle (Stein & Lane, 1996) and higher (Newman, Marks & Gamoran, 1995) grades.

There are many ways to slice through the holistic account of recommended instructional practice discussed above. In the remainder of this chapter, I will analyze this instructional approach in terms of three interactive elements: tasks, discourse, and classroom norms, a typology also used by the National Council in their *Professional Standards for the Teaching of Mathematics* (NCTM, 1991).

At the outset, it is important to note that it is difficult to examine these elements in isolation from one another. For example, attempts to orchestrate classroom discourse that is impassioned, mathematically interesting, and equitable across students will falter unless the task one has selected is rich and allows for multiple solution strategies and points of entry. Attempts to encourage students to think deeply about interconnections between percents and decimals – even with the best of tasks – will falter if students have grown accustomed to algorithmic solutions and rewards based on the number of problems completed in a standard amount of time. Thus, although tasks, discourse, and classroom environment will be discussed separately, it should be understood that they contribute interactively to effective instruction.

Instructional Tasks

The selection of instructional tasks is arguably the most important decision that a teacher makes. Across academic subject areas, tasks have been identified as the essential link between teaching and learning (Doyle, 1983). In mathematics, tasks determine what content students have the opportunity to learn and influence their perceptions of what mathematics is. What are the characteristics of the kinds of tasks with which students should become engaged?

Focus on Important Mathematical Ideas

One of the tenets of the mathematics reform is that students should become engaged with fewer tasks that focus on important mathematics as opposed to many tasks that focus on trivial or tangential mathematics. An important finding from the videotape portion of TIMSS was the general lack of coherence of most American lessons. Lessons in the United States tended to consist of "a mosaic of fragmented, small tasks" (Schmidt et al, 1999, p. 5) while the Japanese lessons were more coherent, primarily due to the fact that they were designed around a single instructional task that focused on an important mathematical idea.

What are the important mathematical ideas on which tasks should focus? The *PSSM* points to the significance of "foundational ideas" in their discussion of important mathematics. Foundational ideas are those which enable students to understand other mathematical ideas and to connect different areas of mathematics. They include place value, equivalence, proportionality, function, and rate of change. Because these ideas reappear throughout the K-12 curriculum, how students come to understand them has far-reaching implications throughout their mathematical careers.

Tasks that focus on important mathematical ideas, more often than not, will also have "longitudinal coherence" (Ma, 1999). When selecting tasks, teachers need to have one eye on where the students are coming from and the other on where they are moving to mathematically. Tasks that have longitudinal coherence are those which build on students' prior understandings and leave behind "residue" (Davis, 1992) – insights, skills, or understandings – that can be accessed and productively used in future learning. Heibert and his colleagues have suggested two types of residue that may be exceptionally robust for future learning: (a) insights into the structure of mathematics (which are likely to be developed when students are invited to explore relationships in the context of solving problems); and (b) strategies or methods for solving problems, including both specific techniques for specific kinds of problems (the aim of most of traditional instruction) and a general approach for how to construct their own methods for solving problems (Hiebert, Carpenter, Fennema, Fuson, Wearne, Murray, Olivier & Human, 1997).

Another line of thinking on important mathematics points to the wisdom of designing instructional tasks based on our knowledge of the big transition points in students' learning of mathematics. We know, for example, that while students develop additive reasoning quite naturally, they typically have difficulty moving from additive to multiplicative forms of reasoning (Lamon, 1995; Simon & Blume, 1994). This suggests that students will need to regularly encounter situations which require multiplicative reasoning in different contexts (i.e. fractions, decimals, ratios, rates, proportions, and percents) and be encouraged to discover parallels in their reasoning processes and emerging insights across these contexts. Situations that provoke students to reconceptualize important concepts related to multiplicative reasoning (e.g. what constitutes a unit, intensive quantities) will also need to be purposively arranged.[2]

From a cognitive psychological perspective, the rationale for tasks that focus on important mathematical ideas is clear. For example, foundational ideas constitute the hubs around which richly interconnected networks of facts, procedures, and other concepts can form. Tasks that focus on transition points in students' understanding are effective because they provoke reorganization of students' internal networks to accommodate new information. If the overall aim is to develop more sophisticated networks of understanding that provide new and more accurate insights into the structure of mathematics, tasks that focus on important mathematical ideas will necessarily form the backbone of effective mathematics instruction.

Utilize Multiple Representations

Effective mathematics instruction also emphasizes students' use of manipulatives, diagrams, sketches, and other representational models, along with oral language, to describe, represent, and make personal meaning out of quantitative situations. Tasks that utilize multiple representations offer students the opportunity to think in diverse ways about important mathematical concepts and provide insight into the structure of mathematics. As students use different representations to think about relationships and patterns, they develop rich, personal, often qualitative understandings of ideas and concepts. These understandings are seen as providing a foundation on which students can later build proficiency in the use of mathematical symbols, definitions, and conventional procedures.

Consider decimals. Students often have difficulty working with the symbolic notation used to represent decimals. For example, Kouba, Carpenter and Swafford (1989) reported that less than half of the seventh graders assessed by NAEP were able to identify the greatest number in the following set of numbers: 0.36, 0.058, 0.375, and 0.4. Resnick, Nesher, Leonard, Magone, Omanson, and Peled (1989) found similar difficulties in basic understanding of decimal place value and the relationship between fractions and decimals, and they argued that these errors result from students applying rules that are not based on a sound understanding of decimal and fraction concepts. For example, many students decide that the number 4.7 is less than the number 4.08 "because the zero does not matter and 8 is bigger than 7" (Resnick et al., p.20).

Tasks that use multiple representations help students to see, talk about, and analyze the inherent structure of the rational number system. For example, the decimal squares representation, shown in Fig. 1, is a particularly powerful representation for building students' understanding of the relationship among the three most commonly used decimal place values: tenths, hundredths, and thousandths.

This model consists of three congruent square-units, each divided into a multiple of 10. The tenths square-unit is divided into 10 strips; the hundredths

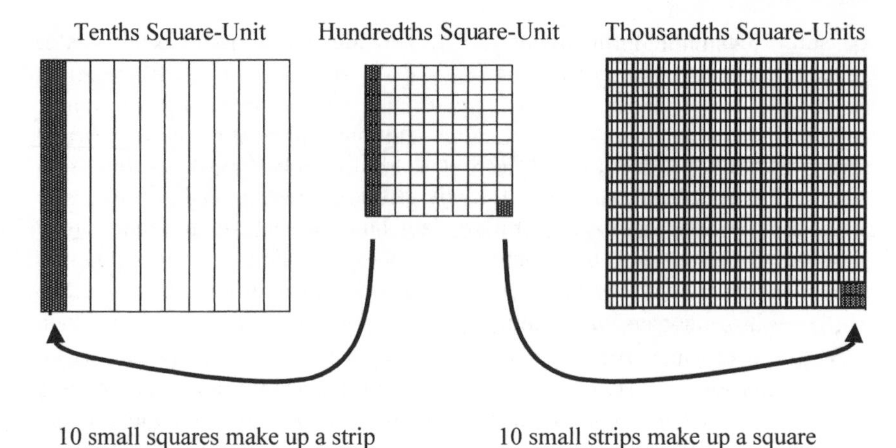
Tenths Square-Unit
Hundredths Square-Unit
Thousandths Square-Units
10 small squares make up a strip
10 small strips make up a square

Fig. 1. The decimal squares representation. (Taken from Silver & Stein, 1996; Reprinted with the permission of Sage Publications.)

square-unit into 100 squares: and the thousandths square-unit into 1000 "small strips." The three square-units are explicitly related to one another, as, for example, 10 small squares make up one strip and 10 small strips make up a square.

As shown in Fig. 2, the decimal squares model makes the differences in the quantities represented by decimals very salient.

As shown in the first row, by connecting multiple representations of the same rational number, the decimal squares representation helps students to see how two tenths is the same quantity as is 20 hundredths, which is the same amount as is 200 thousandths. The decimal squares can also clarify the actual size discrepancy of different decimals that have superficially similar notations (see bottom row, Fig. 2).

Tasks that utilize nonsymbolic representations are effective because they empower students to think in diverse ways about quantity and relationships. In so doing, they aid in the students' development of strong, multiply connected internal networks. The presumption is that students bring with them a storehouse of information already connected to the alternative representations (Hiebert & Carpenter, 1992). In the decimal squares case, the written symbols .2, .02, and .002 become informed by the multiple associations that students have already constructed for squares and fractional quantities. As such, the written symbols become "understood" as links are formed between them and

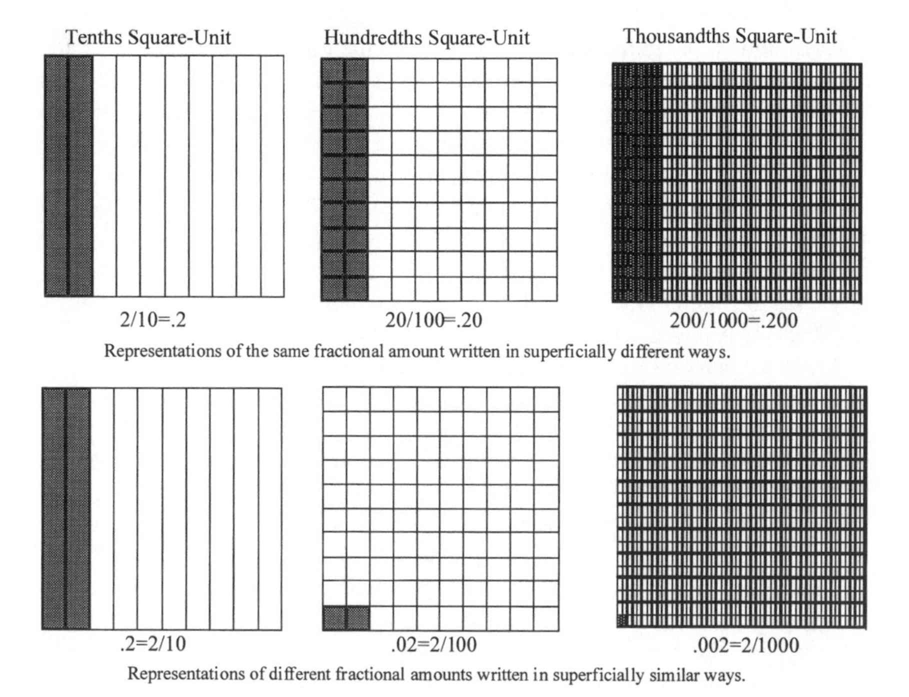

Fig. 2. Using the decimal squares representation to compare and contrast decimal. (Taken from Silver & Stein, 1996; Reprinted with the permission of Sage Publications.)

this much vaster network. When the symbols for decimals are fully integrated into this network, students are much less likely to view 4.7 as less than 4.08.

Can be Solved Using Multiple Strategies

Effective mathematics instruction incorporates the use of tasks that can be solved in multiple ways. Such tasks increase students' flexibility of thinking about mathematical ideas, problem settings and solution methods. Because tasks that can be solved in multiple ways encourage students to invent their own procedures, they help students develop skills in and confidence as mathematical reasoners and problem solvers. Invented strategies also have the advantage of being tied to students' conceptual understandings, rather than simply being procedures that they memorize and apply without much thoughtfulness or connectedness.

A common topic in the middle-school mathematics curriculum is conversion among fractions, decimals, and percents. The conventional form of these tasks consists of scores of similar problems for which students have been trained to use *a* procedure (e.g. to convert from a fraction to a decimal, divide the numerator by the denominator). This approach fails to explore the meaning of fractions, decimals and percents and encourages students to become dependent on poorly understood algorithms. Not surprisingly, when they encounter problems that fall outside the format to which they've become accustomed – problems in which in they have to reason with quantities expressed as fractions, decimals, or percents – they find themselves ill-equipped to do so.

Another approach to converting among fractions, decimals, and percents – one that illustrates the benefit of using multiple strategies – is illustrated in Fig. 3.

This task requires students to think and reason about the actual quantities involved (i.e. what part of the whole rectangle do the six shaded squares represent?) before they consider how to represent those quantities as a fraction, decimal or percent. And, because students chose to shade the six squares in different configurations, it also encourages many different ways of making meaningful partitions of the rectangle, each of which lends different insights into the part-whole relationships embodied by fractions, decimals and percents. For example, one student might shade the six squares as shown in Fig. 4 and

Shade 6 of the small squares in the rectangle shown below.

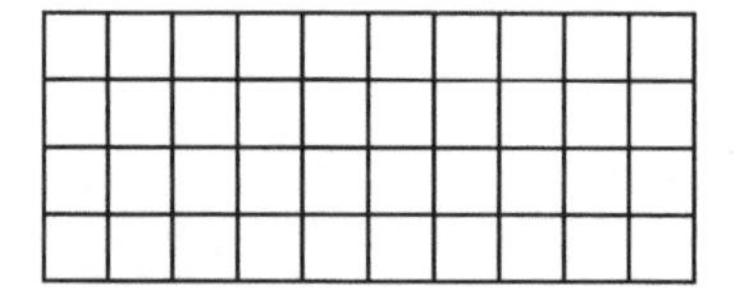

Using the diagram, explain how to determine each of the following:

A) the percent of area that is shaded.

B) the decimal part of area that is shaded.

C) the fractional part of area that is shaded.

Fig. 3. Instructional task that can be solved in multiple ways. (Taken from Stein, Smith, Henningsen & Silver, 2000; Reprinted with the permission of Teachers College Press.)

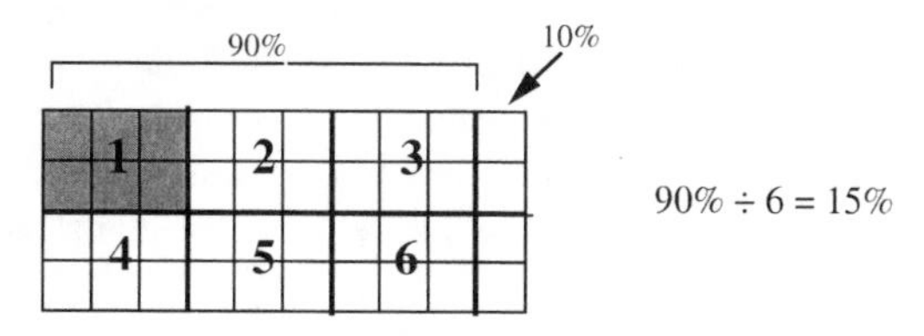

Fig. 4. Possible shading configuration and solution strategy for task shown in Fig. 3. (Taken from Stein, Smith, Henningsen & Silver, 2000; Reprinted with the permission of Teachers College Press.)

begin by reasoning that, since the entire rectangle represents 100%, the "leftover" column on the right represents 1/10 or 10% (i.e. it is one of 10 shaded columns). Thus, the remaining part of the overall rectangle would equal 90% and each of the 2×3 sections would equal 1/6 of the 90% or 15%. The student who reasons her way through the problem in this manner understands a great deal more about the equivalence of 6/40 and 15% than does a student who divides 6 by 40, gets .15, and then moves the decimal point two places to the right.

Another possible strategy begins with the random shading of 6 squares shown in Fig. 5. In this case, the student might begin by trying to figure out the percentage represented by a single square. If the entire rectangle represents 100% and there are 40 blocks, 80% can be distributed across the 40 blocks by giving 2% to each block. That leaves an additional 20% to be distributed evenly across the 40 blocks. Since 20 is half of 40, each block would get another .5%. Thus each block would represent a total of 2.5%. Six blocks would equal $6 \times 2.5\%$ or 15%.

How do students benefit from an instructional practice in which this kind of task is commonplace? First, their own powers of reasoning and sensemaking are continually recruited and validated. Second, when students observe the

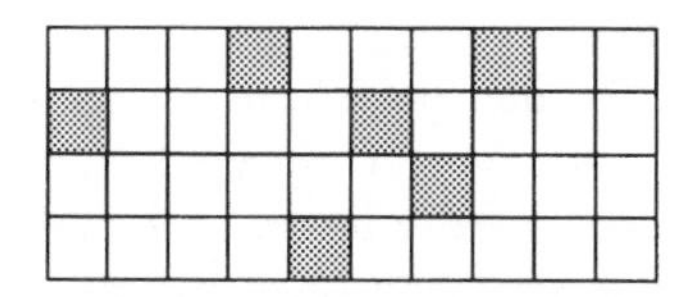

Fig. 5. Possible shading configuration for task shown in Fig. 3. (Taken from Stein, Smith, Henningsen & Silver, 2000; Reprinted with the permission of Teachers College Press.)

variety of approaches that are possible to this one task, they become aware that there is *no one right way* to do *these kinds of problems.* Hence, when confronted with nonroutine problems, they will be less likely to mindlessly try one algorithm after another, and more likely to reason their way through the problem.

Solving the same task in multiple ways helps students develop important mathematical insights as well. By reflecting on what is similar and different about all of the possible approaches to the task in Fig. 3, students can develop insight into what constitutes the key, invariant mathematical idea (the part-whole relationship that sits at the heart of all rational numbers) and where there is room for flexibility of approach and strategy (how the part is configured, the way in which the whole is chunked in order to reason about quantities).

Tasks that can be solved in multiple ways are especially useful when students are learning a new concept or procedure because they encourage students' own sensemaking and serve to further elaborate their networks of understanding. In most cases, the solution methods that students invent will not be as efficient, elegant, or generalizable as the conventional algorithms and it is important that students eventually learn these algorithms as well, preferably in a way that connects to their initial understandings. Beginning with students' own sensemaking, however, ensures that they will possess some conceptual understanding of the related quantities and relationships, thereby giving them something to access and use when confronted with nonroutine problems.

Requires High-level Thinking and Reasoning

Tasks that focus on important mathematics, utilize multiple representations, and can be solved in multiple ways typically encourage students to think and reason in cognitively challenging ways. My colleagues and I have found that the cognitive demands of tasks are another useful lens through which to examine instructional tasks (Stein, Smith, Henningsen & Silver, 2000). By cognitive demands we mean the kind and level of thinking required of students in order to successfully engage with and solve the task.

Tasks that require students to perform a memorized procedure in a routine manner (low-level cognitive demands) lead to one type of opportunity for student learning; tasks that demand engagement with concepts and that stimulate students to make purposeful connections to meaning or relevant mathematical ideas (high-level cognitive demands) lead to a different set of opportunities for student thinking. Day-in and day-out, the cumulative effect of students' experiences with instructional tasks is students' implicit development of ideas about the nature of mathematics – about whether mathematics is something they personally can make sense of, and how long and how hard they should have to work to do so.

Since the tasks with which students become engaged in the classroom form the basis of their opportunities for learning mathematics, it is important for teachers to be clear about their goals for student learning. With learning goals clearly articulated, they can then select or design tasks to match those goals. Awareness of the cognitive demands of different instructional tasks is a central consideration in this matching. For example, if a teacher wants students to learn how to justify or explain their solution processes, she should select a task that is deep and rich enough to afford such opportunities. If, on the other hand, speed and fluency are the primary learning objectives, other types of tasks will be needed.

The example shown in Fig. 6 illustrates four ways in which students can be invited to think about the multiplication of fractions.

As shown on the left side of the figure, tasks with lower-level demands consist of memorizing and executing the rule of multiplying numerator times numerator and denominator times denominator. When these kinds of tasks are used, students typically work 10–30 similar problems within one sitting.

Another way in which students can be asked to think about the multiplication of fractions – one that presents higher-level cognitive demands – might also use general procedures, but do so in a way that builds connections to underlying concepts and meaning. For example, as shown in Fig. 6, students might be asked to use pattern blocks to find 1/6 of 1/2. In so doing, they would be associating the multiplication procedure with specific quantities. In this case, students also must explain their solutions and grapple with the fact that two hexagons constitute the unit whole, an important concept in the transition to multiplicative reasoning.

Another high-level task would entail asking student to generate a real-world situation for a typical problem: $2/3 \times 3/4$. We call this level of task "doing mathematics" because students are challenged to apply their understanding of fractions in novel ways. The student response shown in Fig. 6 shows how, without rules or structure, the student was able to construct a scenario, a representation, and a solution that fit the parameters of the problem. In contrast to the tasks discussed earlier, when working on the kinds of tasks on the right side of Fig. 6, students typically perform far fewer problems (sometimes as few as two or three) in one sitting.

Being aware of the cognitive demands of tasks and matching tasks to one's goals does not in itself, however, guarantee a good lesson. Most teachers with high-level learning goals for their students can successfully identify and set up challenging instructional tasks in their classrooms. Many of these tasks,

Lower-Level Demands	Higher-Level Demands
Memorization What is the rule for multiplying fractions? *Expected Student Response:* *You multiply the numerator times the numerator and the demoninator times the denominator.* OR *You multiply the two top numbers and then the two bottom numbers.*	Procedures With Connections Find 1/6 of 1/2. Use pattern blocks. Draw your answer and explain your solution. *Expected Student Response:* 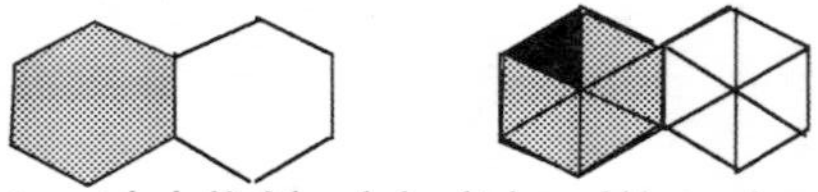 *First you take half of the whole which would be one hexagon. Then you take one-sixth of the half. So I divided the hexagon into six pieces which would be six triangles. I only needed one-sixth so that would be one triangle. Then I needed to figure out what part of the two hexagons one triangle was and it was 1 out of 12. So 1/6 of 1/2 is 1/12.*
Procedures without connections Multiply: $\frac{2}{3} \times \frac{3}{4}$ $\frac{5}{6} \times \frac{7}{8}$ $\frac{4}{9} \times \frac{3}{5}$ *Expected Student Response:* $\frac{2}{3} \times \frac{3}{4} = \frac{2 \times 3}{3 \times 4} = \frac{6}{12}$ $\frac{5}{6} \times \frac{7}{8} = \frac{5 \times 7}{6 \times 8} = \frac{35}{48}$ $\frac{4}{9} \times \frac{3}{5} = \frac{4 \times 3}{9 \times 5} = \frac{12}{45}$	Doing Mathematics Create a real-world situation for the following problem: $\frac{2}{3} \times \frac{3}{4}$. Solve the problem you have created without using the rule and explain your solution. *One Possible Student Response:* *For lunch Mom gave me three-fourths of the pizza that we ordered. I could only finish two-thirds of what she gave me. How much of the whole pizza did I eat?* *I drew a rectangle to show the whole pizza. Then I cut it into fourths and shaded three of them to show the part mom gave me. Since I only ate two thirds of what she gave me, that would be only two of the shaded sections.* *Mom gave me the part I shaded.* 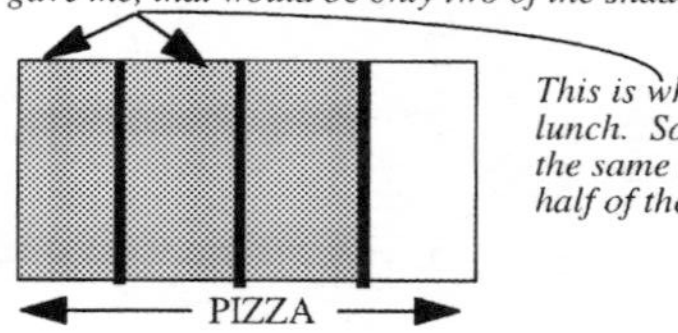 PIZZA *This is what I ate for lunch. So 2/3 of 3/4 is the same thing as half of the pizza.*

Fig. 6. Examples of tasks at each of the four levels of cognitive demand. (Taken from Smith & Stein, 1998; Reprinted with the permission of NCTM.)

however, will fail to engage students in cognitively challenging ways. Instead they will be reduced in the course of their enactment to lower-level or mathematically nonproductive thinking on the part of the students (Stein, Grover & Henningsen, 1996).

A variety of factors conspire to reduce the level of cognitive demand of tasks once they are unleashed into the classroom environment. Because high-level tasks typically are more unstructured and hence anxiety provoking, students often press teachers to show them a series of steps to follow to get to the right answer. In other cases, students might wallow for long periods of time, making little or no progress toward understanding the important mathematical ideas embodied in the task.

We have found that the ways and extent to which the teacher supports students' thinking and reasoning is a crucial ingredient in the ultimate fate of high-level tasks (Henningsen & Stein, 1997). Teachers who successfully encourage their students to think and reason at a high level scaffold students' thinking, provide classroom models of high-level thinking and reasoning, continuously press their students to explain and justify their thinking, provide sufficient time for students to grapple with problems (not too little, but also not too much), and make sure that conceptual connections are drawn between and among ideas.

As tasks are enacted in classroom settings, they become intertwined with the actions and interactions of teachers and students, as well as the history of expectations regarding how work gets done in a particular class. Actions and interactions are ultimately bundled with the language or discourse of the classroom, while the history of expectations of a class constitutes its underlying social and intellectual norms. These are the two topics which constitute the remaining two themes of effective mathematics instruction. I turn now to a discussion of classroom discourse.

Classroom Discourse

At the time the NCTM Standards were first released, the popular vision of mathematics practice was that of a solitary pursuit dominated by numbers and symbols. It is a measure of how much our thinking has shifted that it is now commonly accepted that a productive mathematics lesson is one in which there is a great deal of talk. What kind of talk can be considered the sign of an effective mathematics lesson?

The relationship between language and thinking is complex. The stance taken here is that individuals acquire intellectual skills and dispositions – ways of seeing, styles of reasoning, tools for inferencing, arguing, analyzing, and generalizing – by first using them in social interaction with others. Hence, an effective mathematics classroom is one in which teachers (the experts) gradually socialize students (the novices) into mathematically accepted ways of thinking, reasoning and valuing. By participating in a "microcosm of mathematical practice" (Schoenfeld, 1992), students gradually acquire the habits of mind valued by their classroom community and the larger community of mathematical practitioners. In this view, then, productive forms of talk include classroom discussions in which mathematical ideas are the currency of exchange, and correctness lies in the structure and logic of mathematics (Hiebert et al., 1997).

One of the most frequently cited goals of the mathematics reform is for students to become competent conjecturers, evidence providers, and critics of others' conjectures and evidence. In order to do so, students must learn how to make a claim and present a well-grounded case for why that claim should be accepted. Further, students must learn to listen to, understand, and evaluate the claims of others, both in terms of their accuracy and potential bridges between others' claims and their own.

Given the modal form of mathematics instruction that has prevailed in this country over the past decades, students do not typically arrive at mathematics class prepared to think and reason in these ways. Nor are teachers, most of whom themselves have not learned mathematics in this way, necessarily prepared to support them in learning how to do so. The challenges of incorporating classroom discourse as an integral part of an overall strategy of teaching and learning are great (Lampert, 1998; Hicks, 1998). Doing so successfully involves more than developing a respectful, trusting, and nonthreatening climate for discussion and problem solving. As so aptly described by O'Conner and Michaels:

> The teacher must give each child an opportunity to work through the problem under discussion while simultaneously encouraging each of them to listen to and attend to the solution paths of others, building on each other's thinking. Yet she must also actively take a role in making certain that the class gets to the necessary goal: perhaps a particular solution or a certain formulation that will lead to the next step. . . . Finally, she must find a way to tie together the different approaches to a solution, taking everyone with her. At another level – just as important – she must get them to see themselves and each other as legitimate contributors to the problem at hand (1996, p. 65).

Hence, the teacher can be seen as juggling multiple goals and diverse contributions as she nudges students' developing ideas toward recognizable mathematical goals. In this section, I will focus on the manner in which effective discourse must pay simultaneous attention to how students participate in mathematical discussion and the academic content. In the next section, I will focus on the norms that underlie such discourse.

Few examples of classroom discourse that effectively combine attention to both talk and academic content exist (Hiebert & Wearne, 1993; Hicks, 1998; but see Lampert & Blunk, 1998; Forman, McCormick, & Donato, 1998; and Forman, Larreamendy-Joerns, Stein, & Brown, 1998 for early contributions in this area). They are sorely needed. The professional development literature is filled with exhortations to ask students questions such as: Can you describe how you did that problem? Does anyone have another method for solving that problem? Does everyone agree with ——? Yet the fruitfulness of such questions is largely dependent on the mathematical ideas in play when they are asked, where the class is in the construction of those ideas, and the kind of follow-up that the teacher provides to students' responses. Simply urging teachers to ask more questions like these will not help them to learn how to scaffold student learning. More explicit guidance regarding when and why such questions should be asked is needed.

Michaels has suggested that a promising approach to the identification of effective content- and discourse-based learning approaches may be to ask what kinds of activity types foster the practice of making claims, externalizing one's rationales for the claims, examining the claims and reasoning of others, and comparing frameworks and perspectives offered by different students with one's own (Michaels, 1999). Activity types refer to identifiable, patterned chunks of classroom practice described in terms of goals, participants, and sequences of teacher and student interactions. By refusing to reduce discourse to a decontextualized count of number of utterances per student or cognitive levels of isolated questions, this approach maintains a simultaneous focus on goals, content, and discourse. Below, I identify and discuss two examples of activity types believed to be effective means of fostering classroom discourse that is content-based: whole-class mathematical argumentation and small group collaboration.

Mathematical Argumentation

When students take up positions and defend them, they have the opportunity to develop and practice the skills of conjecturing and mathematical justification in an authentic setting. As students become invested in their point of view, they learn how to garner and present the necessary evidence to convince others that their claim is warranted, inspect others' criticisms against their claim, and examine opposing claims for signs of weaknesses and/or credibility.

Teachers can "set up" opportunities for mathematical argumentation in their classrooms by selecting tasks that have different solutions or that allow differ-

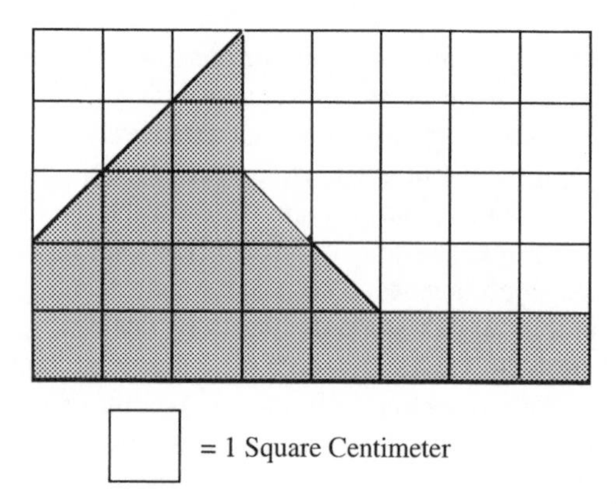

Fig. 7. Instructional task in which students were asked to figure the area of the shaded irregular shape in both square centimeters and square millimeters. (Taken from Silver, Smith & Nelson, 1995; Reprinted with the permission of Cambridge University Press.)

ent positions to be taken and defended. In the following example (see Fig. 7), students were asked to figure the area of the irregular shape shown below both in square centimeters and square millimeters. The teacher's goal for the task was two fold: to have students begin to recognize the proportional relationships that exist between the length and area of two similar figures (two similar figures whose corresponding linear measurements have the ratio $a:x$ will have areas with the ratio $a^2:x^2$) and to give students practice in making and defending claims.

The teacher chose this particular task because she suspected her students would have two different answers – and answers that they would want to defend – for the area of the irregular shape in square millimeters. Based on her past experiences with this task, she expected that some students would think that changing from square centimeters to square millimeters would involve multiplying by 10 (as was done in earlier lessons on linear measurement). Others, however, would have figured out that the conversion factor was 100. Having two possible solutions on the table, she felt, would be useful; by aligning themselves with one or the other, she hoped that her students would become more invested in defending their claims.

She opened the discussion by asking for a volunteer to go to the overhead projector, and provide an answer, in square millimeters, to the area question:

> Okay is there somebody who'd go up with what they did on the first one – for the millimeter squared – because I see a lot of different answers on this and you're going to find out which and prove it.

With this invitation, the teacher not only legitimizes the existence of diverging positions with regard to the task at hand ("different answers") but also sets the norm that students will be expected to express their solution and explain it convincingly ("prove it") to the rest of the class.

The first answer provided (by Larry) is 175, arrived at incorrectly by multiplying 17.5×10. The teacher asked if anyone had a question about it: "Did everybody get 175 millimeter squared for their answer? . . . If you got something else, I don't understand why your hand isn't in the air asking him how he got it."

In the face of critical questioning, Larry sticks by his claim, backing it with the rationale that there are 10 millimeters in 1 centimeter. The teacher restates his original claim and expands upon it: "So Larry is saying that there are 175 *square* millimeters in the figure." This forces Larry to agree or disagree with her, thereby aligning himself with a position that takes the unit of analysis seriously.

Shortly thereafter, the second position is set forth by another student, Eliza:

> It would be, if every centimeter has 10 of those then, the dimensions would by 10 by 10, right? So that makes every centimeter worth 100 millimeters, right?

With this statement, Eliza challenges both Larry's claim and his rationale (i.e. the 1 : 10 ratio for area units) and provides a new claim with a different rationale. As she did with Larry's claim, the teacher restates and expands Eliza's position ("You mean 100 *square* millimeters").

After a few more comments the teacher summarizes the two positions, inviting more students into the debate:

> We still have some more discussion here. Eliza insists that its . . . each one is not 10 [square millimeters] but 100 [square millimeters]. Todd is agreeing [with Larry that it's 10].

With this conversational turn, the teacher clearly articulates her view of the crucial differences in the debate: their contradictory rationales. At this point, another student, Nancy, expands Eliza's initial argument using a visual representation on the overhead projector (a 10×10 grid square which she asks the class to imagine as a blown up version of a square centimeter). Sweeping her hand first horizontally and then vertically across the gird, she indicates how a square centimeter is made up of 10×10 or 100 square millimeters. She completes her argument by noting that each of the square centimeters in the task can be thought about as 10×10 grids, and hence you would need to multiply by 100 to get the answer (not 10). At this point, Eliza spontaneously shouts: "Bravo, bravo, I agree with you 100%".

Forman et al. (1998) label the discourse orchestration moves made by this teacher as a form of revoicing, a strategy for socializing students into mathematical discourse that was first identified by O'Conner and Michaels (1993). They define revoicing as a particular kind of reuttering of a student's contribution by another participant (usually the teacher) in the discussion. In revoicing, "the teacher aligns the student utterance with important academic content ("Eliza insists that each one is not 10, but 100"), she situates the student with other, previous contributions to the discourse ("Todd is agreeing [with Larry] that it's 10"); and she opens up a slot for the student to agree or disagree with the her characterization of the student's contribution, thus ultimately crediting the contents of the reformulation to the student" (T: "A hundred square centimeters?" St: "Yeah, 100 square centimeters.") (O'Conner & Michaels, 1996, p. 71).

Revoicing provides an example of how a teacher can guide the content of classroom discourse by building on students' contributions, as opposed to the more conventional approach of telling students what to think or how to do a particular problem. This strategy has many benefits. By working with student responses, the teacher is more likely to construct a learning experience that is attuned to students' current knowledge and hence more meaningful to them. Moreover, by using their contributions she is helping them develop ownership over their ideas and the self-confidence to see themselves as capable of making sense of the task at hand and contributing to the developing conversation. Finally, in her role as a more expert member of the mathematical community she is modeling the ways in which participants can actively integrate and make sense of multi-voiced conversations.

Heibert and colleagues (1997), while not invoking an underlying socio-cultural frame or participant framework structure, have discussed similar roles for teachers who use classroom discourse as an integral part of their instructional approach. They note that the now-commonly heard admonition of "don't tell" paints an unrealistic and potentially harmful piece of advice for teachers. Depending on one's goals at the time, telling students what to think or do may be in order. Within an inquiry-based approach to mathematics teaching in general, however, certain kinds of teacher talk can be identified as "good." These include articulating the mathematical ideas that can be found in students' methods, sharing with students alternative methods for solving the problem at hand, and, at appropriate times and places, sharing commonly accepted mathematical conventions.

Collaborative Problem Solving

We now know many of the pre-requisites to effective group work. These include selecting a task that is rich enough to allow multiple points of entry and to draw upon different kinds of ability, and taking the time to create and monitor norms of collaboration, such as the explication of the roles and obligations of students in the groups (Noddings, 1989).

Wood and Yackel (1990) have provided examples of how genuine mathematical problems can arise in the course of social interaction in small group problem solving. They claim that in the course of working through problems, students have the opportunity to develop and practice skills of mathematical thinking and communication. More specifically, citing Barnes and Todd (1977), they describe and provide examples of how students must extend their own framework for thinking about a task as they work through it with others. Benefits accrue as students listen to what their peers are saying, and honestly try to make sense of it and coordinate it with their own way of thinking about the situation.

Although the most common situation in which the need to coordinate and reconcile various points of view arises when students have different answers or solution methods, Wood and Yackel point out that such needs also arise when students have the same answer, arrived at through different solution strategies, or when they are trying to explain their own thinking or trying to point out what they perceive to be errors in another students' thinking (see Brown, Stein & Forman, 1996 for an example of students' need to expand their conceptual frameworks based on the need to point out errors in their peer's solution strategy). In classroom observations conducted as part of the QUASAR Project, my colleagues and I observed numerous occasions during which the expectation of a group project (usually a large poster that explained the groups' solution processes to a complex problem) also set up the need for student-to-student communication that fostered genuine collaborative discourse.

Proponents of classroom discourse base claims regarding its effectiveness on underlying theory which purports that students who regularly participate in mathematically rich forms of discourse will, over time, come to internalize the methods and values of that discourse. Hence, advocates tend to rely upon process measures of student engagement and participation as opposed to standardized achievement tests as measures of student learning (Noddings, 1985). Probably the best evidence that higher level forms of discourse result in increased student skill levels and conceptual understanding has been provided by Heibert and Wearne (1993) who found that students in their "experimental" group learned more than their counterparts in a control group. These students participated in classrooms that supported a greater degree of reflective and analytic thought, thought that was encouraged by fewer recall questions (than the control group) and more questions which asked students to describe strategies, generate problems and explain or analyze the underlying features of problems.

Classroom Norms

The undergirding foundation of an effective mathematics lesson includes not only social norms of mutual respect and trust, but also sociomathematical norms of what counts as mathematical activity, what constitutes an adequate explanation, and where the intellectual authority for deciding what is (or is not) acceptable lies. I will discuss each of these in turn and then identify ways in which teachers can set up productive norms in their classrooms.

What Counts as Mathematical Activity

The instructional tasks that teachers select, the resources that they make available to students as they work on those tasks, and the manner in which students' thinking on those tasks is supported, assessed and rewarded have a profound effect on students' developing notions of what mathematics is and how one does it. When students are exposed, year after year, to routine, nonchallenging problem sets (the typical diet of conventional instruction), they form the belief that there is only one right way to solve a mathematics problem (the one the teacher shows them), that the correct method will be provided to them by the teacher (students can't be expected to do something that the teacher has not yet shown them how to do), and that doing mathematics means having a ready method at hand to solve every problem, a method that will produce the correct answer quickly and painlessly (Schoenfeld, 1992, p. 343).

Student beliefs such as these build up over many years and are apt to be reinforced by the adults in their environment (Nickson, 1992). Beliefs that students shouldn't have to figure things out for themselves and that they shouldn't have to work hard or long on mathematics problems constitute one of the most, if not the most, formidable barrier to change in school systems across the country. Having run into such attitudes among administrators and school board members, Cobb and his colleagues found that their initial belief that they could institute change working only at the classroom-level was ill founded. Often, researchers, reformers, and teachers find it necessary to engage parents, administrators and school board members as well (see Dillon, 1990).

What Constitutes an Adequate Explanation

Most students arrive at mathematics class expecting to be sanctioned for use of everyday language and intuitive forms of reasoning. Yet a primary emphasis of the mathematics reform is to build upon and make connections to students' natural ways of thinking and reasoning about the world. For example, it has been suggested that informal and home-based conversational experiences (i.e. arguments outside of school, the provision of justification to parents and siblings) can serve as discourse platforms that students could potentially build upon in the mathematical domain (O'Conner, 1998).

One of the teacher's most important responsibilities is to build bridges between students' intuitive understandings and the mathematical understandings sanctioned by the world-at-large (in Vygotskian terms, bridging from everyday concepts to scientific concepts [1978; 1986]). Under the teacher's guidance, students' everyday language should be gradually replaced by the more precise and generalizable language of mathematics and their situation-based, often inconsistent styles of reasoning replaced with modes of thinking and reasoning characterized by precision, brevity, and logical coherence. In this journey, the teacher is the expert practitioner who must gently guide students without having them lose the ties to their own sensemaking.

Where Intellectual Authority Lies

Mathematics reformers contend that students must learn how to view themselves as the primary authors of their mathematical knowing. Rather than appeal to the authority of the textbook or the teacher, students must learn to justify their own thinking in terms of the logic and rules of evidence of mathematics (NCTM, 1991). How can teachers guide students to "correct mathematics" while, at the same time, helping them to develop the dispositions and intellectual tools required to become their own judges of mathematical quality?

The underlying norms of a conventional mathematics classroom undercut students' attempts to think and reason independently. Consider the traditional sequence of classroom conversational moves known as the I-R-E sequence: [a] teacher initiation, usually in the form of a question; [b] student response, usually a low-level, few-word answer; and [c] teacher follow-up, usually a simple evaluation of right or wrong) (Mehan, 1979). The I-R-E sequence allows teachers to maintain control over the flow of information and the advancement of academic content by giving them complete authority over the topic (the Initiation move) and the decision over what is right and wrong and thus fair game for continued conversation (the Evaluation move) (O'Conner & Michaels, 1996). Students in these kinds of classrooms are not provided with the opportunity to develop their own ideas nor are they encouraged to build their own sense of what constitutes an acceptable mathematical explanation.

As students have been given a more active role in classroom discourse, the potential for their development of a sense of ownership over their own learning has increased. However, in the hands of an inexperienced teacher, classroom discourse can degenerate into a social guessing game, with the teacher steering or funneling students' contributions toward a procedure or answer that she had in mind all along (Voigt, 1985). Students acclimate quickly to such a style of discourse and either come to resent it (if she knows what she wants us to say, why doesn't she just tell us?) or learn to play along (Baxter & Williams, 1998).

Cobb and his colleagues contend that, in order for genuine dialogue and learning to occur, the teacher and students must together constitute a community of co-validators. Describing the second-grade teacher with whom they worked, they note:

> She reformulated their [the students'] explanations and justifications in terms that were more compatible with the mathematical practices of society at large and *yet were accepted by the children as descriptions of what they had actually done*. Thus rather than funneling the children's contributions, the teacher took her lead from their contributions and encouraged them to build on each others' explanations as she guided conversations about mathematics (Cobb, Wood & Yackel, 1993, p. 93, italics added).

The underlying message received by the students – similar to the effect of the revoicing strategy discussed in the prior section – is that mathematics is something about which they personally can think and reason. With this underlying norm, all students are encouraged to see themselves as capable of engaging in a mathematical dialogue.

Although the norms that govern social interactions in the classroom have received extensive treatment in the literature (see Feiman-Nemser & Floden, 1986, for a review), the development of sociomathematical norms for mathematics classrooms has received less attention. Cobb, Wood and Yackel (1993) suggest that teachers begin on the first day of school by deliberately setting up two levels of classroom discourse: one in which teachers and students talk about mathematics and one in which they talk about talking about mathematics. They note that the conversational pattern characterizing these two levels of discourse differs, calling attention to the different goals and purposes of these conversational moves. When the goal is to learn mathematics, teachers and students engage in a very balanced give and take in which the teacher acts as co-constructor of meaning with the students. When the goal is to learn about the mathematical norms that the teacher expects students to follow in the classroom, the conversation shifts into a traditional I-R-E pattern. The following example is provided:

> T: (softly). Oh, okay. Is it okay to make a mistake?
> Andrew: Yes.
> T: Is it okay to make a mistake, Jack?
> Jack: Yes.
> T: You bet it is. As long as you're in my class it is okay to make a mistake. Because I make them all the time, and we learn from our mistakes – a lot. Jack already figured out, "Ooops,

> I didn't have the right answer the first time" (Jack turns and looks at the teacher and smiles), but he kept working at it and he got it.

They note that the teacher didn't plan such occasions of talking about how to talk about mathematics, but she did take advantage of moments that spontaneously presented themselves as ripe for such a discussion. Moreover, she would identify and label these instances – Cobb et al refer to them as paradigm cases – so that she and students could return to them at a later point in time in order to revisit their "rules" for classroom discourse. Cobb and his colleagues report that the manner in which this particular teacher capitalized on unanticipated events by framing them as paradigmatic situations in which to discuss explicitly her expectations and the children's obligations constituted a crucial aspect of her expertise in the classroom" (p. 99).

HOW IS EFFECTIVE MATHEMATICS TEACHING DISTINCT FROM GENERIC TEACHING PRINCIPLES?

Using the core instructional principles and prototypical instructional methods identified in the introductory chapter as my reference point, I will close with a brief discussion of the ways in which I see effective mathematics teaching as both similar to and different from good teaching in general. Let me begin by noting that the research on effective mathematics teaching supports much of what is presented in the initial chapter. The generic principles that were most aligned with principles of good mathematics instruction could be viewed as pre-conditions for effective instruction, many of which were assumed in the present chapter. These include principles such as using instructional time wisely so as to provide students with sufficient opportunity-to-learn and maintaining alignment between instructional goals, curriculum, and assessments.

There are also distinctions that can be made – differences to be drawn – between the characterization of effective teaching practices in the introductory chapter and what mathematics educators consider to be good mathematics instruction. I will group these distinctions into three categories: conflicts, needed elaboration, and questions of focus.

Conflicts

The most serious differences between generic principles of good instruction and what is known about effective methods of teaching mathematics are recommendations regarding how content should be developed. Many of the generic principles cited in the introductory chapter followed the same logic as the Active Mathematics Teaching model, a model developed from a descriptive-correlational-experimental research loop and shown to correlate with large student achievement gains (Good, Grouws & Ebmeier, 1983). Although many useful pedagogical strategies resulted from this model, its broad-based feasibility and atheoretical tenor, as well as its applicability to the teaching of higher order concepts and skills has been questioned (Shulman, 1986).

More specifically, the call to begin lessons with teacher presentation of information, explanation of concepts, and modeling of skills is in conflict with mathematics reform recommendations to allow students to begin lessons by working on their own or in small groups to explore, observe, and invent their own procedures for solving problems. As noted in the present chapter, it is important for students to have the opportunity to make sense of problematic mathematical situations on their own first. It encourages students' own sensemaking and serves to further elaborate their networks of understanding. It also allows students to begin to develop identities as makers and users of mathematics. If teachers take on too much of this burden, students may develop the unhelpful beliefs about mathematics discussed in the section on classroom norms (i.e. that problems always should be solved by a single method and that method should be provided to you).

Beginning with problem situations that students are asked to make sense of does not assure that students will do so. As also discussed in the present chapter, tasks that ask students to impose their own structure and monitor their work often lead to high levels of student anxiety and are especially prone to teachers taking over and doing the task for students, often by showing them a step-by-step solution procedure. Moreover, such tasks are susceptible to degeneration into unsystematic and nonproductive exploration (Stein, Grover & Henningsen, 1996), thereby leaving students with little useable "residue" to take to the next learning situation. Nevertheless, with more intentional efforts on the part of teachers to guide student activity and discourse in productive mathematical directions (such as the strategy of revoicing) and to consolidate learning points before moving onto the next instructional task, project-based pedagogy remains the pedagogy of choice for how to structure meaningful student learning of mathematics.

Needed Elaboration

Generic principles of effective instruction usually begin with the need to set up classroom environments that are conducive to learning. I interpret this as a call for the establishment of expectations regarding how students should act and think and what they can, in turn, expect of their teacher. The explication of effective norms in the introductory chapter deals primarily with what I would label "social norms" (e.g. the need for teachers to have a cheerful and friendly disposition and to create classroom environments in which students feel safe in taking risks). In this chapter, I have argued that teachers also need to take an active role in setting up sociomathematical norms regarding how mathematics gets done, who has the right to question and challenge mathematical solutions, and the basis on which claims and counterclaims will be judged.

These kinds of norms build upon trust and mutual respect, but they must be actively taught *in addition to* purely social norms. In fact, teachers and students who have grown accustomed to a "culture of niceness" in which any criticism or disagreement is regarded as unsocial, often have to overcome fears of being perceived as unfriendly if they disagree with a peer's statement. Mathematics teachers need to become aware that teaching their subject involves not only instilling the content but also socializing their students into a larger social world that honors standards of reason and rules of practice (Popkewitz, 1988).

Questions of Focus

I will use two generic principles to illustrate my concern regarding questions of focus, perhaps the most subtle yet potentially important of what I see as needed distinctions between effective teaching methods, stated generically, and effective teaching methods, stated within a subject-matter specific framework.

As with most lists of effective instruction, this book's introductory chapter awards small group work a place at the table of effective practices. While mathematics educators would agree that group work serves an important function in the mathematics classroom and that many of the features of group work that are mentioned are important considerations, they would advise against leading a discussion of effective teaching methods under the auspices of "small group work."

Featuring small group work as a component of effective mathematics instruction can have the unintended consequence of encouraging workshops that avoid issues of content. As stated by a perceptive staff developer who had witnessed too many sessions devoted to cooperative groups that were devoid of mathematics: (paraphrase)

> The leading issue for any mathematics lesson has to be what ideas should students learn and how should they learn them. Within this framework, small groups *become necessary* when tasks are so rich, solution strategies so numerous, and eagerness to talk so widespread that separating the class into small groups in the only way to accommodate the flood of student thinking and communication.

Also included on the list of generic teaching principles is thoughtful discourse with a focus on preplanning questions related to the content that the teacher wants students to learn. Although it is certainly helpful for teachers to be aware of different forms and cognitive levels of questions and (most certainly) to combine such awareness with the content-to-be-learned, the formulation of actual questions must necessarily be responsive to the unfolding of student contributions *during the lesson*. It is impossible to build on students' thinking without having such a flexible plan from the start. In fact, rather than spending lesson preparation time designing questions, it may be more fruitful for teachers to spend that time thinking about and preparing for the alternative ways of thinking about the instructional task she might expect her students to display.

Many of the features of good questions typically identified in generic teaching principles are difficult to dispute. However, I worry if perhaps it may be dangerous to lead with the features of good questions (e.g. they cannot be answered yes or) because it may provoke a nonproductive mindset in teachers ("If I focus on asking higher questions and leaving enough wait time, I will be teaching effectively). It is much harder to think about questioning – to think about discourse in general – in a manner that is contextualized by the mathematical ideas that are in play at the time. This chapter has provided some early ideas about how this might be done, but it will obviously require more fleshing out, including research on how teachers learn how to lead such discussions.

While generic principles of effective instruction share many similarities with principles of effective mathematics teaching, it may be that we need also to ask ourselves the question: How will such principles be packaged and for whom? My closing remarks are meant to point to certain risks that may arise if teachers and professional developers work off of generic lists, most specifically that these professionals may be deflected from engaging in the hard, but necessary, work of integrating intellectual ideas and teaching methods.

NOTES

1. In the late 1980s and the 1990s, NCTM released a trio of documents calling for profound changes in the curriculum (*Curriculum and Evaluation Standards*), teaching practices (*Professional Standards for the Teaching of Mathematics*) and student assessment (*Standards for Student Assessment)* for K-12 mathematics education. These standards were updated in 2000 with the *Principles and Standards for School Mathematics*.

2. It is important to point out that a nontrivial degree of curricular work is involved in designing tasks that embody foundational ideas, possess longitudinal coherence, and

are attuned to how student thinking develops. In recent years, several curricula have been published that follow NCTM guidelines and that aim to provide a careful selection and sequencing of instructional tasks.

REFERENCES

Barnes, D., & Todd, R. (1977). *Communicating and learning in small groups*. London: Routledge & Kegan Paul.

Bredo, E., & McDermott, R. P. (1992). Teaching, relating, and learning [Review of *The construction zone* and *Rousing minds to life*]. *Educational Researcher*, 31–35.

Brown, C. A., Stein, M. K., & Forman, E. A. (1996). Assisting teachers and students to reform the mathematics classroom. *Educational Studies in Mathematics*, *31*, 63–93.

Carpenter, T., Fennema, E., Peterson, P. L., Chiang, C., & Loef, M. (1989). Using knowledge of children's mathematics thinking in classroom teaching: An experimental study. *American Educational Research Journal*, *26*, 499–532.

Cobb, P. (1994). Where is the mind? Constructivist and sociocultural perspectives on mathematical development, *Educational Researcher*, *23*, 13–20.

Cobb, P., Wood, T., Yackel, E., Nicholls, J., Wheatley, G., Trigatti, B., & Perlwitz, M. (1991). Assessment of a problem-centered second grade mathematics project. *Journal for Research in Mathematics Education*, *22*, 3–29.

Cobb, P., Wood, T., & Yackel, E. (1993). Discourse, mathematical thinking, and classroom practice. In: E. A. Forman, N. Minick & C. A. Stone (Eds), *Contexts for Learning: Sociocultural Dynamics in Children's Development* (pp. 91–119). New York: Oxford.

Davis, R. B. (1992). Understanding "understanding". *Journal of Mathematical Behavior*, *11*, 225–241.

Dillon, D. (1990). *The wider social context of innovation in mathematics education*. Paper presented at the annual meeting of the American Educational Research Association, Boston.

Dossey, J. A., Mullis, I. V. S., Lindquist, M. M., & Chambers, D. L. (1988). *The mathematics report card: Are we measuring up?* Princeton, NJ: Educational Testing Service.

Doyle, W. (1983). Academic work. *Review of Educational Research*, *53*, 159–199.

Epstein, J. L., & MacIver, D. J. (1989). *Education in the middle grades: Overview of a national survey on practices and trends*. Baltimore, MD: Center for Research on Elementary and Middle Schools, Johns Hopkins University.

Feiman-Nemser, S., & Floden, R. E. (1986). The cultures of teaching. In: M. C. Wittrock (Ed.), *Handbook of Research in Teaching* (3rd ed., pp. 505–526). New York: Macmillan.

Forman, E. A., Larreamendy-Joerns, J., Stein, M. K., & Brown, C. A. (1998). "You're going to want to find out which and prove it": Collective argumentation in mathematics classrooms. *Learning and Instruction*, *8*(6), 527–548.

Forman, E. A., McCormick, D. W., & Donato, R. (1998). Learning what counts as a mthematical explanation. *Linguistics and Education*, *9*(4), 313–339.

Good, T. L., Grouws, D. A., & Ebmeier, H. (1983). *Active mathematics teaching*. New York: Longman.

Greeno, J. G. (1997). On claims that answer the wrong questions. *Educational Researcher*, *26*(1), 5–17.

Griffin, S. A., Case, R., & Siegler, R. S. (1994). Rightstart: Providing the central conceptual prerequisites for first formal learning of arithmetic to students at risk for school failure. In:

K. McGilly (Ed.), *Classroom Lessons: Integrating Cognitive Theory and Classroom Practice* (pp. 25–49). Cambridge: MIT Press.

Grouws, D. A., & Smith, M. S. (In Press). Findings from NAEP on the preparation and practices of mathematics teachers. In: E. A. Silver & P. A. Kenney (Eds), *Results from the Seventh Mathematics Assessment of the National Assessment of Education Progress*. Reston, VA: The National Council of Teachers of Mathematics.

Henningsen, M., & Stein, M. K. (1997). Mathematical tasks and student cognition: Classroom-based factors that support and inhibit high-level mathematical thinking and reasoning. *Journal for Research in Mathematics Education, 28*(5), 524–549.

Hicks, D. (1998). Closing reflections on mathematical talk and mathematics teaching. In: M. Lampert & M. Blunk (Eds), *Talking Mathematics in School: Studies of Teaching and Learning* (pp. 241–252). New York: Cambridge University Press.

Hiebert, J., Carpenter, T. P., Fennema, E., Fuson, K. C., Wearne, D., Murray, H., Olivier, A., & Human, P. (1997). *Making sense: Teaching and learning mathematics with understanding*. Portsmouth, NH: Heinemann.

Hiebert, J., & Wearne, D. (1993). Instructional tasks, classroom discourse, and students' learning in second-grade arithmetic. *American Educational Research Journal, 30*(2), 393–425.

Hiebert, J., & Carpenter, T. P. (1992). Learning and teaching with understanding. In: D. A. Grouws (Ed.), *Handbook of Research on Mathematics Teaching and Learning*. New York: Macmillian.

Kouba, V. L., Carpenter, T. P., & Swafford, J. O. (1989). Numbers and operations. In: M. M. Lindquist (Ed.), *Results from the Fourth Mathematics Assessment of the National Assessment of Educational Progress* (pp. 64–93). Reston, VA: National Council of Teachers of Mathematics.

Lamon, S. J. (1995). Ratio and proportion: Elementary didactical phenomenology. In: J. T. Sowder & B. P. Schappelle (Eds), *Providing a Foundation for Teaching Mathematics in the Middle Grades* (pp. 167–198). Albany, NY: SUNY Press.

Lampert, M., & Blunk, M. (Eds) (1998). *Talking mathematics in school: Studies of teaching and learning*. New York: Cambridge University Press.

Ma, Liping. (1999). *Knowing and teaching elementary mathematics: Teachers' understanding of fundamental mathematics in China and the United States*. Mahwah, NJ: Erlbaum.

Mathematical Association of America, Committee of the Mathematical Education of Teachers (1991). *A call for change: Recommendations for the mathematical preparation of teachers of mathematics*. Washington, D.C.: Author.

Mehan, H. (1979). *Learning lessons*. Cambridge, MA: Harvard University Press.

Michaels, S. (1999). Talk given to the Institute for Learning, Learning Research and Development Center, University of Pittsburgh.

National Council of Teachers of Mathematics (2000). *Principles and standards for school mathematics*. Reston, VA: Author.

National Council of Teachers of Mathematics (1995). *Assessment standards for school mathematics*. Reston, VA: Author.

National Council of Teachers of Mathematics (1991). *Professional standards for teaching mathematics*. Reston, VA: Author.

National Council of Teachers of Mathematics (1989). *Curriculum and evaluation standards for school mathematics*. Reston, VA: Author.

National Research Council, Mathematical Sciences Education Board (1989). *Everybody counts: A report to the nation on the future of mathematics education*. Washington, C.C.: National Academy Press.

Newmann, F. M., Marks, H. M., & Gamoran, A. (1995). *Authentic pedagogy and student performance*. Paper presented at the Annual Meeting of the American Educational Research Association, San Francisco.

Nickson, M. (1992). The culture of the mathematics classroom: An unknown quantity? In: D. Grouws (Ed.), *Handbook of Research on Mathematics Teaching and Learning* (pp. 101–114). New York: Macmillan.

Noddings, N. (1985). Small groups as a setting for research on mathematical problem solving. In: E. A. Silver (Ed.), *Teaching and Learning Mathematical Problem Solving: Multiple Research Perspectives* (pp. 345–360). Hillsdale, NJ: Erlbaum.

Noddings, N. (1989). Theoretical and practical concerns about small groups in mathematics. *The Elementary School Journal*, *89*(5), 607–623.

O'Conner, M. C., & Michaels, S. (1993). Aligning academic task and participation status through revoicing: analysis of a classroom discourse strategy. *Anthropology and Education Quarterly*, *24*, 37–72.

O'Conner, M. C., & Michaels, S. (1996). Shifting participant frameworks: Orchestrating thinking practices in group discussion. In: D. Hicks (Ed.), *Discourse, Learning and Schooling* (pp. 63–103). New York: Cambridge University Press.

O'Conner, M. C. (1998). Language socialization in the mathematics classroom: Discourse practices and mathematical thinking. In: M. Lampert & M. Blunk (Eds), *Talking Mathematics in School: Studies of Teaching and Learning* (pp. 17–55). New York: Cambridge University Press.

Popkewitz, T. S. (1988). Institutional issues in the study of school mathematics: Curriculum research. In: A. J. Bishop (Ed.), *Mathematics Education and Culture* (pp. 221–249). Dordrecht, Holland: Kluwer.

Porter, A., Floden, R., Freeman, D., Schmidt, W., & Schwille, J. (1988). Content determinants in elementary school mathematics. In: D. A. Grouws & T. J. Cooney (Eds), *Effective Mathematics Teaching* (pp. 96–113). Reston, VA: National Council of Teachers of Mathematics.

Resnick, L. B., Nesher, P., Leonard, F., Magone, M., Omanson, S., & Peled, I. (1989). Conceptual bases of arithmetic errors. The case of decimal fractions. *Journal for Research in Mathematics Education*, *20*(1), 8–27.

Schoenfeld, A. H. (1992). Learning to think mathematically: Problem solving, metacognition, and sense making in mathematics. In: D. Grouws (Ed.), *Handbook of Research on Mathematics Teaching and Learning* (pp. 334–370). New York: Macmillan.

Schmidt, William H., McKnight, C. C., Cogan, L. S., Jakwerth, P. M., & Houang, R. T. (1999). *Facing the consequences: Using TIMSS for a closer look at U.S. mathematics and science education*. Dordrecht, The Netherlands: Kluwer.

Scribner, S., & Cole, M. (1981). *The psychology of literacy*. Cambridge, MA: Harvard University Press.

Sfard, A. (1998). On two metaphors for learning and the dangers of choosing just one. *Educational Researcher*, *27*(2), 4–13.

Shulman, L. S. (1986). Paradigms and research program sin the study of teaching: A contemporary perspective. In: M. C. Wittrock (Ed.), *Handbook of Research in Teaching* (3rd ed., pp. 3–36). New York: Macmillan.

Silver, E. A. (1997). Learning from NAEP: Looking back and looking ahead. In: P. A. Kenney & E. A. Silver (Eds), *Results from the Sixth Mathematics Assessment of the National Assessment of Educational Progress*. Reston, VA: National Council of Teachers of Mathematics.

Silver, E. A., & Stein, M. K. (1996). The QUASAR Project: The "revolution of the possible" in mathematics instructional reform in urban middle schools. *Urban Education, 30*(4), 476–521.

Silver, E. A., Smith, M. S., & Nelson, B. S. (1995). The QUASAR Project: Equity concerns meet mathematics education reform in the middle school. In: W. Secada, E. Fennema, & L.Byrd Adajian (Eds), *New Directions for Equity in Mathematics Education* (pp. 9–56). New York: Cambridge University Press.

Simon, M. A. & Blume, G. W. (1994). Building and understanding mjuultiplicative relationships: A study of prospective elementary teachers. *Journal for Research in Mathematics Education, 25*, 472–494.

Smith, M. S., & Stein, M. K. (1998). Selecting and creating mathematical tasks: From research to practice. *Mathematics Teaching in the Middle School, 3*(5), 344–350.

Stein, M. K., Smith, M. S., Henningsen, M., & Silver, E. A. (2000). *Implementing standards-based mathematics instruction: A casebook for professional development*. New York: Teachers College Press.

Stein, M. K., Smith, M. S., & Silver, E. A. (1999). The development of professional developers: Learning to assist teachers in new settings in new ways. *Harvard Educational Review, 69*(3), 237–269.

Stein, M. K., Grover, B. W., & Henningsen, M. (1996). Building student capacity for mathematical thinking and reasoning: An analysis of mathematical tasks used in reform classrooms. *American Educational Research Journal, 33*(2), 455–488.

Stein, M. K., & Lane, S. (1996). Instructional tasks and the development of student capacity to think and reason: An analysis of the relationship between teaching and learning in a reform mathematics project. *Educational Research and Evaluation, 2*(1), 50–80.

Stigler, J. W., & Hiebert, J. (1997). Understanding and improving classroom mathematics instruction: An overview of the TIMSS video study. *Phi Delta Kappan, 79*, 14–21.

Stodolsky, S. S. (1988). *The subject matters*. Chicago, IL: University of Chicago Press.

Voigt, J. (1989). Social functions of routines and consequence for subject matter learning. *International Journal of Educational Research, 13*, 647–656.

Vygotsky, L. (1978). *Mind in society: The development of higher psychological processes*. M. Cole, V. John-Steiner, S. Scribner, & E. Souberman (Eds and Trans.). Cambridge, MA: Harvard University Press.

Vygotsky, L. (1986). *Thought and language*. A Kozulin (Ed. And Trans.). Cambridge, MA: MIT Press.

Wertsch. J. (1985). Vy*gotsky and the social formation of mind*. Cambridge, MA: Harvard University Press.

Williams, S. & Baxter, J. (1996). Dilemmas of discourse-oriented teaching in one middle school mathematics class. *Elementary School Journal, 97*(3).

Wood, T., & Yackel, E. (1990). The development of collaborative dialogue in small group interactions. In: L. P. Steffe & T. Wood (Eds), *Transforming Early Childhood Mathematics Education: An International Perspective* (pp. 244–252). Hillsdale, NJ: Erlbaum.

A RESEARCH-BASED PERSPECTIVE ON TEACHING SCHOOL GEOMETRY

Michael T. Battista

INTRODUCTION AND PLAN

Several theoretical and conceptual frameworks are useful for analyzing geometry teaching. I discuss these frameworks in increasing level of specificity. First, I summarize the relevance of the general teaching principles described by Brophy in this volume. Then I briefly examine research-based principles for mathematics learning and teaching. Finally, I discuss research and principles for learning and teaching geometry.

BROPHY'S PRINCIPLES

In general, Brophy's principles are consistent with accepted tenets in mathematics education. But there are a few differences. In particular, Brophy's principles seem to emphasize direct instruction more than the inquiry-based problem-solving approach supported by recent research in mathematics education. Even so, most of Brophy's principles can be adapted to make important points about effective mathematics teaching. Below, I briefly rephrase Brophy's principles so that they are consistent with current principles in mathematics education.

Supportive classroom climates (B1), appropriate opportunities to learn (B2), properly aligned and cohesive curricula components that focus on core ideas and principles (B3), high teacher expectations for student learning that are appropriately communicated to students (B12), instruction that engages

Subject-Specific Instructional Methods and Activities, Volume 8, pages 145–185.

ISBN: 0-7623-0615-7

students in sustained discourse structured around powerful ideas (B6), and appropriate assistance to enable students to productively engage in learning activities (B8) are essential for effective mathematics instruction. Establishing appropriate learning orientations (B4) is one of the key principles in reformed mathematics instruction. Such learning orientations focus on sense making and understanding, not rule following (which is the norm in traditional mathematics education).

Although current best practice in mathematics education calls for curricula to consist of coherent content emphasizing structure and connections (B5), it is not necessary that content be *explained* by teachers. Instead, teachers should structure activities so that students – through problem solving, inquiry, and reflection – construct their own personal networks of connected knowledge. Coherence of content "for students" (as opposed to knowledgeable adults) is absolutely vital.

Unthoughtful practice, a mainstay of traditional mathematics instruction is inconsistent with current research in mathematics education. However, instructionally supporting students development of fluency with core ideas and procedures is essential. Thus, students need sufficient opportunities to practice and apply what they are learning, and to receive improvement-oriented feedback (B7), but this must be done in appropriate contexts. Quite frequently, it can be effectively accomplished by revisiting core ideas and skills in many different contexts or in games. It can also be accomplished through targeted thoughtful practice embedded in problem-solving situations.

Helping students develop self-regulation (metacognitive) strategies (B9) is an important component of effective mathematics teaching. For instance, classroom discussions should deal with what it means to make sense of a mathematical idea, how to make sense of ideas, and how to know when you have made sense of an idea. Highlighting problem-solving strategies like Polya's heuristics is also valuable (1945).

Collaborative learning (B10) is a critical component of mathematics instruction. I use the phrase "collaborative learning" to distinguish small group activity in which all students are fully engaged as partners in learning from activity in which each student in a group performs a specific subtask (so individual students may not truly understand the large problem or its solution). Only the former type of collaboration is consistent with research and reform-based mathematics instruction.

Using a variety of formal and informal assessment methods to monitor learning progress (B11) is a critical component of effective mathematics instruction. It is especially important that assessment focus on students' cognition, not merely their behavior.

LEARNING AND TEACHING MATHEMATICS: RESEARCH AND PRINCIPLES

> Students must learn mathematics with understanding, actively building new knowledge from experience and prior knowledge. . . . In recent decades, psychological and educational research on the learning of complex subjects such as mathematics has solidly established the important role of conceptual understanding in the knowledge and activity of persons who are proficient. . . . Mathematics makes more sense and is easier to remember and to apply when students connect new knowledge to existing knowledge in meaningful ways (NCTM, 2000, p. 20).
>
> Because students learn by connecting new ideas to prior knowledge, teachers must understand what their students already know. Effective teachers know how to ask questions and plan lessons that reveal students' prior knowledge; they can then design experiences and lessons that respond to, and build on, this knowledge (NCTM, 2000, p. 17).

All current major scientific theories describing how students learn mathematics with understanding (instead of by rote) agree that (a) mathematical ideas must be personally constructed by students as they intentionally try to make sense of situations, (b) how students construct new ideas is heavily dependent on the cognitive structures students have previously developed, and (c) to be effective, mathematics teaching must carefully guide and support students' cognition as they construct mathematical ideas (Bransford, Brown & Cocking, 1999; De Corte, Greer & Verschaffel, 1996; Goldin, 1992; Greeno, Collins & Resnick, 1996; Lesh & Lamon, 1992; NRC, 1989; Lester, 1994; Hiebert & Carpenter, 1992; Prawat, 1999; Schoenfeld, 1994; Steffe & Kieren,1994; Romberg, 1992). Accordingly, an abundance of empirical research has shown that mathematics teaching that genuinely supports students' construction of personally meaningful mathematical knowledge through problem- or inquiry-based teaching produces powerful mathematical thinkers who not only can compute but have strong mathematical conceptions and problem-solving skills (Ben-Chaim et al., 1998; Boaler, 1998; Carpenter et al., 1998; Cobb et al., 1991; Fennema et al., 1996; Hiebert, 1999; Muthukrishna & Borkowski, 1996; Quinn, 1997; Wood & Sellers, 1996, 1997; Silver et al., 1996).

Critically important in the studies cited above is that teaching was based on careful attention to research on students' mathematical thinking. For example, Cognitively Guided Instruction (CGI) projects helped teachers acquire "understanding of the general stages that students pass through in acquiring the concepts and procedures in the domain, the processes that are used to solve different problems at each stage, and the nature of the knowledge that underlies these processes" (Carpenter & Fennema, 1991, p. 11). *Thus, teaching in ways that are consistent with research on mathematics learning requires use of substantial knowledge about how students construct the particular mathematical ideas and processes targeted by instruction.*

RESEARCH ON TEACHING AND LEARNING GEOMETRY

Traditional Geometry Teaching in the U.S.: Poor Performance

An abundant amount of research shows that the great majority of students in the United States have inadequate understanding of geometric concepts and poorly developed skills in geometric reasoning, problem-solving, and proof (Clements & Battista, 1992; Beaton et al., 1996; Mullis et al., 1997; Mullis et al., 1998). The primary cause for this poor performance is both *what* and *how* geometry is taught (Clements & Battista, 1992). Most U.S. geometry curricula consist of a hodgepodge of superficially covered concepts with no systematic support for students' progression to higher levels of geometric thinking. These curricula even lack careful attention to the choice of instructional examples, with students often developing limited conceptualizations for geometric concepts due to presentation of overly-constrained sets of examples (Clements & Battista, 1992; Driscoll, 1983; Vinner & Hershkowitz, 1980). For instance, a student might recognize a right triangle only if its legs are vertical and horizontal because all instructionally presented examples were positioned that way.

Conceptual Frameworks on Geometry Learning

Piaget

According to Piaget, mental representations of spatial objects and relationships are not "read off" objects and the environment, but are constructed from the mental actions an individual performs on perceptual input and images (Clements & Battista, 1992). Geometric concept development follows a sequence from acting on objects, to internalizing these actions so that they can be imagined, to representing the actions with formal geometric concepts. Consistent with this theory, a majority of studies indicate that tactile-kinesthetic experiences such as body movement and manipulating shapes can facilitate students' construction of geometric concepts (Clements & Battista, 1992; Fuys, 1988). Unfortunately, the instructional approach implemented in most U.S. textbooks rarely incorporates the proper use of geometric manipulatives (Fuys, 1988; Stigler, Lee & Stevenson, 1990). In contrast, more successful Japanese instruction features greater use of manipulatives (Mitchelmore, 1980; Stigler, Lee & Stevenson, 1990).

van Hiele

According to the van Hiele theory, as students are exposed to appropriate instruction, they progress sequentially through several levels of qualitatively different and increasingly sophisticated and powerful levels of geometric thought (Clements & Battista, 1992; van Hiele, 1986).

Level 1: Visual

Students reason about shape according to appearance; their thinking is dominated by perception not conception. Students recognize and mentally represent shapes such as squares and triangles as visual wholes. When identifying shapes, students often use visual prototypes, saying that a figure is a rectangle, for instance, because "it looks like a door". They might judge that two shapes are congruent because they look the same or because they can be turned to look the same. Students at the visual level do not explicitly attend to geometric properties of shapes.

Level 2: Descriptive/Analytic

Students recognize and characterize shapes by their geometric properties, that is, by explicitly conceptualized spatial relationships between their parts. For instance, students might think of a rectangle as a figure that has opposite sides equal and four right angles. While still important, the holistic appearance of shapes becomes secondary because students identify shapes by their properties rather than by simply matching visual prototypes. Students do not, however, see relationships between classes of shapes (e.g. a student might contend that a figure is not a rectangle "because it is a square").

Level 3: Abstract/Relational

Students can form abstract definitions, distinguish between necessary and sufficient sets of properties for a class of shapes, and understand and sometimes even provide logical arguments in the geometric domain. They can meaningfully classify shapes hierarchically and give arguments to justify their classifications (e.g. a square is identified as a rhombus because it has the defining property of a rhombus, all sides congruent). Because students see that some properties imply others, they no longer feel a need to list all the properties of a class of shapes. However, students do not grasp that logical deduction is the method for establishing geometric truths.

Level 4: Formal Deduction and Proof
Students can formally prove theorems within an axiomatic system. That is, they can produce a sequence of statements that logically justifies a conclusion as a consequence of the "givens". Thinking at Level 4 is required for a proof-oriented high school geometry course.

Teaching Based on the van Hiele Levels
Research supports that the van Hiele levels are useful in describing students' geometric concept development (Burger & Shaughnessy, 1986; Clements & Battista, 1992; Fuys, Geddes & Tischler, 1988). Unfortunately, most U.S. geometry curricula do not support students' sequential attainment of the levels. In fact, over 70% of U.S. students begin high school geometry below level 2, and almost 40% of students end the year there (Clements & Battista, 1992).

Proof and Justification in Geometry
Traditionally, formal proof has been a major component of high school geometry. Unfortunately, teaching of this topic has been extremely ineffective; only about 30% of students in courses that teach proof reach a 75% mastery level in proof writing (Clements & Battista, 1992). Furthermore, formal proof does not, as it should, *convince* high school geometry students of the validity of ideas (Burger & Shaughnessy, 1986; Fischbein & Kedem, 1982; Hoyles & Jones, 1998; Martin & Harel, 1989; Schoenfeld, 1986, 1988; Usiskin, 1982). To understand students' failure with proof, we must understand both the nature of mathematical proof and relevant research on students' learning.

The Nature of Mathematical Proof
A major and critical component of doing mathematics at all levels, from elementary school to front-line research, is intuitive and empirical/inductive thinking (Eves, 1972; Polya, 1954). In doing mathematics, problems are posed, examples analyzed, conjectures made, counterexamples offered, conjectures revised. Unfortunately, these mathematical processes are hidden by the deductive-proof format in which mathematics is recorded and traditionally presented (Lakatos, 1976). That is, because advanced mathematics is presented in a strict sequence of theorems and proofs, this format is mistakenly seen by many teachers and most students as the core of mathematical practice (Hanna, 1989). It is believed that "learning mathematics must involve training in the ability to create this form" (Hanna, 1989, pp. 22–23).

Piaget on the Development of Proof
According to Piaget, the ability to construct proofs as logical necessity develops through several stages, with progress through the stages caused by

age-related maturation and social interaction (Clements & Battista, 1992). At stage 1, the student's thinking is not reflective, systematic, or logical. Each example is treated separately and not integrated with other examples. Exploration proceeds randomly, without a plan. Conclusions may be contradictory. At stage 2, there is an anticipatory character and purposefulness to students' examination of examples. Students make predictions based on empirical results and begin trying to justify their predictions. Students at this stage, however, do not establish logical necessity.

Only at stage 3 does the student progress beyond empirical beliefs to deductions that are true by logical necessity. The student is capable of formal deductive reasoning based on any assumptions (and so is capable of operating explicitly within an axiomatic system). Thought experiments in which conjectures are tested by imagining phenomena can be replaced by deductive arguments.

van Hiele on the Development of Proof

Research by Senk (1989) indicates that formal proof in geometry requires thinking at least at van Hiele level 3, with mastery coming at level 4. Researchers explain that at van Hiele level 1, students' conclusions are restricted to the specific example for which a justification is given (e.g. a particular rectangle), while at level 2, justifications and conclusions refer to collections of similar objects (e.g. sets of rectangles) (Clements & Battista, 1992). Only at van Hiele level 3 and beyond can students justify statements by forming logical arguments because it is at level 3 that students establish a network of logical relations between properties of shapes. Moreover, research indicates that only students who enter a proof-oriented high school geometry course at van Hiele level 2 or higher have a good chance of becoming competent with proof by the end of the course (Senk, 1989; Shaughnessy & Burger, 1985).

Given that over 70% of students begin high school geometry at level 1 or below, it should not be surprising that almost all attempts to teach formal proof to high school students have been unsuccessful (Clements & Battista, 1992). Because students cannot bypass levels and achieve understanding, prematurely dealing with formal proof leads students only to attempts at memorization and to confusion about the purpose of proof.

Summary on Proof

Before high school students can effectively and meaningfully learn geometric proof, they must successfully pass through the first three van Hiele levels and the first two stages of proof as described by Piaget, with such passage dependent on both developmental factors and appropriateness of curricula. Furthermore, to move students' toward genuine mastery of geometric proof, instruction must focus on students' meaningful justification of ideas. For example, Fawcett (1938) conducted a two-year experiment in geometry in which students were challenged to develop their own axioms, definitions, and theorems, and to examine, debate, and justify their conjectures. At the end of the two years, the experimental students outperformed traditional students on several measures of geometric knowledge and reasoning.

Geometry and Computers

Use of computers can greatly facilitate students' construction of geometric knowledge because there are pedagogical functions computers can perform that cannot be duplicated in other situations.

Logo

In the Logo computer programming language, students enter commands to control the screen movement of a graphic "turtle" (Battista & Clements, 1991). For instance, the string of commands FD 50 RT 90 FD 50 makes the turtle move forward 50 units, turn right 90 degrees, then move forward another fifty units, causing it to draw a right angle.

Because Logo commands can create parts of shapes and specify how those parts are put together, appropriate Logo activities can facilitate students' progression to higher levels of geometric reasoning (Clements & Battista, 1989, 1990, 1992, in press). However, as with all instructional computing environments, appropriate teacher guidance is necessary for successful student construction of geometric concepts. Such guidance includes helping students link their Logo experiences with more traditional conceptual knowledge (Battista & Clements, 1991). For instance, before asking students to write Logo commands that make the turtle draw a rectangle, a teacher might have students find physical examples of rectangles, or to make rectangles on a geoboard, then ask students how they know these examples are rectangles and what all the examples have in common. As students say things like, "They all have two long sides and two short sides", the teacher might ask, "How will you make the turtle draw two long sides and two short sides?"

As another example, consider the task of giving Logo commands to make the turtle draw a rectangle. Many young students use trial-and-error visual approaches, especially if the turtle starts off heading obliquely rather than horizontally or vertically. So they might give the set of commands – FD 50 RT

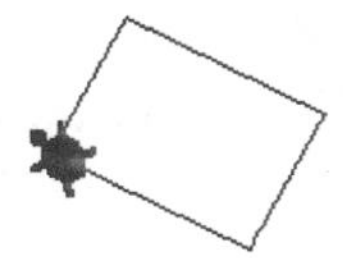

Fig. 1. Student's attempt at drawing a rectangle in Logo.

75 RT 5 RT 5 FD 80 RT 80 RT 5 RT 5 RT 5 FD 40 FD 5 FD 5 FD 5 RT 75 RT 10 FD 75 – resulting in the "rectangle" in Fig. 1.

In such situations, the teacher must find ways to encourage students to progress to higher levels of analysis. For instance, she could ask students to how to make the turtle do the same thing, but with the fewest possible commands (FD 50 RT 85 FD 80 RT 95 FD 55 RT 85 FD 75). Then she could guide students to set up an explicit correspondence between Logo commands and shape components (forward commands correspond to sides, turn commands correspond to angles). Setting up this correspondence will cause some of the students to see that what was made by their Logo commands does not have certain geometric properties they attribute to rectangles (e.g. opposite sides congruent).

Computer Construction Programs

Consistent with the alternatives to axiomatic approaches to geometry teaching, the focus of computer construction programs such as the *Geometric Supposer* software series (Schwartz, Yersushalmy & Wilson, 1993) is to facilitate students making, testing, and justifying geometric conjectures. These programs enable students to choose a shape such as a triangle or quadrilateral and perform measurement operations and geometric constructions on it. The program can record sequences of constructions performed on shapes and can automatically perform them again on other like shapes, thus permitting and encouraging students to explore the generality of their geometric conjectures. Research has demonstrated the effectiveness of such programs (Wiske & Houde, 1988; Yerushalmy, Chazan & Gordon, 1987; Yerushalmy & Chazan, 1993).

Computer-Based Dynamic Geometry

Dynamically variable and manipulable geometric computer screen shapes can be intriguing and convincing to students. Thus, dynamic geometry software such as the *Geometer's Sketchpad* and *Cabri Geometry* is strongly recommended by NCTM (2000). Furthermore, use of such software has been found to increase the geometric knowledge and reasoning of students (Hoyles &

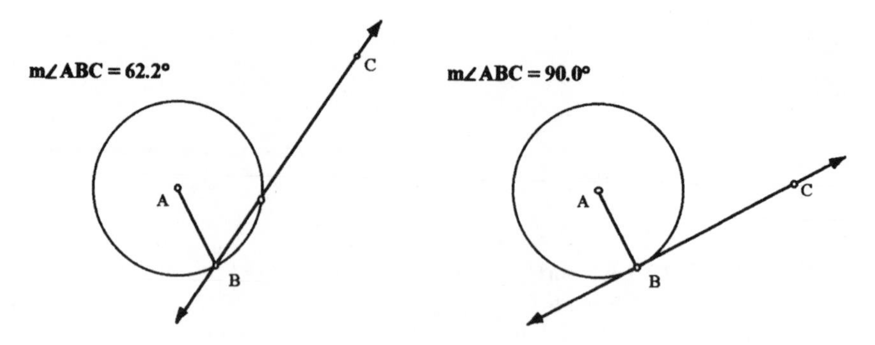

Fig. 2. Tangent to a circle in a dynamic geometry program.

Jones, 1998; Laborde, 1998; Olive, 1998), and has been found to move students to increasingly sophisticated methods of justification (Battista & Clements, 1995; De Villiers, 1998; Olive, 1998). That is, with proper teacher questioning, such software can motivate students to use more formal proof: "Why does this always happen? How do we know this is always going to be true?"

However, good curricula, instructional tasks, and instruction are necessary components for proper pedagogic use of dynamic geometry software. For instance, teachers and students often use such programs only as a drawing tool instead of as a means to embed screen drawings with geometric properties (Laborde, 1998; Olive, 1998). Suppose, for example, a teacher asks students to draw a tangent to a circle in a dynamic geometry program. Some students might draw a circle with a line through a point on it, with the line visually positioned so that it looks tangent to the circle. Unless a teacher asks questions such as, "How can we know if your line is really tangent to the circle?" such a problem promotes little deep conceptual understanding. The instructional effectiveness of the task can be improved by constructing another point of intersection of the line with the circle, along with the radius through the first point, and the measure of the angle between the radius and the line (see Fig. 2). Then, in answering the teacher's questions, the student could discover the conceptually rich notion that there is only one point of intersection between the line and the circle (i.e. the line is tangent) only if the measured angle is 90°.

Another approach to engendering students' development of conceptual knowledge and connections in dynamic geometry environments is to require that students' drawings are "draggable" (i.e. their geometric properties stay intact as points on the figure are dragged). For instance, students could construct, through a point on a circle, a line perpendicular to a radius through this point. Then, no matter how the figure is changed by dragging, the line remains tangent to the circle. This "draggability" requirement forces students to make a geometric connection between circles and tangents via the concept of perpendicularity to radii.

There are other ways that teachers and curriculum designers have tried to ensure that students' work with the dynamic geometry is conceptually useful. For instance, Laborde and her colleagues have given students interesting draggable figures and asked students to construct draggable figures with the same properties (1998). Students can make such constructions only by establishing connections between the geometric concepts that are implemented in the dynamic geometry programs. The *Shape Maker* environment described below further illustrates this point.

SUGGESTIONS AND ISSUES FOR CLASSROOM PRACTICE

Principles for Geometry Curricula

K-12 Geometry Curricula

Unquestioned Importance, But Lack of Content Consensus
There is widespread agreement that geometry learning is a critical goal throughout the mathematics curriculum (Mammana & Villani, 1998; NCTM, 1989, 2000). "However, as soon as one tries to enter into details, opinions diverge on how to accomplish the task. There have been in the past (and there persist even now) strong disagreements about the aims, contents, and methods for teaching of geometry" (Mammana & Villani, 1998, p. 337). Geometric content varies by country as well as era, and is often, especially in the U.S., a hodgepodge of unconnected ideas. Perhaps one of the main reasons for this situation is that geometry has so many aspects and mathematical approaches that educators cannot agree on what to teach. A coherent view of school geometry curricula is needed.

Euclid's Legacy
Around 300 B.C., Euclid created an axiomatic system that enabled him to deductively describe the entire body of geometric knowledge that was known at the time (Kline, 1953). According to many, the impact of Euclid's monumental achievement has been profound. The *Elements* "taught the Greeks and later civilizations the power of reason and gave them confidence in what could be achieved by this faculty. Encouraged by this evidence, Western man was inspired to apply reason elsewhere. . . . Greek geometry served as the

progenitor of the science of logic". (Kline, 1953, pp. 54–55). It is little wonder that ever since the *Elements* appeared, generations of students have studied them to learn how to reason (Kline, 1953).

However, despite the historical importance of Euclid's deductive-logical approach to geometry, past and current educators have called for a different form of geometry curriculum (Chazan & Yerushulmy, 1998; Richards, 1988). These calls for change have occurred both because of the belief that a more practical and usable approach to geometry is needed by scientists and technologically literate workers, and, more recently, because research shows that the traditional Euclidean approach is ineffective for all but a small minority of students (Chazan & Yerushulmy, 1998; Clements & Battista, 1992; Richards, 1988).

An Improved Framework for Geometry Instruction and Curricula

The framework presented below can be used by both teachers and curriculum developers to understand the purpose and extent of school geometry instruction. It is based on how students think about geometric and spatial ideas, a careful analysis of school geometry, and the types of reasoning that are essential for surviving in, managing, and interpreting spatial environments. The central concepts in this framework are structuring and certain related processes (described below). Crossing these structure/process categories with core geometry strands provides an overall organization to geometry curricula and instruction. The four core geometry strands are: (a) measuring spatial entities; (b) analyzing shape via properties; (c) indicating position, including coordinate systems and maps; and (d) analyzing motion, including transformations and paths (See Fig. 3).

Spatial Structuring

The most fundamental process in geometric thinking is spatial structuring. To *spatially structure* an object, set of objects, or spatial environment is to mentally construct an organization or form for it. Spatial structuring determines how an object's shape is perceived by identifying its spatial components,

	Spatial Structuring	Geometric Structuring	Axiomatic Structuring	Processes (describing, representing/modeling, analyzing, justifying, spatial imaging)
Measuring				
Analyzing Properties				
Indicating Position				
Analyzing Motion				

Fig. 3. Suggested organization for school geometry curriculum.

combining components into spatial composites, and establishing interrelationships between and among components and composites (such as how they are placed in relation to each other). It results in mental models of environments and objects that enable us to move about in them, manipulate or operate on them, and reflect on and analyze them so as to understand their spatial nature.

Geometric Structuring
A geometric structuring is created by utilizing formal geometric concepts to mentally construct or analyze a spatial structuring. Geometric structurings involve defining or using specific formal concepts such as angles, slope, parallelism, length, rectangle, coordinate systems, and geometric transformations to conceptualize and operate on spatial situations. In order for a geometric structuring to make sense to a student, its description must create or be connected to an appropriate spatial structuring.

Axiomatic Structuring
Axiomatic structuring formally organizes geometric concepts into a system and specifies that interrelationships must be described and established through logical deductions.

Related Mathematical Processes
Interlaced with the structuring processes are the related processes of describing, representing/modeling, analyzing, and justifying. And underlying almost all spatial and geometric thinking is spatial imaging, which consists of imagining operations on spatial phenomena.

The processes of describing and representing are performed for two different purposes, communication and reflection. They enable individuals to communicate their geometric thinking to others and to mentally "fix" or solidify that thinking so that they can better reflect on it.

Initially, analysis involves only visual decomposition and comparison (van Hiele level 1); later it involves examination of geometric objects by their formal geometric properties (level 2 and higher). Justification involves making analyses and inferences explicit and believable. Both justification and analysis include inference, which is the drawing of conclusions from known or assumed facts or statements.

Illustrations of Types of Structuring

Spatial structuring
An example from elementary school geometry illustrates a fundamental spatial structuring concept that is almost always neglected in traditional instruction

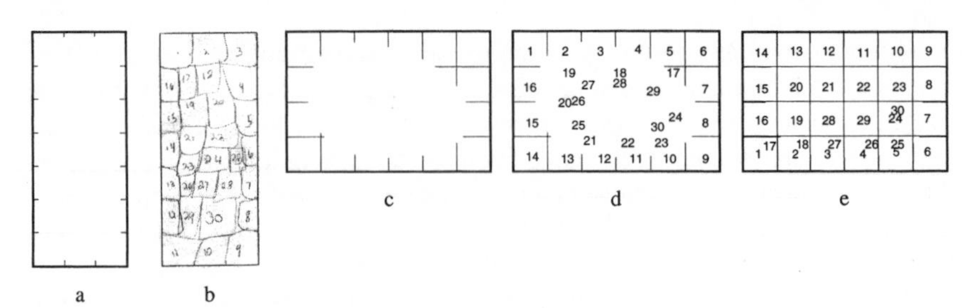

Fig. 4. CS' rectangular array conceptions.

(Battista et al., 1998). Second grader, CS, was shown a plastic inch square and the 7 inch by 3 inch rectangle shown in Fig. 4a. She was also shown that the plastic square was the same size as one of the squares suggested by the tick marks on the rectangle. She was then asked to predict how many of the plastic squares it would take to completely cover the rectangle. CS drew squares where she thought they would go and counted 30 as shown in Fig. 4b.

On a similar problem, CS was asked to predict how many squares would cover the rectangle shown in Fig. 4c (this time, however, she was asked to make a prediction without drawing). CS pointed and counted as shown in Fig. 4d, predicting 30. When checking her answer with plastic squares, she pointed to and counted squares as shown in Fig. 4e, getting 30. But she got confused, so she counted again, getting 24, then 27.

Clearly, CS was not imagining the row-by-column organization that most people "see" in these rectangular arrays of squares. Although, as educated adults, we immediately "see" how rows and columns of squares cover rectangles A and C, CS had not yet mentally constructed this organization. *For CS, this row-by-column organization was not there.* To construct a proper row-by-column structuring of such arrays, CS would have to spatially coordinate the squares in the rows and columns, something that is quite difficult for many students (Battista et al., 1998). In fact, my research indicates that only 19% of second graders, 31% of third graders, 54% of fourth graders, and 78% of fifth graders make correct predictions for problem 4c. Thus, this example illustrates the fundamental nature of students' spatial structuring in their construction of the concept of area.

Geometric Structuring

Seeing the rotation of an object in terms of rotation about three perpendicular coordinate axes is a simple but extremely powerful conceptual device that enables one to describe and compute movements with much more precision than is possible without it.

Axiomatic Structuring

Different axiomatic structurings of geometry focus and organize geometric analysis in particular ways. For instance, taking triangles as the basic shape for which congruence axioms hold encourages decomposition of shapes into triangles. This is an extremely powerful way of looking at the geometric world. It is not the only way, but it is a way that mathematicians have found extremely powerful. It provides a way of reasoning about shape that makes description and justification manageable.

Summary

In essence, all of geometry consists of ways of structuring space and studying the consequences of these structurings. Focusing on students' structuring of various spatial environments offers us a new and powerful perspective for analyzing geometric thinking, learning, and curricula. It helps us better understand the relationship between spatial reasoning and formal geometric concepts. It suggests different instructional emphases. For instance, students in traditional curricula get very few opportunities to develop their own personally meaningful spatial structurings. They simply apply formal structurings invented by mathematicians, usually without a proper foundation of personally meaningful spatial structurings. The structuring framework, along with current research, suggests that students need many opportunities to pass through the structuring sequence described above. This framework can be used to construct geometry curricula that not only are consistent with research on students' geometric learning, but that lead students to Euclid's monumental achievement, but in a psychologically sound way.

PRINCIPLES FOR GEOMETRY TEACHING

Geometry teaching that is consistent with research on mathematics learning and professional recommendations requires a deep understanding of the ways that students' make sense of the particular geometric topics that are being taught (Battista, in press a; Lehrer et al., 1998). Much of the discussion so far has described principles of such student sense making. I now turn to general principles for geometry teaching.

Establishing a Culture of Inquiry

Both professional standards and scientific theories of mathematics learning suggest that mathematics instruction focus on inquiry, sense making, and problem solving (Battista, in press a; NCTM 1989, 2000). In such instruction, teachers provide students with numerous opportunities to solve substantive and interesting problems; to read, represent, discuss, and communicate mathematics; and to formulate and test the validity of personally constructed mathematical ideas. Students use demonstrations, drawings, and physical objects – as well as formal mathematical arguments – to convince themselves and their peers of the validity of their problem solutions.

To foster meaningful learning in such classrooms, teachers must establish a culture of inquiry in which students are not only involved in inquiry, problem solving, and invention, but in classroom discourse that establishes ideas and truths collaboratively. Students' participation in such a culture promotes their personal construction of ideas as they

- attempt to elaborate and clarify personally-developed ideas so that they can communicate them to others;
- reflect on, evaluate, and justify their personally-developed ideas in response to conflicts with others' ideas or challenges posed by classmates;
- attempt to make sense of and sometimes utilize new ideas offered by classmates.

Critical to success in establishing a classroom culture of inquiry is choosing instructional tasks based on knowledge of *students'* mathematics (Steffe & D'Ambrosia, 1995). The choice of tasks must be "grounded in detailed analyses of children's mathematical experiences and the processes by which they construct mathematical knowledge" (Cobb, Wood & Yackel, 1990, p. 130). If instructional tasks are inappropriate, students will be unable to make personal sense of mathematics, so a culture of inquiry will be impossible to establish.

Responsibilities in a Culture of Inquiry

The major responsibilities of teachers and students in a classroom culture of inquiry are described below.

Students are Responsible For

(1) Attempting to solve and make sense of all problems given to them.
(2) Explaining their mathematical thinking to other members of the class and justifying problem solutions in response to challenges.

(3) Listening to, as well as attempting to make sense of, other students' mathematical explanations and problem solutions. This includes asking for clarification if an explanation is not understood, and challenging strategies and problem solutions that do not seem reasonable.
(4) Working collaboratively with other students. This includes attempting to reach consensus on problem solutions while respecting the rights of others to derive or justify solutions differently.

Teachers are Responsible For

(1) Selecting instructional tasks and guiding students' work on these tasks so that students' thinking becomes increasingly more sophisticated. This includes:
   - Choosing appropriate sequences of problematic tasks based on detailed knowledge of how students construct meanings for specific mathematical topics, as well as the conceptual advances that students can make with those topics.
   - Continually assessing students' learning progress and adjusting instruction accordingly.
   - Encouraging students to reflect on their mathematical experiences.
(2) Establishing a social environment that supports a spirit of inquiry and collaborative small-group work. This includes explaining, illustrating with classroom examples, and regularly reminding students of their responsibilities.
(3) Encouraging productive dialogue among students by
   - Having students explain and justify their mathematical ideas.
   - Highlighting conflicts between alternative student interpretations or solutions.
   - Unobtrusively encouraging potentially fruitful student contributions.
   - Redescribing student ideas in more sophisticated ways that students can still comprehend. This includes introducing, at appropriate times, formal mathematical concepts, symbolism, and terminology to redescribe student-invented ideas, and encouraging student use of these formal ideas.

Helping Students Meet Their Responsibilities
If students are not accustomed to participating in a culture of inquiry in mathematics class, it may take them several weeks to become comfortable and competent with such an instructional environment. They need regular and explicit discussion of this new way of learning. Posting student responsibilities

on a poster board in class and regularly asking students what their responsibilities are and how they are implementing them can help.

But students also need to see classroom examples of their responsibilities in action. For instance, to learn how to explain and justify their strategies and problem solutions, students need to talk about the processes of explanation and justification. If a teacher sees that students are having difficulty explaining their thinking about a particular problem, the teacher should make such explanations an explicit topic of class discussion. The discussion might begin with the teacher asking students who are having difficulty articulating an idea to explain it to the class as best as they can. The teacher can then ask students with similar solutions to explain their work. "How do you think we should talk about these ideas? What words should we use to refer to what you are talking about?" Such discussions can help students develop a language and set of conceptualizations for describing their developing ideas.

Supporting Students' Personal Construction of Knowledge

The goal of mathematics instruction should be to help *each* student build mathematical ideas and theories that are more complex, abstract, and powerful than he or she currently possesses. The major instructional mechanism for encouraging students' construction of knowledge is the presentation of properly chosen problematic tasks. These tasks guide students' knowledge construction by properly focusing their attention, encouraging reflection and abstraction, and promoting perturbations that require reorganization of current knowledge structures.

However, to be effective, instructional tasks must fall within students' current *zones of construction*. That is, students' construction of new concepts required to complete instructional tasks must be possible given their current conceptual structures and mental processes. In fact, because students' existing structures determine how they think about all new tasks, teachers must constantly monitor the development of these structures and adjust instruction accordingly.

Whenever teachers ignore students' current ways of thinking and attempt to impose methods on them, students' sense-making activity is stifled. Students parrot the procedures they are shown. Their beliefs about the nature of mathematics change from viewing mathematics as sense making to viewing it as learning set procedures that make little sense. Students change from intellectually autonomous thinkers to rule followers.

Teaching According to the van Hiele Theory

For many topics in geometry, the van Hiele levels provide the best analysis of students' constructive itineraries during learning, and therefore, are essential for guiding instruction. These levels provide teachers and curriculum developers with a framework for (a) setting instructional goals, (b) choosing appropriate sequences of instructional tasks and questions, (c) guiding classroom discussions, (d) encouraging student reflection and focusing their attention, and (e) assessing student work.

In addition to the levels, the van Hiele theory also suggests a set of instructional phases that can be used to encourage students' movement from one level to the next (Clements & Battista, 1992; van Hiele, 1986). The essence of these phases follows.

Phases of Geometry Learning

Phase 1: Intuitive Orientation
Teachers choose instructional activities that encourage students to actively engage in exploring, through manipulation and measurement, objects and phenomena from which targeted geometric concepts can be abstracted.

Phase 2: Explicitation
Teachers guide students to explicitly formulate conceptualizations of instructionally targeted geometric concepts and to describe these conceptualizations in their own language.

Phase 3: Formal Orientation
Teachers introduce relevant formal mathematical terminology and conceptual formulation for the intuitive ideas previously elaborated. They select problematic tasks that require students to interrelate their intuitive and formal concepts, orienting themselves within the network of relations they are building.

Phase 4: Integration and Consolidation
Teachers encourage students to consolidate their geometric knowledge using formal mathematical concepts as a framework. Students synthesize all they have learned about the objects of study, integrating their knowledge into a coherent network that they can fluently apply and describe in traditional mathematical language.

Examples of Inquiry-Based Geometry Instruction

The picture of high quality geometry teaching that emerges from modern research on learning is that geometry instruction should encourage and support students' movement from intuitive informal ideas of space and shape to progressively more sophisticated and formal geometric concepts, all within the context of sense making, inquiry, and problem solving. Within this progressive development, students personally construct increasingly sophisticated conceptual and reasoning tools that increase the power of their geometric analysis of spatial situations.

I now illustrate the nature of high-quality geometry instruction with two examples. Both instructional treatments are firmly grounded in knowledge of how students' thinking about particular geometric concepts develops. Both support student inquiry, sense making, and conjecturing, as well as testing of ideas. Furthermore, both have been shown to be effective in engendering students' acquisition of powerful geometric ideas (Battista, in press b; Battista, 1999; Battista & Borrow, 1997).

Shape Makers: A Dynamic Geometry Microworld

Consider two pairs of equal length rods hinged together at their ends (see Fig. 5). As we manipulate this "parallelogram maker", the visual and kinesthetic experiences we abstract from our actions, along with our reflections on those actions, are integrated to form a mental model that can be used in reasoning about the geometric concept of parallelogram. The *Shape Makers* computer microworld, a special add-on to the dynamic geometry computer program *The Geometer's Sketchpad,* provides students with screen manipulable shape-making objects similar to, but more versatile than, this physical parallelogram maker (Battista, 1998). For instance, the computer Parallelogram Maker can be used to make any desired parallelogram that fits on the computer screen, no matter what its shape, size, or orientation – but only parallelograms. Its shape can be changed by using the mouse to drag its vertices (which are indicated by small circles in Fig. 6).

The *Shape Makers* environment provides a powerful representation of shapes and their properties that supports students' geometric reflection and

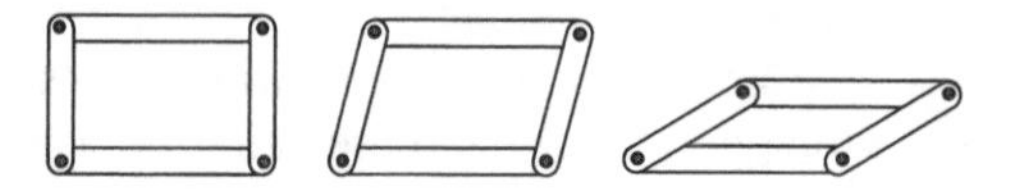

Fig. 5. Hinged rod "rectangle maker".

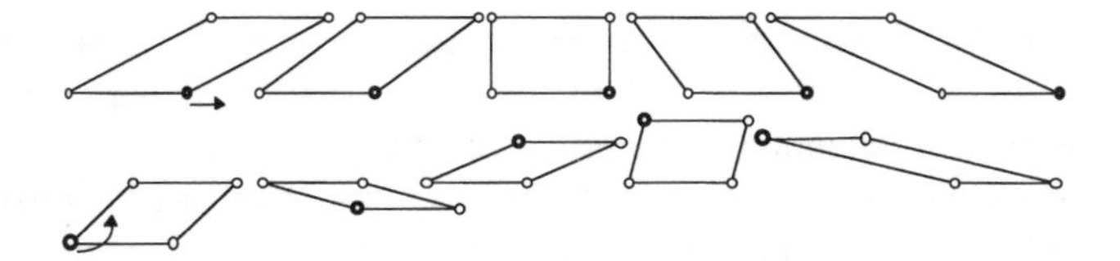

Fig. 6. Computer Parallelogram Maker.

Fig. 7. Three students' work on Parallelogram Maker.

analysis. It presents students with interesting manipulable screen objects whose behavior can be understood using geometric concepts, encouraging students' creation and use of important geometric structurings. That is, each Shape Maker embodies the defining properties of a class of geometric shapes. Because these properties are invariant as the Shape Maker is altered by dragging, its properties are more perceptually and conceptually salient then if one or several static examples are examined. Moreover, the *Shape Makers* microworld and activities encourage students' passage from van Hiele level 1 to level 3.

In initial activities, students use the Shape Makers to make their own pictures, then to duplicate given pictures. These activities encourage students to create mental models for the movement possibilities of the Shape Makers and to formulate informal conceptions of geometric properties. For example, three students were considering whether the Parallelogram Maker could be used to make a trapezoidal target figure (Fig. 7). Their knowledge of the Parallelogram Maker was insufficient to predict that this was impossible. However, as they manipulated the Parallelogram Maker on the computer screen, one of the students discovered something that enabled her to solve the problem.

> ST: No, it won't work. [Pointing to the non-horizontal sides in the Parallelogram Maker] See this one and this one stay the same, you know, together. If you push this one [a non-horizontal side] out, this one [the opposite side] goes out. . . . This side moves along with this side.

As ST manipulated the Parallelogram Maker in her attempts to make the target figure, she detected a pattern or regularity in its movement. As she abstracted this movement pattern, incorporated it into her mental model for the Parallelogram Maker, and described it in terms of parts of the Shape Maker, she was able to infer that the target figure was impossible to make. Although ST had not yet developed a geometric structuring that enabled her to precisely describe and completely understand the relevant spatial relationships she was observing, she had formed a spatially structured mental model of the Parallelogram Maker that could, with further elaboration, serve as the basis for conceptualizing the formal mathematical property "in a parallelogram, opposites sides are parallel and congruent". Thus, this experience with the Shape Makers helped ST move toward van Hiele level 2 thinking and it became part of a foundation for her later geometric structuring.

The next example shows students at three different points along the path toward geometric structuring. The students were investigating the Square Maker.

MT: I think maybe you could have made a rectangle.
JD: No; because when you change one side, they all change.
ER: All the sides are equal.

MT, JD, and ER abstracted different things from their Shape Maker manipulations. MT noticed the visual similarity between squares and rectangles, causing him to conjecture that the Square Maker could make a rectangle. JD took an initial, imprecise, step toward geometric structuring as he abstracted a movement regularity in terms of side length – when one side length changes, all side lengths change. Only ER, however, conceptualized the movement regularity with complete precision by expressing it in terms of a geometric property. Thus, JD began a geometric structuring of this property as he gave a van Hiele level 1 response, whereas ER completed a geometric structuring of the property, giving a level 2 response.

Subsequent *Shape Maker* activities involve students in tasks that require more careful analysis of shapes, further supporting their formulation and use of geometric properties. Unmeasured Shape Makers are replaced by Measured Shape Makers that display measures of angles and side lengths which are instantaneously updated when the Shape Makers are manipulated. The following episode shows how a teacher supported two students as they were struggling to develop a proper geometric conceptualization of their vague intuitive ideas.

The task was to determine which of Shapes 1–7 could be made by the Rectangle Maker (explaining and justifying each conclusion). See Fig. 8. Fifth graders M and T predicted that the Rectangle Maker could make shapes 1–3, but not 4–6. They are now checking and

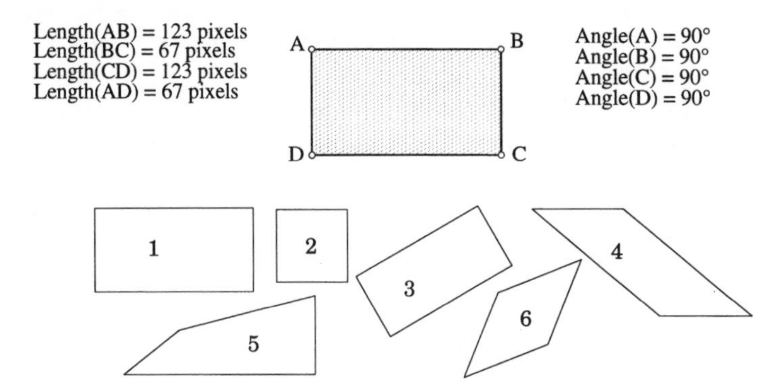

Fig. 8. Rectangle Maker activity.

discussing their results. (These students were participating in a class taught by a teacher who was highly skilled in creating a classroom culture of inquiry, problem-solving, and sense making.)

After using the Rectangle Maker to check Shapes 1–3, M and T move on to Shape 4.

T: I'm positive it can't do this one.

M: It [the Rectangle Maker] has no slants. We had enough experience with number 3, that it can't make a slant.

T: Yes it can, it has a slant in the other one [Shape 3]. It's has a slant right now. . . .

Tchr: What do you mean by slant?

M: Like this. See how this is shaped like a parallelogram [motioning along the perimeter of Shape 4].

T: This is in a slant right now [points at the Rectangle Maker, which is rotated from the horizontal]. [Note that "slant" means non-perpendicular sides for M, but rotated from the horizontal for T.] . . .

M: It can't make that kind of a shape [pointing to Shape 4] . . . . It can't make something that has a slant at the top and stuff.

T: Do you mean it has to have a straight line right here [pointing to Shape 6], like coming across? [T now uses "straight" to mean horizontal.]

M: I know they are straight, but they are at a slant and it [the Rectangle Maker] always has lines that aren't at a slant. . . .

Tchr: When you put two sides together, what do they make?

T: A right angle, if you put them right.

Tchr: OK. You get a right angle and what effect does that have on the sides?

M: It makes them all be, like, none of them will be at a slant.

Tchr: What do you think of that T?

T: So are you saying if there is one right angle.

M: None of the two will be at a slant [drawing an acute angle]. . . .

Tchr: What's different about [Shapes] 3 and 4?

M: The top lines on these are not exactly on top of each other [pointing to the nearly horizontal sides on Shape 4]; these ones are [motioning to two opposite sides on shape 3]

Tchr: OK. What else?

T: This side is pushed up farther up than this side [motioning along the two long sides of Shape 4] . . . .

Tchr: Keep on really looking at what makes these different. And maybe some of the information on the screen will help you. Watch the numbers up there and see if that will help you. . . .

T: [Manipulating the Rectangle Maker after the teacher leaves] Oh, this [the Rectangle Maker] always has to be a 90° angle. And that one [Shape 4] does not have 90° angles. And so this one [Shape 3] has to have a 90° angle, because we made this with that [Rectangle Maker]. So there is one thing different. A 90° angle is a right angle, and this [Shape 4] does not have any right angles.

M and T were trying to find a way to conceptualize and describe the spatial relationship that the sides of the Rectangle Maker are perpendicular (i.e. that they intersect in right angles). However, they had not yet constructed the formal concept of perpendicularity. So they used terminology and concepts that they had available, like "slanted" and "straight". Unfortunately, these terms did not adequately describe the new idea they were grappling with. Furthermore, M and T used the terms differently, making communication difficult.

The teacher did several things to try to help the boys: She constantly asked for clarification, "What exactly do you mean by this?" She tried to expose differences and inadequacies in the boys' use of language. When the boys brought up the idea of right angle, she attempted to get them to use this concept to analyze the differences in shapes that could and could not be made by the Rectangle Maker. However, when the teacher subsequently asked the boys what was different about Shapes 3 and 4, they did not use the concept of right angle to describe the differences. Recognizing that the boys needed more time to manipulate and reflect on the issues, she left them. But before leaving, she tried to focus their attention on a productive way of thinking by suggesting that they look at the Rectangle Maker's measurements.

After the teacher left, T manipulated the Rectangle Maker, focusing on its measurements. Through this manipulation, he discovered and abstracted that the Rectangle Maker always has four 90° or right angles. Furthermore, he abstracted this property sufficiently so that he was able to use it to analyze the differences between Shapes 3 and 4. He subsequently saw that the Rectangle Maker could not make Shapes 4, 5, and 6 because they do not have four right angles. In fact, by the end of the class period, the boys saw that the spatial relationship they were attending to could be described in terms of the formal mathematical concept of right angle. They had constructed a geometric structuring that enabled them to solve the problem they had embraced.

Inquiry on the Concept of Volume

The following example of how an inquiry-based unit should be structured and implemented has been synthesized from research and curriculum development projects that I conducted in the fifth-grade classroom of Linda Hallenbeck, a teacher and colleague who is highly skilled in inquiry-based teaching. On the surface, the goal of the unit was for students to develop an understanding of how to enumerate cubes in 3D arrays. At a deeper level, however, because research suggests that many students are unable to correctly enumerate the cubes in such arrays because their spatial structuring of the arrays is incorrect (Battista & Clements, 1996), the unit's goal was to develop students' ability to imagine correct spatial structurings of 3D cube arrays.

Initially, the teacher distributed the *How Many Cubes?* activity sheet (Fig. 9) to each student and explained that the students' goal was to find a way to correctly predict the number of cubes that would fill boxes described by pictures, patterns, or words. (So the goal was not to "complete the worksheet".) Students worked collaboratively in pairs, predicting how many cubes would fit in a graphically represented box, then checking their answers by making the box out of grid paper and filling it with cubes. Importantly, students predicted then checked their results for one problem before going on to the next. The teacher circulated about the room, asking questions to promote student sense making and reflection, inquiring about student thinking, and encouraging within-pair communication and collaboration. The nature of student work on this activity, which was completed in two-and-a-half one-hour periods, will be illustrated by excerpts from one pair of students. Class discussion of student findings, which occurred during the second half of the third day, follows the description of student pair work.

Small Group Work

Episode 1, Day 1. For Box A, N counts the 12 outermost squares on the 4 side flaps of the pattern picture [see Fig. 10a], then multiplies by 2: "There's 2 little squares going up on each side, so you times them." P counts the 12 visible cube faces on Box Picture A, then doubles that for the hidden lateral faces of the box. The boys agree on 24 as their prediction.

P: [After putting 4 rows of 4 cubes into the box] We're wrong. It's 4 sets of 4 equals 16.

N: What are we doing wrong? [Neither student has an answer, so they move on to Box B.]

P: What do you think we should do? [Pause.] [Pointing at 2 visible faces of the cube at the bottom right front corner of Box Picture B] This is 1 box [cube], those 2.

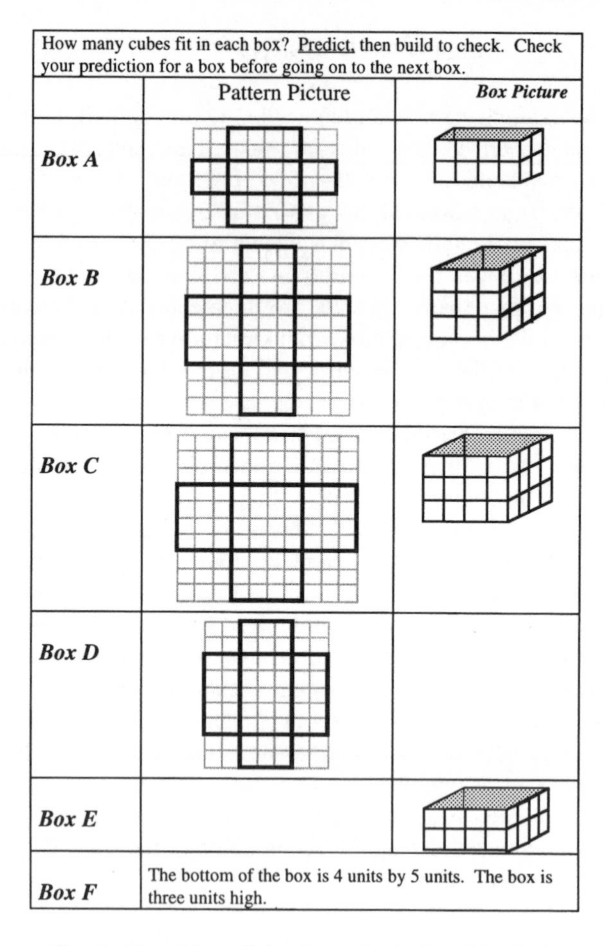

How many cubes fit in each box? Predict, then build to check. Check your prediction for a box before going on to the next box.
Pattern Picture
Box Picture
Box A
Box B
Box C
Box D
Box E
Box F
The bottom of the box is 4 units by 5 units. The box is three units high.

Fig. 9. How Many Cubes? activity (reduced in size).

N: Oh, I know what we did wrong. We counted this [pointing to the front face of the bottom right front cube] and then the side over there [pointing to the right face of that cube].

P: So we'll have to take away 4 [pointing to the 4 vertical edges of Box Picture A], no wait, we have to take away 8. [P subtracts 8 from their prediction of 24 and tells N that this subtraction would have made their prediction correct.]

In their prediction for Box B, P counts 21 visible cube faces on the box picture, then doubles it for the box's hidden lateral faces. He subtracts 8 for double-counting (not taking into account that this box is 3 cubes high, not 2, like Box A), predicting $42 - 8 = 34$. N adds

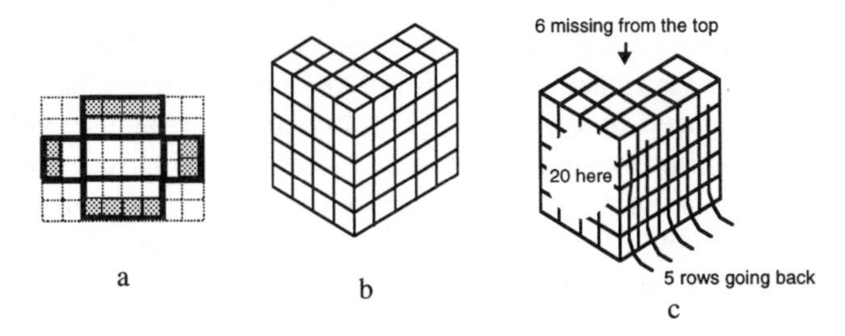

Fig. 10. Work of N and P.

12 and 12 for the right and left lateral sides of Box Picture B, then 3 and 3 for the middle column of both the front and back, explaining that the outer columns of 3 on the front and back were counted when he enumerated the right and left faces. He predicts 30.

Although N and P agreed on a prediction of 24 for Box A, the discrepancy between their predicted and actual answers caused them to reflect on and reevaluate their reasoning. As P reflected on his original conception of the two visible faces of the bottom right corner cube of Box Picture B, he realized that they were, in fact, the front and right faces of the same cube. When N and P extended this new interpretation of adjacent edge faces to other vertical-edge cubes, they recognized that P's current enumeration strategy double-counted such cubes. P compensated for the error by subtracting the number of cubes he thought he had double-counted. N attempted to imagine the cubes so he would not double-count them.

Episode 2. After Box B is constructed, the boys determine that 36 cubes fill it, which is a different answer than either boy predicted. This puzzles the boys.

P: What are you thinking we should do?

N: Well, I was thinking I forgot some in the middle.

P: I got it! We were doing minus 8 [pointing to Box Picture A] but really there are 1, 2, 3–4, 5, 6–7, 8, 9–10, 11, 12 [pointing successively to the 4 vertical edges of Box Picture B]. [But since this adjustment does not give the correct answer, he reflects further.] I got it. We missed 2 in the middle. There's 2 more blocks in the middle, and that would make 36.

The discrepancy between N and P's predicted and actual answers for Box B lead them to reflect further on the situation. At first, P thought that his prediction error arose from failing to account for the building height in his subtract-to-compensate strategy. However, when P's adjustment for this error did not give the correct answer, he reflected on N's comment about failing to count interior cubes, deciding that the prediction of 34 missed 2 cubes in the middle.

Although neither P nor N had yet developed a correct spatial structuring of 3D arrays, they were making progress. As they reflected on, analyzed, discussed, and revised their work, they abstracted important aspects of the spatial organization of the arrays that would later help them make the needed restructuring. Importantly, however, P focused on a numerical strategy that was based on somewhat fuzzy spatial structuring, whereas N focused on constructing a spatial structuring that enabled him to locate all the cubes in the array.

Episode 3. N and P jointly count 21 outside cube faces for Box Picture C, not double-counting cubes on the right front vertical edge. They then multiply by 2 for the hidden lateral sides and add 2 for the middle cubes (which is how many cubes they concluded they had missed in the middle of Box B). Their prediction is 44 cubes. The boys make and fill the box and find that it contains 48 cubes. They are puzzled.

Episode 4. On Day 2, N and P reflect on their incorrect predictions for Box C. P concludes that they failed to count some of the cubes in the four vertical edges. However, as in his previous adjustments, P derives the number to be added to compensate for these omitted cubes by comparing their predicted and actual answers, not by finding an error in his spatial structuring.

N: I think I know Box D; I think it's going to be 30; 5 plus 5 plus 5 [pointing to the columns in the pattern's middle], 15. And it's 2 high. Then you need to do 3 more rows of that because you need to do the top; 20, 25, 30 [pointing to middle columns again].

P: I don't know – that'd probably work, I guess.

After the boys make the box and find that it contains 30 cubes as N predicted, both boys smile. P then asks N to explain his strategy again, which N does.

P: So there's 15 right here [pointing to the bottom of the box].

N: Yeah, on the bottom. And in the top part here [motioning] there'd be another 15.

In Episode 4, N made a correct prediction by structuring the array into 2 layers of 3 columns of 5 cubes. Only after P saw that N's strategy gave the correct answer did P become interested in really listening to and making sense of N's strategy description.

Episode 5. On the next problem [Box Picture E], because neither boy is able to employ their new layering strategy in this different graphic context, both return to variants of their old strategies. However, after the boys complete the pattern for Box E, N applies his layering strategy, predicting "16 times 2", but he is not yet confident in its validity. When the boys find that 32 cubes actually fit in the box, N comments, "It is 32," seemingly realizing that his new layering strategy is, in fact, valid in this situation.

Episode 6. The boys are unwilling to make a prediction for Box F until they draw the pattern. Once the pattern is drawn, N silently points to and counts the squares in its middle section, 1–20 for the first layer, 21–40 for the second, and 41–60 for the third. The boys build the box, fill it cubes, then count the cubes by fives to 60. They do not seem relieved that their count matches their prediction but expect their answer to be correct.

Episode 7, Day 3. The teacher asks N and P how many cubes would be in a box that has the same bottom as Box A but is 3 cubes high.

P: Eight times 3 equals 24.
N: Yeah, 8, 16, 24. I'm not too good at my multiplication facts.

The boys also correctly solve several other problems, including the one shown in Fig. 10b. For instance, [see Fig. 10c], P explained his solution as, "There's 20 on the left side, times 5 rows going back, equals 100. There's 6 cubes missing from the top, and it's 5 down, so we have to subtract 6 five times; 100 minus 6 equals 94, 94 minus 6 equals 88 . . . 76 minus 6 equals 70, 70 is the answer.

During their collaborative small group work, N and P developed a layer-based enumeration strategy that they verified, refined, and became confident in on subsequent problems. They also successfully adapted their layer-based strategy to apply it in new situations. N and P, like almost all of their classmates, had developed a powerful and general way of reasoning about cube enumeration problems (Battista, 1999).

Factors that Contributed to Students' Success

There are several important factors that contributed to the overall success of this activity. First, having students predict, then check their predictions with cubes was an essential component in their establishing the viability of their mental models and enumeration strategies. Because students' predictions were based on their mental models, making predictions encouraged them to reflect on and refine those mental models. Strategies were refined as students reorganized their models and strategies to better fit their experiences. Indeed, it is students' mental models that the instructional unit was attempting to develop. Having students merely make boxes and determine how many cubes fill them would have been unlikely to have promoted nearly as much student reflection as having students make and check predictions because (a) opportunities for reflection arising from discrepancies between predicted and actual answers would have been greatly reduced and (b) students' attention would have been focused on physical activity instead of on their own thinking.

Second, the effectiveness of this approach depends heavily on students' ability to independently check the viability of their predictions. It is highly unlikely that the current instructional treatment would have been so effective if students had been unable to reliably determine the number of cubes that fill actual boxes. For instance, this unit would have been inappropriate for third graders because, at that level, students still have difficulty reliably enumerating cubes in actual 3D arrays (Battista & Clements,1998).

Third, not shown in the episode, but an essential element of the classroom environment is the work the teacher did as she circulated around the classroom interacting with student pairs. Especially at the end of the second day and on the beginning of the third, she was alert for students who still seemed to have major misconceptions. In fact, as is shown below, the teacher had to recognize and deal individually with several students' lack of proper mental models for cube arrays (see Battista, 1999).

Episode 8, Day 3. L is working with M and K because her usual partner is absent. The teacher, who suspects from previous observations that L does not have a proper mental model of the cube arrays, asks the students to explain their enumeration strategies.

Tchr: L, could you explain your strategy to M and K? Try it on Box F.

L: [Pointing successively to the front, right side, back, and left side on a previously completed Box F] I took 3 times 4, 5 times 3, 3 times 4, and 5 times 3. [The girls calculate that this gives a total of 54.]

Tchr: You started out saying that this [motioning to a 5-by-3 lateral side of Box F] was 5 times 3 and that is how you got 15. Where would these 15 blocks be when I put them in the box? [L places 5 columns of 3 cubes inside the box as shown in Fig. 11a.] If I cut your pattern [cutting the vertical edges] and fold this [the right side] down, what should I be able to see?

L: Cubes everywhere.

Tchr: What would I have to be able to do so that you could see them "everywhere"?

L: Put more in?

Tchr: Okay. Do you want to do that? [L does not do anything.] M, do you have an idea?

Tchr: [M rearranges the cubes as shown in Fig. 11b.] If I wanted to look at this side [rotating the box 180° so that the opposite 5-by-3 side is facing L], what would I see?

L: Fifteen.

Tchr: Where would they be? [L motions along a column of 5 in the bottom of the box. The teacher places a 5-by-3 array on this column; see Fig. 11c] How many would be in the middle [rotating the box so that a 4-by-3 lateral side is facing L]?

L: [Motioning to the 4-by-3 face] 12; 12 [pointing to the opposite 4-by-3 face].

Tchr: That's 24 and 30, and that's 54. Is that how you got your 54?

L: Yeah. But I don't think that's the right answer. There would be less because there's not enough room for all these [indicating the 4-by-3 flap] to fit in there.

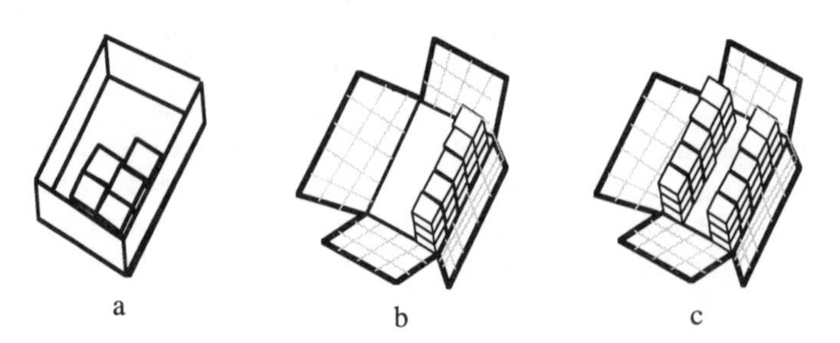

Fig. 11. L's work with teacher.

Tchr: Okay. But M thinks there's going to be more [than 54], right? M tell us how you got 60.

M: Five, 10, 15, 20 [counting columns of 5 in the top layer of the partially filled box].

Tchr: Okay, we have 20. Now what?

L: If there's 20 on the first layer [motioning to the top layer of cubes, then taking a stack of 3 cubes out of the box], then there is 20 on the second layer [pointing to the middle cube in the stack of 3], and 20 on the bottom layer [pointing to the bottom cube in the stack]. [L continues to use a layer structuring on subsequent problems.]

As L explained her enumeration strategy, the teacher recognized that L was making a substantive *structuring* error – she knew that L did not have a proper mental model of how the cubes filled the box. To aid L, the teacher focused on getting L to connect her enumeration scheme with how the cubes had to be arranged to fill the box. She tried to help L see exactly where the 15 cubes that L had enumerated for the right side fit in the box, providing similar guidance for the other three lateral sides. By comparing her lateral-sides strategy with her newly acquired understanding of how the cubes fit into the box, L saw that her strategy was flawed. She then used her improved spatial structuring of the cube array to adopt a layer-based mental model and enumeration strategy.

Subsequent Class Discussion

Midway through Day 3, the teacher showed a transparency of the activity sheet on an overhead projector at the front of the classroom. She explained that pairs of students would come to the front of the classroom and describe their enumeration strategies.

Tchr: If you are not explaining [at the front of the room], your job is to listen to and think about how the strategy being explained is the same or different than the one you used. Do you agree or disagree with the person and why? . . .

K: [Box C]. We got 16 going around the bottom [pointing to the middle area of Pattern Picture C on the overhead].

Tchr: Time out. I'm confused. Where did you get the 16 going around?

L [K's partner]: Like the whole thing [making a circular motion in the air].

Tchr: How do you mean there are 16 around? Around the outside?

L: Well, no; I mean in the bottom layer, the whole thing.

Tchr: The whole thing. I just wanted to clarify that. So it's not just the cubes around the edges, it's the whole bottom side of the box. [L nods yes.] Now, K, tell us exactly how you got your 16.

K: I went 4 times 4.

Tchr: Where did you get those two 4s?

K: Right here was 4 and right here was 4 [pointing to the bottom row and left column in the middle section of Pattern Picture C; see Fig. 12a]. And then there is 3 right here that fold up [pointing to one of the side flaps]. OK, 16 times 3 [pause]. So there is 48 cubes in that box.

Tchr: Who's next?

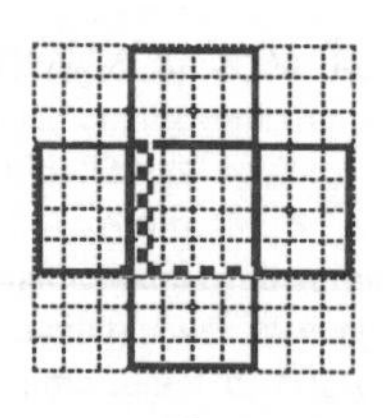
a

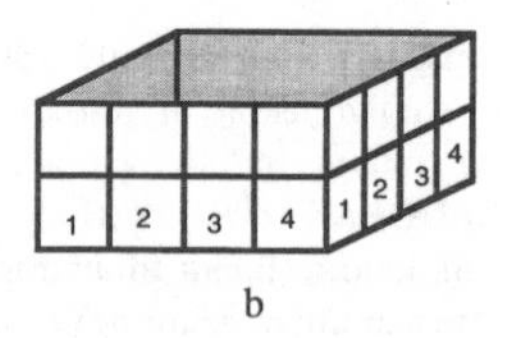

b

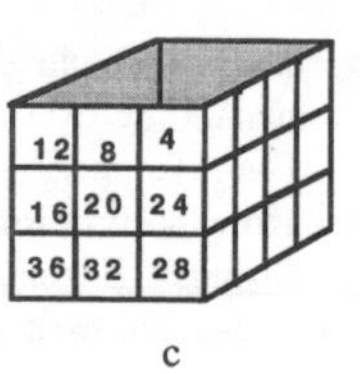

c

Fig. 12. Class discussion.

H: Well we did it a little bit differently. [Pointing to each square in the middle section of Pattern Picture B] there's 12 in the center. Then you just see how many go up and how many go that way [pointing to the top and left side flaps]. And there's 3 that goes each way. So 12 times 3 equals [pause], 36.

Tchr: Another group?

L: We are doing kind of the same thing that they are doing. But we don't have the floor plan, so we are going to use the box thing [Box Picture E].

M: And so we took these four, and these four [writes numerals on the Box Picture E; see Fig. 12b] and we timesed them together; and we got 16 times 2, 32.

Tchr: N?

N: [Box F]. If the bottom of the box is 4 units by 5 units, there would be 4 blocks going like this way and then there would be 5 like going this way [N draws two adjacent sides of a rectangle.] And then you times 4 by 5 because that's how many boxes [cubes] you need. And you would get 20. And then there's 20 times 3 because there's 3 blocks high.

Tchr: Nice job. Did anybody use a different method?

Q: [Box B]. Well, if there's four strips going back here [bottom right side row]. That's 4, 8, 12, 16, 20, 24, 28, 32, 36 [points to squares on the front face; see Fig. 12c].

Tchr: Is this procedure the same or different than the ones we've already seen?

L: Everyone else counted the bottom layer and timesed it by the side. They [Q's pair] kind of added rows going back.

Tchr: OK, they didn't do horizontal layers. [Connecting four cubes in a row and placing them on the overhead] They called this a strip. So this 4 going back was one strip. They counted or added 4 how many times?

K: Three.

Tchr: If they counted 4, three times what would they get?

M: Twelve.

Tchr: Did they end up with 12? How many times did they count 4?

X: Nine.

Tchr: How do you know they did it 9 times?

W: [Goes to the overhead] Because of back and back and back [motioning along horizontal rows of four cubes on Box Picture B]. So I just counted 3, 6, 9 [pointing to the columns on the front face of Box Picture B].

Tchr: So what multiplication would this be?
W: Nine times 4.
R: We did sort of what W is saying, but with layers. We counted 4, 8, 12 on the right side [pointing to rows on the right side of Box Picture B]. Then we took 12 and timesed it by 3.
Tchr: Why did you multiply by 3?
R: Cause there's like 3 up-and-down layers [motioning vertically along the three columns in Box Picture B].
Tchr: Is it okay to have vertical layers instead of horizontal layers?
R: Yeah.
Tchr: Why?
R: Because you're just counting all the cubes either way.
W: It's the same because the up-and-down layers would be flat [motioning horizontal] if you turned the box on its side.
Tchr: So, what do you think is the best way to find out how many cubes are in a box?
Q: Use layers. That's faster than the way we did it [skip counting].
Tchr: Okay. And what do we do with the layers?
R: Find out how many cubes are in a layer, then you times that by how many layers there are in the box.

This class discussion encouraged students to consolidate and deepen their mathematical thinking in several ways. First, students had to be sure they understood their ideas well enough so that they could communicate them. Often, students in pairs could be seen talking to each other before their presentations, making sure that they knew what they were going to say. (True collaboration among pairs was encouraged by the teacher because she often asked one partner to finish an explanation that the other partner started.) Second, understanding other students' strategy descriptions often required a student to construct a new mental model of a 3D array, or view an old one from a new perspective, thus deepening and enriching the student's spatial conception of arrays. Third, and to ensure that students would reflect on the methods being presented, the teacher asked students to evaluate and compare presented strategies. This required students to reflect on and analyze their spatial structuring and enumeration of arrays. Finally, as when the direction of layers was discussed, seeing different strategies helped students interrelate and generalize ideas, strengthening their relevant conceptual networks and helping them form more abstract concepts.

From the outset, the teacher solicited explanations and justifications, not mere answers. The focus was on sense making, understanding, and connecting ideas. For instance, near the end of the episode, she took the opportunity to help students see how multiplication is connected to skip counting, a fundamental idea for elementary school students. The teacher also pushed students to clarify

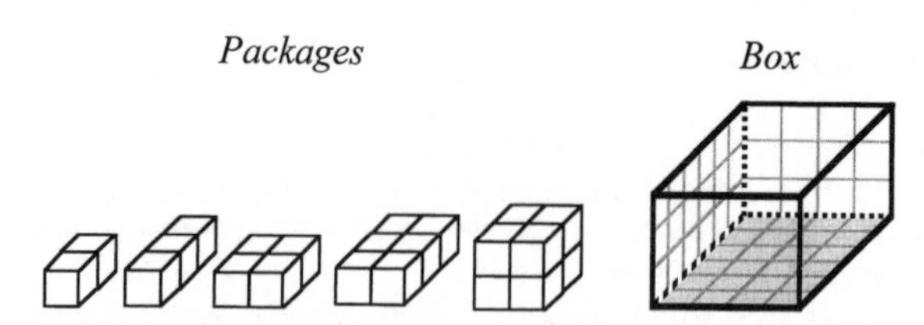

- How many of each package do you predict will fit in the box at the right? The bottom of the box is 6 cubes long and 4 cubes wide. The box is 3 cubes high.
- Use only one type of package at a time. You may not break packages apart.
- Predict, then make the box and fill it with cube packages to check your predictions.

Fig. 13. Additional 3D spatial structuring problems.

their language, having them point to objects or pictures or refine their explanations until what was being referred to was clear. For instance, at the beginning, the teacher was alert to the conceptual confusion that could be caused by using the word "around". She knew that students often consider only perimeter squares when thinking about area-like situations.

Extensions

It is important to note that this inquiry-based instructional unit did not stop with the lesson shown (Battista & Berle-Carman, 1996). In the following days, students moved on to problems like those shown in Fig. 13. Here, the goal is to ensure that students actually spatially structure the set of volume units rather than merely use a numerical procedure in an inappropriate situation. Rotely applying a procedure such as $V = L \times W \times H$ will not work on all of these problems. Neither will applying division, reasoning that because 72 cubes fill the box and the 2-by-2-by-2 package is made up of 8 cubes, $72 \div 8 = 9$ of these package will fill the box.

Additional problems such as asking students to find a box that holds twice as many cubes as a 3-by-2-by-5 box further encourage students' thinking about the spatial structure of rectangular cube arrays and strengthen students' fluency with their newly acquired concepts and strategies. Students then extend their thinking to the general concept of volume by addressing problems such as, "How much room for air is there in our classroom? Whose classroom holds more air, ours or Teacher B's?" Because it is surprisingly difficult for students to move from the small-scale-box context to the large-scale context of classrooms, students need appropriate opportunities to restructure their knowledge in this new situation.

Summary for 3D Example

The 3D cube arrays example illustrates how powerful mathematics learning can occur in problem-centered inquiry-based instruction. In this classroom culture of inquiry, students, like scientists, constructed, revised, and refined theories to solve and make sense of problems. Unlike scientists exploring uncharted intellectual territories, however, students' theory building was guided by the instructional materials and teacher. During the small group work, guidance of students' theory building came mainly through the sequence of problems on the activity sheet. Within the culture of inquiry that had been established in this classroom, the gradual fading of perceptual scaffolding in the sequence of problems, along with the combination of making predictions then checking those predictions with cubes, induced students to construct increasingly more sophisticated and viable structurings and enumeration procedures.

The teacher's role was more subtle than, but just as important as, that of the instructional materials. Prior to this unit, she had orchestrated the creation of a classroom culture of inquiry. During the unit, as she interacted with students in their small group work, she not only maintained the inquiry and sense-making culture, she used her knowledge of how students construct mathematics in general, along with her knowledge of how students construct the concept of volume, to monitor, guide, and support students' mathematical activity and cognition. In essence, she was the engineer who controlled and maintained the inquiry-learning machine operating in her class.

This example also illustrates that in problem-centered inquiry-based learning – as in scientific endeavor in general – the development of viable theories and concepts is often confused, irregular, and untidy. Students, like scientists, make mistakes and struggle to make sense of ideas. Consequently, teachers attempting to implement problem-centered inquiry-based instruction must understand that grappling with ideas, and even becoming confused (within reason), are natural components of students' sense making. This situation is unlike that of traditional instruction, in which student struggles are generally interpreted as deficiencies in the students, curriculum, or teacher. Indeed, one of the most difficult things for teachers to learn in implementing problem-centered inquiry-based instruction is when to intervene with students and when to let them struggle. Only by thoroughly understanding the usual paths students take in learning particular mathematical ideas – including stumbling blocks and learning plateaus – can teachers know when to intervene. Indeed, the design and implementation of this inquiry-based activity depended heavily not only on understanding the general nature of students' mathematics learning, but understanding the constructive itineraries students follow in learning this particular geometric idea.

CONCLUSION

Research indicates that effective geometry teaching has many components. Included are the general principles discussed by Brophy, along with the principles for effective mathematics and geometry teaching derived from research and professional standards. Among all these components, however, two critical principles stand out. First, instruction must focus on developing students' understanding and supporting their personal sense making. Second, to support students' geometric sense making and understanding, all aspects of geometry instruction must be based on a detailed knowledge of how students construct geometric ideas. Geometry teaching, curricula, and assessment must be thoroughly based on analysis of students' mathematical and geometric cognitions.

ACKNOWLEDGMENT

Support for some of the work described in this chapter was provided by Grant RED 8954664 from the National Science Foundation. I would like to thank Linda Hallenbeck, not only for participating as a teacher in the classroom research described in this chapter, but for our numerous rich discussions about the learning and teaching of geometry. The opinions expressed, however, are mine and do not necessarily reflect the views of that foundation.

REFERENCES

Battista, M. T. (1998). *Shape makers: Developing geometric reasoning with the Geometer's Sketchpad*. Berkeley, CA: Key Curriculum Press.

Battista, M. T. (1999). Fifth graders' enumeration of cubes in 3d arrays: Conceptual progress in an inquiry-based classroom. *Journal for Research in Mathematics Education*, 417–448.

Battista, M. T. (in press a) How Do Children Learn Mathematics? Research and Reform in Mathematics Education. In: T. Loveless (Ed.), *Curriculum Wars: Alternative Approaches to Reading and Mathematics*. Brookings Press.

Battista, M. T. (in press b). Shape Makers: A computer environment that engenders students' construction of geometric ideas and reasoning. *Computer in the Schools*.

Battista, M. T. & Berle-Carman, M. (1996). *Containers and cubes*. Palo Alto, CA: Dale Seymour Publications.

Battista, M. T., & Borrow, C. V. A. (1997). *Shape Makers:* A computer microworld for promoting dynamic imagery in support of geometric reasoning. In: J. Dossey (Ed.), *Proceedings of the Nineteenth Annual Meeting of the North American Chapter of the International Group Psychology in Mathematics Education*.

Battista, M. T., & Clements, D. H. (1991). *Logo geometry*. Morristown, NJ: Silver Burdett & Ginn.

Battista, M. T., & Clements, D. H. (1995). Geometry and proof. (Connecting Research to Teaching) *Mathematics Teacher, 88*(1), 48–54.

Battista, M. T. & Clements, D. H. (May 1996). Students' understanding of three-dimensional rectangular arrays of cubes. *Journal for Research in Mathematics Education, 27*(3), 258–292.

Battista, M. T., & Clements, D. H. (1998). Students' understanding of 3D cube arrays: Findings from a research and curriculum development project. In: R. Lehrer & D. Chazan (Eds), *Designing Learning Environments for Developing Understanding of Geometry and Space* (pp. 227–248). Mahwah, NJ: Lawrence Erlbaum Associates.

Battista, M. T., Clements, D. H., Arnoff, J., Battista, K., & Borrow, C. V. A. (1998). Students' spatial structuring and enumeration of 2D arrays of squares. *Journal for Research in Mathematics Education, 29*(5), 503–532.

Beaton, A. E., Mullis, I. V. S., Martin, M. O., Gonzalez, E. J., Kelly, D. L., & Smith, T. A. (1996). *Mathematics achievement in the middle school years: IEA's third international mathematics and science study (TIMSS)*. Chestnut Hill, MA: Boston College.

Ben-Chaim, D., Fey, J. T., & Fitzgerald, W. M. (1998). Proportional reasoning among 7th grade students with different curricular experiences. *Educational Studies in Mathematics, 36*(3), 247–273.

Boaler, J. (1998). Open and closed mathematics: Student experiences and understandings. *Journal for Research in Mathematics Education, 29*(1), 41–62.

Bransford, J. D., Brown, A. L., & Cocking, R. R. (1999). *How people learn: Brain, mind, experience, and school*. Washington, D.C.: National Research Council.

Burger, W., & Shaughnessy, J. M. (1986). Characterizing the van Hiele levels of development in geometry. *Journal for Research in Mathematics Education, 17*, 31–48.

Carpenter, T. P., & Fennema, E. (1991). Research and cognitively guided instruction. In: E. Fennema, T. P. Carpenter & S. J. Lamon (Eds), *Integrating Research on Teaching and Learning Mathematics* (pp. 1–16). Albany: State University of New York Press.

Carpenter, T. P., Franke, M. L., Jacobs, V. R., Fennema, E., & Empson, S. B. (1998). A longitudinal study of invention and understanding in children's multidigit addition and subtraction. *Journal for Research in Mathematics Education, 29*(1), 3–20.

Chazan, D., & Yerushalmy, M. (1998). Charting a course for secondary geometry. In: R. Lehrer & D. Chazan (Eds), *Designing Learning Environments for Developing Understanding of Geometry and Space* (pp. 67–90). Mahwah, NJ: Lawrence Erlbaum Associates.

Clements, D. H., & Battista, M. T. (1989). Learning of geometric concepts in a Logo environment. *Journal for Research in Mathematics Education, 20*, 450–467.

Clements, D. H., & Battista, M. T. (1990). The effects of Logo on children's conceptualizations of angle and polygons. *Journal for Research in Mathematics Education, 21*, 356–371.

Clements, D. H., & Battista, M. T. (1992). Geometry and spatial reasoning. In: D. A. Grouws (Ed.), *Handbook of Research on Mathematics Teaching* (pp. 420–464). Reston, VA: National Council of Teachers of Mathematics/Macmillan.

Clements, D. H., & Battista, M. T. (in press). *Logo and Geometry*. Reston, VA: NCTM: JRME Monograph Series.

Cobb, P., Wood, T., & Yackel, E. (1990). Classrooms as learning environments for teachers and researchers. In: R. B. Davis, C. A. Maher & N. Noddings (Eds), *Constructivist Views on the Teaching and Learning of Mathematics. Journal for Research in Mathematics Education Monograph Number 4* (pp. 125–146). Reston, VA: National Council of Teachers of Mathematics.

Cobb, P., Wood, T., Yackel , E., Nicholls, J., Wheatley, G., Trigatti, B., & Perlwitz, M. (1991). Assessment of a problem-centered second-grade mathematics project. *Journal for Research in Mathematics Education*, *22*(1), 3–29.

De Corte, E., Greer, B., & Verschaffel, L. (1996). Mathematics teaching and learning. In: D. C. Berliner & R. C. Calfee (Eds), *Handbook of Educational Psychology* (pp. 491–549). New York: Simon & Schuster Macmillan.

de Villiers, M. (1998). An alternative approach to proof in dynamic geometry. In: R. Lehrer & D. Chazan (Eds), *Designing Learning Environments for Developing Understanding of Geometry and Space* (pp. 369–394). Mahwah, NJ: Lawrence Erlbaum Associates.

Driscoll, M. J. (1983). *Research within reach: Elementary school mathematics and reading*. St. Louis: CEMREL, Inc.

Eves, H. (1972). *A survey of geometry*. Boston: Allyn and Bacon, Inc.

Fawcett, H. P. (1938). The nature of proof, *Yearbook, The National Council of Teachers of Mathematics* (Vol. 13). New York: Teachers College.

Fennema, E., Carpenter, T. P., Franke, M. L., Levi , L., Jacobs, V. R., & Empson, S. B. (1996). A longitudinal study of learning to use children's thinking in mathematics instruction. *Journal for Research in Mathematics Education*, *27*(4), 403–434.

Fischbein, E., & Kedem, I. (1982). Proof and certitude in the development of mathematical thinking. In: A. Vermandel (Ed.), *Proceedings of the Sixth International Conference for the Psychological of Mathematics Education* (pp. 128–131). Antwerp, Belgium: Universitaire Instelling Antwerpen.

Fuys, D., Geddes, D., & Tischler, R. (1988). *The van Hiele model of thinking in geometry among adolescents*.

Goldenberg, E. P., & Cuoco, A. A. (1998). Students' understanding of 3D cube arrays: Findings from a research and curriculum development project. In: R. Lehrer & D. Chazan (Eds), *What is Dynamic Geometry?* (pp. 351–368). Mahwah, NJ: Lawrence Erlbaum Associates.

Goldin, G. A. (1992). Toward an assessment framework for school mathematics. In: R. Lesh & S. J. Lamon (Eds), *Authentic Assessment Performance in School Mathematics* (pp. 63–88). Washington, D. C.: AAAS Press.

Greeno, J. G., Collins, A. M., & Resnick, L. (1996). Cognition and learning. In: D. C. Berliner & R. C. Calfee (Eds), *Handbook of Educational Psychology* (pp. 15–46). New York: Simon & Schuster Macmillan.

Hanna, G. (1989). More than formal proof. *For the Learning of Mathematics*, *9*, 20–23.

Hiebert, J. (1999). Relationships between research and the NCTM standards. *Journal for Research in Mathematics Education*, *30*(1), 3–19.

Hiebert, J., & Carpenter, T. P. (1992). Learning and teaching with understanding. In: D. A. Grouws (Ed.), *Handbook of Research on Mathematics Teaching* (pp. 65–97). Reston, VA: National Council of Teachers of Mathematics/Macmillan.

Hoyles, C., & Jones, K. (1998). Proof in dynamic geometry contexts. In: C. Mammana & V. Villani (Eds), *Perspectives on the Teaching of Geometry for the 21st Century* (pp. 121–128). Dordrecht: Kluwer.

Jackiw, N. (1991). *Geometer's Sketchpad [Computer software]*. Berkeley, CA: Key Curriculum Press.

Kline, M. (1953). *Mathematics in western culture*. London: Oxford University Press.

Laborde, C. (1998). Visual phenomena in the teaching/learning of geometry in a computer-based environment. In: C. Mammana & V. Villani (Eds), *Perspectives on the Teaching of Geometry for the 21st Century* (pp. 113–121). Dordrecht: Kluwer.

Lakatos, I. (1976). *Proofs and refutations: The logic of mathematical discovery.* NY: Cambridge University Press.

Lehrer, R., Jacobson, C., Thoyre, G., Kemeny, V., Strom, D., Horvath, J., Gance, S., & Koehler, M. (1998). Developing understanding of geometry and space in the primary grades. In: R. Lehrer & D. Chazan (Eds), *Designing Learning Environments for Developing Understanding of Geometry and Space* (pp. 169–200). Mahwah, NJ: Lawrence Erlbaum Associates.

Lesh, R., & Lamon, S. J. (1992). Assessing authentic mathematical performance. In: R. Lesh & S. J. Lamon (Eds), *Authentic Assessment Performance in School Mathematics* (pp. 17–62). Washington, D. C.: AAAS Press.

Lester, F. K. (1994). Musing about mathematical problem-solving research: 1970–1994. *Journal for Research in Mathematics Education, 25*(6), 660–675.

Mammana, C., & Villani, V. (1998). *Perspectives on the teaching of geometry for the 21st century.* Dordrecht: Kluwer.

Martin, W. G., & Harel, G. (1989). Proof frames of preservice elementary teachers. *Journal for Research in Mathematics Education, 20*, 41–51.

Mitchelmore, M. C. (1980). Three-dimensional geometrical drawing in three cultures. *Educational Studies in Mathematics, 11*, 205–216.

Mullis, I. V. S., Martin, M. O., Beaton, A. E., Gonzalez, E. J., Kelly, D. L., & Smith, T. A. (1997). *Mathematics achievement in the primary school years: IEA's third international mathematics and science study (TIMSS).* Chestnut Hill, MA: Boston College.

Mullis, I. V. S., Martin, M. O., Beaton, A. E., Gonzalez, E. J., Kelly, D. L., & Smith, T. A. (1998). *Mathematics and science achievement in the final year of secondary school: IEA's third international mathematics and science study (TIMSS).* Chestnut Hill, MA: Boston College.

Muthukrishna, N., & Borkowski, J. G. (1996). Constructivism and the motivated transfer of skills. In: M. Carr (Ed.), *Motivation in Mathematics* (pp. 40–63). Cresskill, New Jersey: Hampton Press.

National Council of Teachers of Mathematics (1989). *Professional standards for teaching mathematics [draft].* Reston, VA: Author.

National Council of Teachers of Mathematics (2000). *Principles and standards for school mathematics.* Reston, VA: Author.

National Research Council (1989). *Everybody counts.* Washington, D. C.: National Academy Press.

Olive, J. (1998). Opportunities to explore and integrate mathematics with the Geometer's Sketchpad. In: R. Lehrer & D. Chazan (Eds), *Designing Learning Environments for Developing Understanding of Geometry and Space* (pp. 395–418). Mahwah, NJ: Lawrence Erlbaum Associates.

Polya, G. (1945). *How to Solve It.* Princeton, NJ: Princeton University Press.

Polya, G. (1954). *Mathematics and Plausible Reasoning, Volumes 1 & 2.* Princeton, NJ: Princeton University Press.

Porter, A. (1989). A curriculum out of balance: The case of elementary school mathematics. *Educational Researcher, 18*, 9–15.

Prawat, R. S. (1999). Dewey, Peirce, and the learning paradox. *American Educational Research Journal, 36*(1), 47–76.

Quinn, A. L. (1997). *Justifications, argumentations, and sense making of preservice elementary education teachers in a constructivist matheamatics classroom.* Unpublished doctoral dissertation, Kent State University, Kent, OH.

Richards, J. L. (1988). *Mathematical visions: The pursuit of geometry in victorian England.* Boston: Academic Press/Harcourt Brace Jovanovich.

Romberg, T. A. (1992). Further thoughts on the standards: A reaction to Apple. *Journal for Research in Mathematics Education, 23*(November), 432–437.

Schoenfeld, A. H. (1986). On having and using geometric knowledge. In: J. Hiebert (Ed.), *Conceptual and Procedural Knowledge: The Case of Mathematics* (pp. 225–264). Hillsdale, NJ: Lawrence Erlbaum Associates.

Schoenfeld, A. C. (1994). What do we know about mathematics curricula. *Journal of Mathematical Behavior, 13*, 55–80.

Schoenfeld, A. H. (1988). When good teaching leads to bad results: The disasters of "well-taught" mathematics courses. Special Issue: Learning mathematics from instruction. *Educational Psychologist, 23*(2), 145–166.

Schwartz, J. L., Yersushalmy, M., & Wilson, B. (1993). *The Geometric Supposer: What is it a case of?* Hillsdale, NJ: Lawrence Erlbaum.

Senk, S. L. (1989). Van Hiele levels and achievement in writing geometry proofs. *Journal for Research in Mathematics Education, 20*, 309–321.

Shaughnessy, J. M., & Burger, W. F. (1985). Spadework prior to deduction in geometry. *78*, 419–428.

Silver, E. A., & Stein, M. K. (1996). The QUASAR project: The "revolution of the possible" in mathematics instructional reform in urban middle schools. *Urban Education, 30*, 476–521.

Steffe, L. P., & D'Ambrosia, B. (1995). Toward a working model of constructivist teaching: A reaction to Simon. *Journal for Research in Mathematics Education, 26*(2), 146–159.

Steffe, L. P., & Kieren, T. (1994). Radical constructivism and mathematics education. *Journal for Research in Mathematics Education, 25*(6), 711–733.

Stigler, J. W., Lee, S. Y., & Stevenson, H. W. (1990). *Mathematical knowledge of Japanese, Chinese, and American elementary school children.* Reston, VA: National Council of Teachers of Mathematics.

Suydam, M. N. (1985). The shape of instruction in geometry: Some highlights from research. *Mathematics Teacher, 78*, 481–486.

Thomas, B. (1982). *An abstract of kindergarten teachers' elicitation and utilization of children's prior knowledge in the teaching of shape concepts*: Unpublished manuscript, School of Education, Health, Nursing, and Arts Professions, New York University.

Usiskin, Z. (1982). Van Hiele levels and achievement in secondary school geometry (Final report of the Cognitive Development and Achievement in Secondary School Geometry Project) . Chicago: University of Chicago, Department of Education.

van Hiele, P. M. (1986). *Structure and insight.* Orlando: Academic Press, Inc.

Vinner, S., & Hershkowitz, R. (1980). Concept images and common cognitive paths in the development of some simple geometrical concepts. In: R. Karplus (Ed.), *Proceedings of the Fourth International Conference for the Psychology of Mathematics Education* (pp. 177–184). Berkeley, CA: Lawrence Hall of Science, University of California.

Wiske, M. S., & Houde, R. (1988). *From recitation to construction: Teachers change with new technologies. Technical Report.* Cambridge, MA: Educational Technology Center, Harvard Graduate School of Education.

Wood, T., & Sellers, P. (1996). Assessment of a problem-centered mathematics program: third grade. *Journal for Research in Mathematics Education, 27*(3), 337–353.

Wood, T., & Sellers, P. (1997). Deepening the analysis: longitudinal assessment of a problem-centered mathematics program. *Journal for Research in Mathematics Education*, *28*(2), 163–186.

Yerushalmy, M., & Chazan, D. (1993). Overcoming visual obstacles with the aid of the Supposer. In: J. L. Schwartz, M. Yersushalmy, & B. Wilson (Eds), *The Geometric Supposer: What is it a case of?* (pp. 25–56). Hillsdale, NJ: Lawrence Erlbaum.

Yerushalmy, M., Chazan, D., & Gordon, M. (1987). *Guided inquiry and technology: A year long study of children and teachers using the Geometric Supposer: ETC Final Report*. Newton, MA: Educational Development Center.

HIGH SCHOOL BIOLOGY INSTRUCTION: TARGETING DEEPER UNDERSTANDING FOR BIOLOGICAL LITERACY

James H. Wandersee

THE CHAPTER'S GOAL, INTENDED AUDIENCE, FOCUS, AND CONTEXT

This chapter's focus is on ways of enhancing high school students' knowledge and understandings of biology – within the frames of a constructivist model of learning and the unique nature of biology as a domain. Everything that follows should be interpreted within this context (see Fig. 1 for a concept map of this chapter's basic design).

Constructivism and Biology Teaching

For the purposes of this chapter, *constructivism* may be defined as a recently proposed theory of science pedagogy derived from a theory of knowing (NRC, 1999, pp. 10–11). The latter assumes that humans are goal-driven agents who actively pursue information, and who cognitively "construct new knowledge and understandings based on what they already know and believe." This view is based, in part, on thousands of educational research studies, worldwide, investigating pupils' alternative conceptions in science (previously called science misconceptions; cf. Chapter 5 in Gabel, 1994). This ever-expanding,

Subject-Specific Instructional Methods and Activities, Volume 8, pages 187–214.

ISBN: 0-7623-0615-7

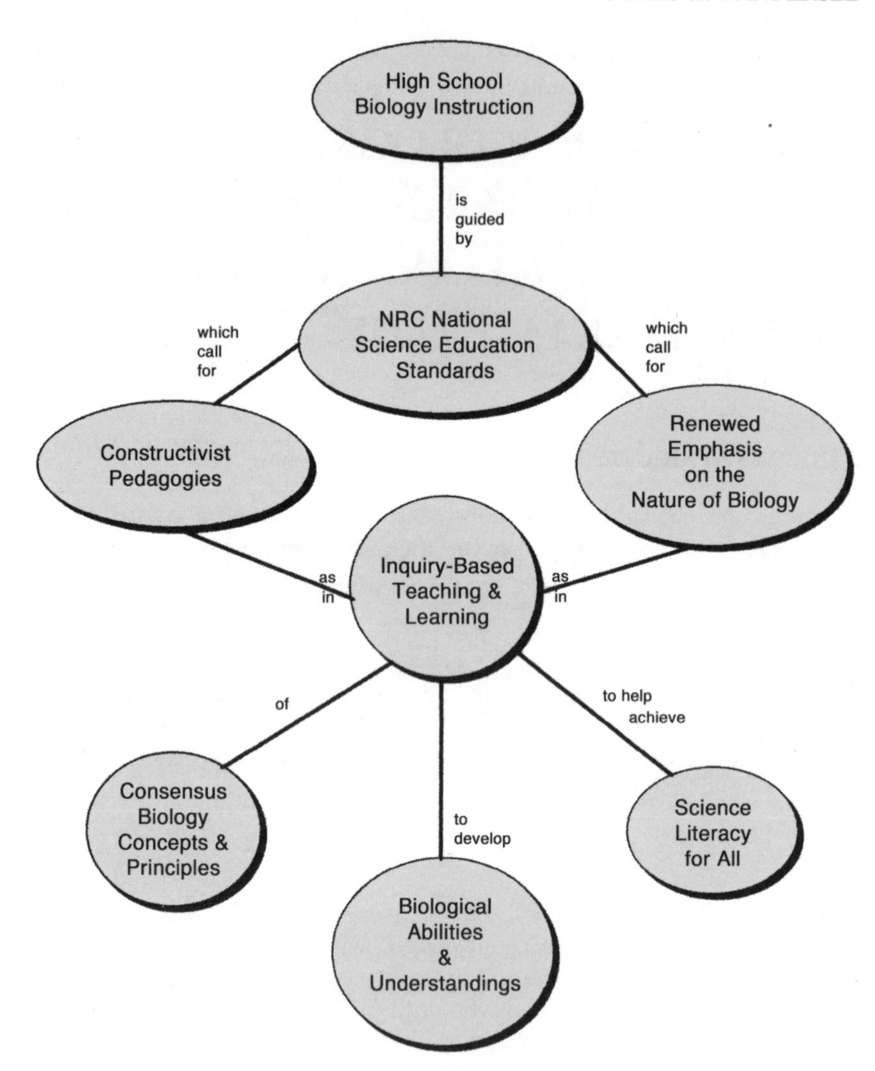

Fig. 1. Concept Map of Biology Chapter's Design.

global research corpus indicates that a person's cognitive understanding of science is primarily self-constructed within a social setting, and varies in knowledge content and structure (along with associated attitudes and skills) from student to student, in response to the individual's unique set of prior knowledge about, experiences with, and expectations for the particular scientific phenomena of interest. It has been shown that the biology learner's nascent, emergent, and progressive understanding often deviates markedly from the contemporary, consensual scientific view of that topic, and thus, from the intended learning outcomes within the science classroom. Yet, there also seem to be small sets of anticipatable categories of alternative conceptions (cf. Chapter 5 in Gabel, 1994) for almost every biology topic – from photosynthesis to animal classification to cellular respiration to natural selection – a research input to pedagogical content knowledge in biology.

So, if we know that students enter the high school biology classroom already knowing and believing certain things about the living world (both scientifically valid and invalid), it seems reasonable to think that their subsequent learning will be enhanced if biology instructors attend to, rather than ignore, this "cognitive baggage" and actually use it as starting point for new learning.

Therefore, it is assumed, for purposes of this chapter, that the goal of high school biology teaching is to help the learner actively undergo conceptual change – constructing a more empowering scientific understanding of life. Metaphorically, biology teachers begin with their students "somewhere in the middle of the stream" conceptually, not standing pristine and dry on the near bank (Gowin, 1981). Constructivist pedagogy in biology focuses on promoting students' biological knowledge construction, on emphasizing and valuing the connection of ideas, on catalyzing the negotiation of and reflection upon the meaning of the biology curriculum, on helping students self-monitor their progress in personally understanding targeted biological concepts and principles, and on letting students grapple with and interpret the new ideas of the biology curriculum using what they already know – so that transferable, meaningful, and mindful learning may eventually result (Langer, 1997). Langer defines mindful learning as having three facets: (a) "continuous creation of new categories;" (b) "openness to new information;" and (c) "implicit awareness of more than one perspective" (p. 4).

The commonplace teaching expressions "I/we covered that" are antithetical to constructivist pedagogy, in that they suggest that biological knowledge is finite, discrete, superficial, easily packaged for learning, disconnected, and need not be revisited once "covered." Those 20th-century expressions also ignore the facts that, (a) at best, any high school biology course is but a small sample of what the world's biologists know about life, and (b) that this biological knowledge both grows and changes with each passing year. Replacing the coverage-driven teaching goals (which are not supported by contemporary cognitive science research) is the idea that fewer, but carefully chosen, biology topics studied in greater depth are more likely to produce biologically literate high school graduates.

An Example of a Constructivist Biology Unit Emphasizing Inquiry

In *Inquiry and the National Science Education Standards* (National Research Council [NRC], 2000, pp. 66–75), we read about a biology teacher named Ms. Idoni who allocates a block of class time in the spring of each school year for her students to conduct a full-inquiry field study. Students who take her biology course know that she does this and they look forward to it as a rite of passage. It is reported, via the school grapevine, to be hard work, but also very satisfying.

This year Ms. Idoni chooses a lake in a city park as her field site. First, she leads her classes there on a field trip to engage their interest. When they arrive, she asks them to simply walk the perimeter of the lake, carefully observe the biological and physical features of the lake, and think about questions they might be interested in answering. This taps into their prior knowledge of biology as well.

Having employed a series of the classic, thought-provoking, biology teaching technique known as "Invitations to Inquiry" (Mayer, 1978) prior to this day – these are compact teaching units designed to focus on developing a particular skill of biological investigation – she knows she can now ask the students, during the course of this new teaching unit, to identify important questions that they can investigate at the field site, design an inquiry around one of them, choose appropriate data they can gather, plan how they will analyze these data, consult relevant and reliable sources on the World Wide Web, and draw defensible conclusions based on the evidence they collect.

Formative assessments are as important to constructivist biology teaching as summative assessments are. One formative assessment she has designed asks each student to keep a personal research journal, and she directs them to organize it using categories such as these: (a) questions and scientific ideas to guide the research, (b) design of the research, (c) technology and mathematics to be used in the research, (d) use of evidence to support the explanations, (e) alternative explanations considered, and (f) conclusions and defense of the explanations. Her categories are designed to be slight modifications of the fundamental abilities of inquiry described in the National Science Education Standards (1996, pp. 98–100) and the students' common journal format also allows her to give more frequent feedback.

Her first journal-writing assignment asks them to record their lakeside observations and questions. Subsequently, she uses the students' journal entries, along with whole class discussion (and later, small-group work, research progress reports, and cooperative learning activities) to deftly balance the tension between sustaining students' interests and insuring that they consider and reflect upon all the critical elements of a scientific inquiry – including the relevant science concepts and principles that undergird any successful scientific inquiry.

Gradually, via in-class discussions and homework assignments, students' questions narrow from water quality (Is the water safe for drinking? For swimming?), organism surveys (What lives in the lake?), and human impacts (How have humans changed the lake?) to focus upon pollution (Why does the lake stink in some places along the shoreline?). Therefore, she decides that it is important for her students to clarify and elaborate their understandings of water pollution and its sources. This takes small-group work and several class periods of discussion to accomplish. Ms. Idoni lets her students grapple with the question of how to differentiate human-influenced pollution from normal change – namely, What counts as pollution?

Students eventually come to realize that they need to understand the natural ecology of this lake before they can pursue the question of human impact. Now, silently, Ms. Idoni realizes she may have just uncovered her summative assessment task: She will propose that something has polluted the lake and the students will have to apply what they learn in their field study to this new problem that she'll pose. But, at the present time, she will have to remain a "guide on the side" – letting the students pose their own questions and helping to catalyze their own research studies. She brings the small groups together and facilitates a whole-class review of the ideas to date. Ultimately, the class decides on these two general questions: "Is city park lake polluted?" and, "If so, how have humans influenced the pollution?" (NRC, 2000, p. 68).

The students conclude that they first need to establish baseline data for the lake. But what shall they describe, monitor, and measure? Initial consultations with some water quality literature and inspections of scientific web sites have shown them that many factors can impact water quality. In consultation with Ms. Idoni, the class decides to zero-in on three types of factors: (a) physical, (b) chemical, and (c) biological. With much discussion, this initial problem typology eventually leads to the proposing of studies by various class teams of, respectively: (a) lake temperature, water color, the limits of penetration of sun light, and suspended particulate amounts and types; (b) water pH (something that they had learned to measure quite competently in a previous science course), and the concentrations of oxygen, carbon dioxide, phosphates, and nitrates (some tests they had not done before); and (c) the numbers and kinds of organisms present in the lake.

The students decide each group will gather data for two months and will share their data regularly with the other groups. They agree that they will also share their emergent hypotheses and accumulated computer search files about the pollution factors they are studying for this city lake. Meanwhile, Ms. Idoni keeps track of the 9–12 inquiry abilities being used as the research project unfolds, monitors the science backgrounds students have and need for this research, and checks the *National Science Education Standards* (1996, pp. 186, 198) to maximize the educational yield of the project. She also makes sure that the students have access to the necessary investigative technologies (from pH meters to database computers, to Hach dissolved oxygen test kits). She meets with each team on a regular basis and asks stimulating questions (Are there any surprising results? How confident are you that your data are accurate and reliable? What explanation do you expect to give for the data you collected?). She also conducts periodic update sessions where research teams share their emerging results.

Together, her students act as a scientific community, weighing the current evidence, debating interpretations, analyzing competing hypotheses and explanations, challenging each other's findings, and slowly coming to the conclusion that all the factors being studied must be integrated in explaining the lake's pollution – and that new factors must even be added! She helps them cope with emergent barriers to scientific thought, such as uncontrolled variables, anomalous data, and uninformed opinions masquerading as fact. New vocabulary is learned spontaneously by the students and on a "need to know" basis and students incorporate the terms into their conversations quite naturally. The students also learn about and ultimately decide to use coliform bacteria testing as their operational definition of human pollution. After all, they reason, these bacteria make water unsafe for swimming and drinking, but live longer and are easier to detect than other human disease-causing bacteria. And, after lots of self-conducted bacteriological tests, compiled reports and a major summary document were produced and presented. The students conclude that, by their self-established standard, city park lake is *not* polluted by humans – even though they now know that the lake undergoes many changes over time.

As a final (summative) assessment, Ms. Idoni poses a new problem for them about the lake and asks them to propose an inquiry that might be conducted for the City Council in order to solve the problem. This is the scenario she poses: A major fish kill has occurred at the lake and everyone suspects pollution. The one thing they do know is that no coliform bacteria have been found in the lake water. She typically finds that the new knowledge her students have constructed during the course of this research unit is deeper and more likely to transfer to solve such a novel problem as the fish kill because they have learned it doing science. She has demonstrated that, with respect to freshwater ecology, it is possible for her high school biology students to learn the abilities of scientific inquiry as well as scientific concepts and principles – important milestones on the road to scientific literacy.

Yet, inquiry is not the only teaching approach she uses. She knows that there are simply not enough hours in the school year to teach every lesson this way – and that there is not a single constructivist approach to biology teaching. Constructivism, as used in science education, also serves as a "knowledge-building" metaphor for how humans learn science – and many teaching strategies can be modified to incorporate it. Key elements of it include: active student orientation to the learning task; student engagement with a scientific question; direct (hands-on) experiences with natural phenomena; elicitation of students' existing ideas; exchange of, dialog about, clarification of, and testing of student ideas; contemporary scientific content input and negotiated restructuring of students' existing ideas; evaluation of alternative explanations by the classroom learning community, testing the application of current explanations in new situations, and continuous feedback of results to the learner (with subsequent modification of explanations as indicated), culminating in teacher-student review and assessment of what they have learned and how they have learned it (NRC, 2000, p. 35). She does all this in an ever-changing social context; as new societal issues come to the fore, they require a well-considered instructional response following established professional guidelines. As a cautionary note, it is also important to quote Harlen (1999a, p. 39), "To date, however, there is no research to show that one approach to changing pupils' ideas is more effective than another nor that any has long-term effects on the development of concepts." That's why using a variety of pedagogical strategies seems prudent at this point in the history of biology education.

It seems that, at present, we may actually know more about what's wrong with biology learning than we do about how to prevent or fix it via improved instruction. We think that the new ideas we intend to teach must be made to appear "plausible, intelligible, and fruitful" (Strike & Posner, 1985) if learners are to change their science conceptions. Those researchers proposed that conceptual change is most likely to occur when students (a) are dissatisfied with the status quo of their understanding, (b) understand how the new ideas would work for them, (c) find the new ideas plausible, and (d) concede that the new ideas will indeed be more fruitful than the old. The science education research community also thinks students need to test their ideas in order to come to see the discrepancy between their current limited or inappropriate scientific explanations and how the natural world actually works.

However, Fensham and Kass (1988) have observed that what may seem obviously discrepant to the teacher may not even be noticed by the student, and that students can be remarkably tolerant of exceptions and oblivious to the need to change their ideas just because of a demonstrated exception. One thing we do know is that students cannot simply be told what to think via didactic instruction; they need to be guided to take ownership of the biological concepts and principles we want them to learn. We now know that biology teaching is much more than just telling students about life.

The Nature of Biology

A recent National Academy of Science publication is unique in that it focuses on *Teaching about Evolution and the Nature of Science* (1998). Indeed, the nature of science has received increasing emphasis in U.S. high school biology classrooms during the past decade. Biology is the scientific study of life. The biological sciences differ from the physical sciences in many ways. Because living things not only have a current state, but also have an evolutionary history, which is reflected in their genes, causality is attributed at two levels in biology – *proximate causality* (in terms of presently acting mechanisms of structure and function) and *ultimate causality* (in terms of phylogeny and mechanisms of evolutionary change – genetic, developmental, and ecological).

Thus, no monolithic "nature of science" exists. The nature of the biological sciences differs in important ways from that of the physical sciences (Mayr, 1982). In contrast, while the physical sciences emphasize experimentation and elevate prediction, the long evolutionary history and complexity of living organisms, as well as their myriad interactions at multiple levels of organization explain why many aspects of life in the biosphere are not amenable to such research approaches. Descriptive, observational, and integrative studies such as Charles Darwin conducted during the voyage of the *H.M.S. Beagle* are also highly valued in the biological sciences. Biology is not "temporarily delayed en route to becoming a science like physics," as earlier histories of science have implied, but it has its own unique suite of methods, limits, explanations, and knowledge products which together constitute the nature of biology. The central theory of biology is evolution and its key subconcepts are variation, selection, divergence, and biodiversity.

> Evolutionary biology allows us to determine not only how and why organisms have become the way they are, but also what processes are currently acting to modify or change them (American Society of Naturalists, 1999, p. 3).

The National Research Council, in a book that focused on improving biology teaching in the nation's schools, (1990, p. 23) stated that evolution is a natural process "that is as fundamental and important in the living world as any basic concept of physics one can name."

It seems important to remember these two mediating contextual frameworks – constructivism and the nature of biology – in carrying out instructional methods and activities in today's high school biology classroom.

An Example of a 5E-Model Biology Lesson Emphasizing the Nature of Biology

The 1998 National Academy of Science publication, *Teaching About Evolution and the Nature of Science*, offers eight exemplary learning activities. One is entitled: "Proposing the Theory of Biological Evolution: Historical Perspective" (pp. 93–95). Since 1989, a growing international movement (sparked by a sequence of five influential international meetings) has been advocating increased use of history and philosophy of science in science teaching. The following lesson shows how the history of biology can help students understand the nature of biology. It also illustrates the use of the 5E model of instruction, a 5-phase elaboration of the learning cycle model, developed for teaching biology at the middle school and high school levels by the BSCS in 1990. It is an easy-to-remember teaching sequence, because the first letter of each phase begins with an "E" and the E-words follow in logical order.

In this high school biology lesson (NAS, 1998, pp. 93–99), students are given copies of 1- or 2-page excerpts of primary-source historical writings by three renowned 19th-century biologists Jean Lamarck (1744–1829), Alfred Russel Wallace (1823–1913), and Charles Darwin (1809–1882) on their original concepts of evolution. Thereby, students have the opportunity to read and talk about actual statements by key contributors to the development of the theory of biological evolution. They begin to understand the people behind the science and begin to value original intellectual documents.

The teacher first ENGAGES the learners by asking pertinent questions based on the intended learning outcomes. The teacher might say: "I want you to think about these questions as you go about reading these famous writings – What influence do you think society had on their views? What makes an explanation a scientific one? Can scientific explanations ever change; and if so, how? Can you name some major theories in science? In biology?" The first homework assignment is to read an excerpt from Lamarck's Zoological Philosophy written in 1809.

The teacher next asks the students, who have now completed the first assigned reading, to EXPLORE Lamarck's ideas by working in groups of four and discussing Lamarck's explanations for why organisms change – considering the role he thinks the environment plays in change, his emphasis on the observation of facts, whether or not his explanation was scientific, and what plausible alternatives there might be to his use-disuse observations. The next homework assignment is to read an excerpt entitled "The Struggle for Existence" from Wallace's 1858 publication *On the Tendency of Varieties to Depart Indefinitely from the Original Type*.

The following day, the teacher asks the students, who have now completed their second assigned reading, to EXPLAIN (in a teacher-guided, full-group discussion) why Wallace thinks wild animals' lives are struggles to exist, how his view is scientific, why he thinks useful variations will tend to increase, how his explanation of biological changes differs from Lamarck's, and what they think about Wallace's critique of Lamarck's hypothesis. This oral comparing and contrasting helps students come to see biology as a human endeavor, see the value of history of biology for understanding biology, and see how biologists reason. Teacher feedback and introduction of associated current biological terms is important here. The next homework assignment is to read an excerpt entitled "Introduction" from Darwin's 1859 publication *On the Origin of Species by Means of Natural Selection*.

On the next day, the teacher asks students to ELABORATE their current ideas about changes in organisms (in a teacher-facilitated, full-group discussion) by reacting to the following questions: What stimulated Charles Darwin to develop his ideas about how species originate? What did Darwin suggest was the origin of species? How does Lamarck's and Wallace's work relate to Darwin's? Was Darwin's explanation scientific? How did Darwin go about trying to find out how changes in species occurred? How did Darwin explain the fact that his ideas were incomplete? Greater understanding of how the concept of evolution developed should be evident in students' answers. Students should begin to speak with greater confidence in describing scientific thought and to grasp that biology is question-driven.

Finally, the teacher EVALUATES the week's work by asking the students to write a short essay on the nature of biological knowledge, as demonstrated in the three historical readings on evolution. They are advised that they should use at least two quotations to support their assertions. The essays should also include the concepts of *descent from common ancestors, adaptation*, and *natural selection*. Students then receive feedback from the teacher about their progress in understanding the nature of biology and biological evolution. New questions captured in some students' essays may lead to a new 5E cycle, with appropriate excerpts from more contemporary evolutionary biologists.

As Hagen, Allchin and Singer (1996) aptly observe, "Without understanding the background of [biological] discoveries we have little to justify our scientific beliefs. We must [then] rely on the authority of the scientists themselves" (p. v). The focus is not on learning the history of biology; instead, we look to the past to understand contemporary biological ideas better.

THE PROBLEM OF REACHING CONSENSUS ON U.S. BIOLOGY TEACHING GOALS

Unlike many other nations, the United States has no Ministry of Education; this means there is no official national agency to determine what should be taught (as well as, how and when) in U.S. high school biology courses. This has resulted in high school biology courses that vary dramatically in funding, content, and modes of instruction.

In addition, American high school biology instruction has long sought to resolve the pulls of two powerful forces. One force comes from the nation's scientific and technical communities – expecting, as an outcome, students who are well prepared for future college coursework and possible careers in related scientific and technical fields. The other force springs from the nation's citizenry – expecting, as an outcome, more civic-minded youth who possess improved biological knowledge and understanding that enables them to function as better decision makers in their local, national, and global communities – making sensible choices about current biological issues that touch their own lives, and others' lives as well. While these two forces are not necessarily antagonistic, they do act in substantially different directions – changing in magnitude and direction over time, chiefly in response to perceived threats, from Sputnik to global economic competition to inferior test results on international surveys of scientific understanding.

Returning to the two fundamental forces impinging on high school biology instruction – Is high school biology to be taught and learned primarily "for college and career" or "for living?" Examine all of the issues of *The American Biology Teacher* journal that were published by the National Association of Biology Teachers between 1938 to the early 1990s and you will see a curricular and instructional oscillation that mirrors the changing relative influence of the two camps. Fortunately, within the last decade, the *National Science Education Standards* (National Research Council, 1996, p. ix) has provided a welcome and consensus-driven solution to that problem for the nation – a solution that tapped the pooled wisdom of U.S. scientists, university science educators, and those who teach K-12 science, heralding a "Call to Action" which describes a "vision of [K-12] science education that will make *scientific literacy* for all a reality in the 21st century (p. ix)."

The Standards as Central to the Design of this Chapter

Not only does the *Standards* (NRC, 1996) address high school biology content, but it also explains how to teach it, and what science-reform-appropriate, classroom activities should "look like" if scientific literacy is to be attained. The *Standards* (p. 12) "provide(s) criteria to judge progress toward a national vision of learning and teaching science . . . that promotes excellence. . . .

It therefore seems reasonable to locate this increasingly influential, consensus document at the core of this chapter – using it and its new supplement, *Inquiry and the National Science Education Standards: A Guide for Teaching and Learning* (NRC, 2000), along with selected documents in the biology education literature, as the primary alignment template for today's high school biology instruction and activities. Although this chapter spotlights some important ideas from these two vanguard NRC volumes, it is intended primarily as a pointer to the 464 pages that comprise those two texts. Both are available in print and are accessible on-line, from the National Academy Press web site at http://www.nap.edu.

What is Scientific Literacy?

The goal of the K-12 science education standards (NRC, 1996) is scientific literacy for all Americans. "Scientific literacy is the knowledge and understanding of scientific concepts and processes required for personal decision making, participation in civic and cultural affairs, and economic productivity" (NRC, 1996, p. 22). Not only are the *Standards* (1996) built on the groundwork laid by the American Association for the Advancement of Science's long-term, school science literacy reform initiative which is called *Project 2061* (AAAS, 1989, 1993), but they also carry a federal agency's imprimatur to guide science instruction in the nation's schools, seeking to reconnect "learning science, learning to do science, and learning *about* science" (National Research Council, 2000, p. xv). Operationally, this means that (a) understanding some important, carefully chosen biological concepts and principles in-depth – as well as key biological theories (Duschl, 1990), (b) learning the kinds of scientific thinking that lead to such biological knowledge, and (c) knowing "how biology knows what it knows" (the nature of biology) are the three most helpful navigational way-points that teachers can use when sailing the instructional-strategy-and-activity-selection waters for high school biology instruction.

Scientific Literacy and Biology

Achieving science literacy is, by national consensus, the übergoal of every U.S. science class from K-12 – including high school biology, where it is frequently recast as the subset called "biological literacy" (BSCS, 1993). The essence of biological literacy is understanding a small number of pervasive biological principles and theories well, and applying them in appropriate ways to activities in personal and social spheres.

The *Standards* (1996, p. 181) specifically recommends that high school biology content focus on helping students first understand six core topics well: (a) the cell; (b) the molecular basis of heredity; (c) biological evolution; (d) interdependence of organisms; (e) matter, energy, and organization in living systems; and (f) behavior of organisms. It is assumed that students in grades K-8 have already developed some fundamental understandings of biology by meeting the K-4 and the 5–8 life science standards.

The bull's-eye of this target, however, is conceptual understanding of core biological concepts, principles, processes, and theories (Bybee, 1986). What is conceptual understanding? To a biology educator, concepts are patterns or regularities in nature that we have given a name – for example, *photosynthesis* (a process), *plant cell* (an object), *fertilization* (an event), and *alive* (a property) are all concepts of importance to achieving biological literacy. Concepts are what we think with, and are what we link together semantically to form networks of individual propositions such as "plants make food" or "red blood cells carry oxygen." Higher order concepts, called constructs – such as the construct of *carrying capacity* in ecology – are invented by biologists to encapsulate an important set of relationships among key subconcepts. For this chapter, learning biology is therefore considered to be the constructing of an ever-more-elaborate and well-integrated prepositional hierarchy of knowledge consisting of concepts, constructs, linking words or relations, as well as biological *images* – according to the dual-coding theory of memory (Paivio, 1971). Also to be fluently integrated with this knowledge are associated attitudes and skills that are congruent with prevailing contemporary biological thought.

The BSCS, the nation's leading biology curriculum development organization established via funding from the National Science Foundation in 1958, has greatly influenced U.S. bioeducational thought. It describes conceptual understanding as a kind of knowing that is much deeper and better integrated than the superficial understanding which typically results from didactic presentation of biological facts and memorization of terms. "When students understand biology content conceptually, they can apply their knowledge in new situations, organize information from different sources into a coherent explanation, and retain the major ideas of biology long after they have left high school" (cf. the FAQ's at http://www.bscs.org).

The BSCS Model of Biological Literacy and its Application to Teaching

In 1993, BSCS (formerly called the Biological Sciences Curriculum Study) published a very helpful biological literacy model that established and provisionally defined four (increasing) levels of biological literacy: (a) Nominal, (b) Functional, (c) Structural, and (d) Multidimensional (pp. 18–24). To provide an operational definition of the model's levels, BSCS (p. 18) describes them as follows. At the Nominal level, "Students identify terms and concepts as biological in nature, possess misconceptions, and provide nave explanations of biological concepts." At the Functional level, "Students use biological vocabulary, define terms correctly, but memorize responses." At the Structural level, "Students understand the conceptual schemes of biology, understand procedural knowledge and skills, and can explain biological concepts in their own words." At the highest level, the Multidimensional level, "Students understand the place of biology among other disciplines, know the history and nature of biology, and understand the interactions between biology and society." They possess appropriate knowledge, attitudes, and skills that reflect the depth of their understandings of and interest in biology. They are capable of investigating a biology-related problem or issue on their own. They are well on their way to becoming a lifelong biology learner. Interestingly, BSCS (p. 23) notes that few professional biologists, much less students, will attain this level in all areas of the discipline. Thus, the Multidimensional level is like a distant destination on a map that may serve as a life-long challenge to attain.

Pivotal Aspects of the NRC Teaching Standards

The *Standards* (NRC, 1996) are based on the premise that learning science is something that students themselves *do*, not something that *is done to them*. The *Standards* call for active student involvement and for stimulating learning environments in which students describe objects and events, ask questions, formulate explanations, test those explanations, communicate their ideas to others, and build critical and logical thinking skills. A resonant document, the

Benchmarks for Science Literacy (AAAS, 1993) proposes minima for what all students should know and be able to do in science, mathematics, and technology by the end of grades 2, 5, 8 and 12. It is a companion volume to *Science for All Americans* (AAAS, 1989), the initial and highly influential handbook for AAAS's *Project 2061*, which spearheaded the current reform in U.S. science education, and which sought to answer the question (p. 14), What should the content and characteristics of science, mathematics and technology education be for today's children who will enter tomorrow's world? (The promise of tomorrow's world is symbolized by the year, 2061, in which Halley's comet will return and the children who were just starting school when the comet last appeared and the reform project started will be senior citizens, p. 11).

HOW RESEARCH ON LEARNING INFORMS BIOLOGY TEACHING

The National Research Council's recent book *How People Learn* (1999) suggests that we begin by assuming that high school biology teaching ought to be based upon and align itself with what research says about how humans learn. For purposes of this chapter, learning can be defined as a change in the meaning of experience which involves "storing" knowledge – declarative, procedural, and episodic – within long-term memory (LTM), in a reflective, hierarchical, and well-indexed manner, so that it is retrievable and transferable when needed months or years later. It is, by exclusion, *not* arbitrary, poorly integrated, verbatim, rote learning that takes place over a short time ("cramming," in students' vernacular). Rote learning rarely transfers well, has a "shelf life" of hours to weeks, and, in the past, quite conveniently gave both students and teacher the illusion that deeper understanding had actually occurred. The Private Universe Project's probing of biological understanding, in recent video interviews of new Harvard University graduates about where the biomass of a log comes from, demonstrates that rote learning can occurs at even our best institutions of learning.

The opposite, "meaningful learning," involves building on one's prior knowledge, connecting the new knowledge to be learned to one's existing knowledge in a substantive, nonarbitrary way, and distributed, reflective practice, as well as monitoring the trajectory of one's own understanding (via metacognitive skills and tools, such as concept mapping). Meaningful learning has shown to be more stable, more transferable, better integrated, and more easily accessible – especially for science content that does not need to be remembered verbatim. Another important consideration for sound instructional design is the recent research about the importance of promoting "mindfulness" during instruction – defined as "openness to new information, continuous creation of new categories, and an awareness that there is more than one perspective – and its positive effects on human learning (Langer, 1997, p. 4).

Thus, we may safely assume that the focus of high school biology education is to effect conceptual change leading toward a more meaningful scientific understanding of life and how biologists have come to know what they claim to know about life, so as to ultimately help create a scientifically literate high school graduate. Research on learning (and teaching) influences classroom practice via professional organizations, educational materials, teacher education programs, educational policy, and the media (Donovan, Bransford, & Pellegrino, 1999).

Students come to the secondary school biology classroom already knowing something about life – and research allows us to predict that their prior knowledge for many biology topics has strengths and weaknesses, misconceptions and gaps. However, if that knowledge, problematic as it may be, is not activated during instruction, they may partition (in memory) what they learn in high school biology class as simply "school science," reverting to their pre-existing "street science" whenever they leave the school building.

Some Learning Principles Worth Noting

Assuming that brain function, nutrition, family support, security, and other preconditions are met, the National Research Council has scoured the research on how people learn (Donovan, Bransford & Pellegrino, 1999) and has concluded that three learning research findings have a very solid base and strong indications for teaching, constituting the foundation of sound instructional design:

(1) "Drawing out and working with existing understandings is important for learners of all ages" (p. 11).
(2) "To develop competence in an area of inquiry, students must (a) have a deep foundation of factual knowledge, (b) understand facts and ideas in the context of a conceptual framework, and, (c) organize knowledge in ways that facilitate retrieval and application" (p. 12).
(3) "A metacognitive approach to instruction can help students learn to take control of their own learning by defining learning goals and monitoring their progress in achieving them" (p. 13).

These align well with Brophy's principles in Chapter One of this volume and use of these findings is evident in this chapter's vignettes.

An Example Illustrating How Metacognitive Tools May be Applied in Biology Teaching

Metacognitive tools used in biology education include concept maps, flow charts, semantic networks, Vee diagrams, K-W-L charts, and so forth. These are types of self-constructed graphics (see Fig. 1) taught to, and then completed by, biology students in order to help them monitor the state of their own thinking/ understanding with respect to the subject matter they are studying – and these visible constructions are also useful to the biology teacher in negotiating a more complete scientific understanding with each student (Hyerle, 1996).

Because such visual tools spark reflection upon, dialoguing about, and restructuring of one's own understanding, they are well worth the extra class time they take to construct and discuss. Spiegel and Barufaldi (1994) tested the use of student-constructed graphic organizers following science instruction and found that they increased retention and recall of science textbook information. Generally, it takes at least two months and 10 constructions before a student feels comfortable in representing his/her knowledge this way. After a sufficient period of training and practice, these tools can also be used in items on assessments.

The following abridged example of a high school biology lesson using metacognitive tools was inspired by Hodson's "teaching/learning strategies for conceptual development in science" (1993, p. 108). Mr. Brown is teaching a unit on the interdependence of life. He begins by having his students write about any two organisms they know about which are interdependent or have a long-standing interaction – such as the shark and the remora. One student writes about the relationship between an alga and a fungus comprising the lichen on the rock by the school flagpole, another about the relationship between the yucca moth and the yucca plant she saw at a desert museum when she vacationed in Arizona, yet another about an old-fashioned weight-loss scheme for humans involving the unwitting ingestion of tapeworm eggs, and so forth.

Subsequently, Mr. Brown reads each of the papers himself, marks and attaches a small rubric sheet to each for concise feedback to the student, returning them promptly the next school day. He then uses the prior knowledge they contained as a benchmark and referent for facilitating a whole-class exploration and discussion of those ideas – so that his students can compare their own knowledge and experiences with that of others. The class ultimately decides that there are various kinds of relationships between organisms – relationships that they will call beneficial, harmful, and neutral. Mr. Brown challenges them to find evidence and support for their ideas during the course of the unit – using not just exotic examples, but local examples too. He also has them construct and complete a K-W-L chart (Carr & Ogle, 1987) probing what the students already know, want to know, and have learned so far in the new unit about ecological relationships between organisms. He posts these around the room as stimuli for research.

He draws a 3 meter-wide, blank, spectrum diagram on the wall-mounted marker board at the front of the classroom and then he labels it sequentially from *harmful* pole to the *neutral* midpoint to the *beneficial* pole. Their own words now stand before them. He tells them that by the end of the unit they will be able to fill-in a full range of examples of interacting organisms, position them at the proper place on the diagram and supply reasons for the positions of each. They will also have learned some of the technical terms that scientists use in talking about these relationships from their explorations of science resources and their "talking science" in the classroom.

He encourages them to investigate the topic using the broad array of high school biology texts he has placed on a classroom bookshelf, to conduct "advanced searches" on the World Wide Web via the classroom's computers, to interview an aquarist and a local farmer who raises alfalfa, and also to read some accessible trade books with ecological and organismal relationship themes, such as *Islands of Hope* (Manning, 1999; a book on the research findings on what ecological factors make today's nature preserves successful).

Student pairs work rather independently using a variety of resources in different parts of the classroom – talking to and consulting with Mr. Brown as problems in their research arise. He roams the room, asking probing questions, "pollinating" the inquiry, and encouraging the students as they discuss and record their findings.

Midweek progress reports are required of each group. These are to be in the form of large concept maps showing their key findings and identifying their model organisms. Some teams use diagramming software, a World Wide Web imagebase, presentation software, and one of the high school's LCD projectors to show the class their concept maps. Others use alternative media, and even role-play their animal-animal and plant-animal interactions! Gradually, the classroom spectrum chart fills in too, and student teams explain the entries they posted to the whole class. Spontaneously, the students add the appropriate technical terms to the chart and use them in their group presentation. Mr. Brown sometimes gently interrupts the speakers to correct their pronunciation or to tell them what the component word forms mean in Greek or Latin. Sometimes they ask him questions, so as to be scientifically correct. All of this happens as free-flowing and open conversation, rather than a feared inquisition run by the biology teacher.

With their co-constructed spectrum chart in full view, Mr. Brown makes a short computer slide-show presentation on how the unit connects to a major theme in biology – namely, evolution. He uses the examples that they themselves generated in order to operationally define coevolution as a series of reciprocal adaptations in two species, with changes in one species acting as new selective forces on another species. He then asks them to reinterpret and represent the process of coevolution, via a self-constructed standard flow chart, tracking scientific evidence for coevolution of the relationship between a selected flowering plant and its pollinator – including the energy costs and benefits at each stage. A take-home test follows the activity.

On the last day of the unit, after handing in their take-home exams, students translate some actual scientific research journal articles on coevolution supplied by Mr. Brown into Gowin's Vee diagrams (diagrams used to represent how scientists come to know something via inquiry). Their diagrams must reveal the driving questions, the natural events of interest, the epistemological and methodological framework of the investigation, and the resulting knowledge claims and value claims for the article they received. This metacognitive construction project also serves as a summative assessment for the unit. The walls of the room were now full of metacognitive artifacts – Mr. Brown calls it their "museum of biology thinking." Students linger to observe their cognitive art exhibition.

SOME FEATURES OF AN EXEMPLARY BIOLOGY COURSE

From multiple readings of the biology education literature, it seems prudent, therefore, to assume that a contemporary high school biology course should integrate conceptual development activities, investigative activities that involve problem posing, problem solving, and persuasion of peers (Peterson & Jungck, 1988); issue-driven activities, and interactive presentation activities (Udovic, 1997).

In addition, a reading of the mainstream biology teaching literature suggests to your author that a contemporary high school biology course needs to focus its instructional activities on elucidating these three themes

(1) understanding the evolution of life on earth – via biology's central theory and others (NRC, 1998),
(2) conducting biological inquiry processes (NRC, 2000), and
(3) understanding the nature of biological science (AAAS, 1989, 1993; Mayr, 1982, NRC, 1998).

Detailed Vignettes and Accounts That Model Good Teaching Practices

First of all, for biology topics that involve understanding aspects of inorganic, organic, and biochemistry (e.g. soil composition, human nutrition, digestion, photosynthesis), Treagust and Chittleborough's Chapter 8 on Chemistry, in this volume, offers not only an exemplary exposition of research findings, but also valuable and appropriate insights for practice. Second, the *Inquiry* book (NRC, 2000), the *Standards* book (1996), and the *Evolution and Nature of Science* book (NRC, 1998) present carefully constructed and extensively reviewed vignettes of the planning and unfolding of several high school biology lessons and inquiry activities that operationalize the Standards (1996) and model good practice for inquiry teaching. Rather than attempt to replicate additional full and rich descriptions (such as the one about Ms. Idoni) and their accompanying photographic images here, since they are only a few mouse clicks away, it is suggested that the reader examine: (a) Ms. M's provocative history-of-photosynthesis lesson (NRC, 1996, pp. 194–196; (b) Mr. D's hands-on fossil investigation (NRC, 1996, pp. 182–184), and (c) the set of eight activity lesson narratives which apply the 5-E learning (previously introduced) model to guide instruction on concepts such as natural selection, the magnitude of geological time, and fossil evidence (NAS, 1998, pp. 61–103). Lists of some recommended inquiry-based teaching materials are found at the back of the *Inquiry* book. Note that, although the *Standards* (1996) emphasize inquiry teaching, not all topics lend themselves to such an approach, and biology teachers should use a variety of approaches, applying pedagogical content knowledge in line with the topic, learning goals, and motivation factors.

SOME ADDITIONAL RESEARCH INSIGHTS

Harlen (1999a, p. 4) recently conducted a review of the research on effective science teaching in eight different areas: doing practical work [laboratory learning]; using computers; varying approaches to constructivism; changing pupils' ideas; practicing reflection and metacognition; using assessment to help learning; planning, questioning, and using language; changing the curriculum; and, "improving teachers' own understanding." Her criteria for inclusion of a study in her review were: intergroup comparison, involvement of participants from middle school to high school levels, achievement-based comparison, and sustainability of the innovation in normal classrooms – not just under favorable conditions. Here are some of her and others' findings, of high relevance to contemporary biology teaching.

Laboratory Work

Increasingly, high biology instruction has involved more and more laboratory experiences and field work. Yet, from a science learning perspective, Harlen (1999a) cites studies that show the following: there is no evidence that increased amounts of laboratory work increase pupils' interest in science or motivation to learn science (cf. Gardner & Gauld, 1990). They found that often what students like most about laboratory work (handling materials, freedom, sociality) is quite unrelated to what is purportedly being learned. In fact, only science laboratories that provide the right level of cognitive challenge, have a clear purpose, "work," and allow students' to have sufficient amounts of control and independence were valued by young teenagers (Hodson, 1993).

Harlen (1999b) winnows the research on laboratory teaching and concludes that laboratory work has three important functions: (a) letting students see a phenomenon or effect for themselves, (b) letting them decide what to test, and then try it to test their own or scientists' ideas, and (c) letting them carry out an open-ended investigation – thus trying out relevant science process skills and gaining some firsthand understanding of the nature of science. She reminds us that laboratories can be supplemented by demonstrations or computer programs, but never replaced by them.

Computers

While 54% of the American general public thinks computers have improved learning a great amount, only 31% of U.S. teachers agree (Wurman, 1999). Although using computers to simulate natural phenomena or to data-log and graph data can, indeed, help take the drudgery out of laboratory work, it can also give pupils a false sense of reality, and of what can actually be done in an authentic scientific investigation (Harlen, 1999b). She points out some studies which confirm that when the teacher actively directs students' attention and probes their scientific thinking by eliciting prediction and hypothesizing while they are using the computer to visualize their data, then students will make greater efforts to interpret and explain their data. Yet even the best biology simulation software can seed alternative conceptions (misconceptions) unless students' thinking is artfully probed during the experience and students are debriefed after it. Perhaps the best rule of thumb, based on practical experience, is to teach biology with the actual objects and phenomena of nature – unless this is impossible, due to economic, time, safety constraints. For example, ADAM™ software allows students to study selected aspects of human anatomy when work with cadavers is inappropriate or impossible. Simulations and models are simplified versions of reality – and are not full replacements for studying nature directly. Use the computer in biology teaching to word process, data base, image base, data log, graph, present, visualize, analyze data, simulate, search the World Wide Web, and model in support of biological inquiry activities. It is an extension of the human mind and a tool for thinking. However, the teacher's roles of motivator, guide, subject matter specialist, meaning negotiator, and learning diagnostician appear to remain central to effective biology teaching.

High School Biology Textbooks

Biology differs from physics and chemistry in its heavy use of polysyllabic Greek and Latinate terms – complete with multiple synonyms! In the past decade, biology education has tried to reduce the thicket of terminology that can seem so daunting to a high school biology student. The latest research study of high school biology textbooks, conducted by AAAS's *Project 2061* (Nelson, 2000, p. 639), found that they are still "filled with pages of vocabulary and unnecessary detail." Sadly, they "fail to make important biology ideas comprehensible and meaningful to students. . . ." and they show "serious shortcomings in both content coverage and instructional design." Their biggest defects are the failure to tell coherent stories about biological phenomena and the prevalence of a writing style characterized by piecemeal treatment, unconnected details, and obscurantism. AAAS spokespersons suggested that high school biology teachers address some of the weaknesses of their textbooks by (a) using some of the trade books published on biology topics, (b) changing their teaching practices to align with the recent research on student learning, and (c) avoiding an instructional focus on technical terms and trivial details. None of the 10 leading textbooks received high marks by the AAAS review teams. "Few kids will learn much about biology by using these textbooks as intended," said AAAS's Project 2061 Director, George Nelson (Nelson, 2000, p. 639).

Constructing and Reconstructing Students' Ideas

Learners often know quite a bit about the topics studied in biology before taking high school biology, even though we can predict that some of that knowledge consists of misconceptions (Treagust, 1988), which out of respect to the learner, are now preferentially called *alternative conceptions* (ACs) in the biology education research literature. These ACs are pervasive, topic-specific, tenacious, actually "work" to some degree in everyday life, often reinvoked once the student leaves the science classroom, and frequently held by science teachers as well. More than 3,200 research studies worldwide have been conducted and published on pupil's alternative conceptions in science, and hundreds of such studies address science topics in the high school biology curriculum. The majority of science educators around the world seems convinced that science teaching works best when envisioned as a process of active knowledge construction by students – because the "active-teacher-transmitter" broadcasting to "passive-student-receivers" model does not coincide with what we now know about how people learn meaningfully and mindfully. This new approach is known as "constructivist" science teaching, and stacks of worthy academic books, science teaching journal articles, and science education research journal articles exist to explore the meaning and implementation of this construct in the science classroom – although, now, the essential elements may easily be found in the NRC publications mentioned earlier in this chapter.

LIMITATIONS

No course in biology is truly representative of the total knowledge and activities of today's biologists in all of their fields. Similarly, no chapter on the state of high school biology instruction dare claim to have captured the collective, extant, instructional wisdom derived from research and practice, or the advances in instruction being made in various schools, cities, states, and funded curriculum projects. Appendix A constitutes but a small sample of innovative high school biology teaching activities being carried out in the U.S. at this writing; its purpose for inclusion is merely to underscore the existence of instructional diversity and experimentation. Having carefully assembled more than 200 relevant books, reports, and articles before beginning to write this chapter, your author has agonized over the omission of many topics and many published research studies of obvious merit. Instead, the focus has been on framing a cognitive infrastructure independent of the author's and his collaborators' work – on building an intellectual- and exemplar-based framework which can not only support further exploration and understanding of instruction in high school biology, but also afford insightful perspectives for those scholars who work in other domains of instruction.

Using the History of U.S. Biology Education as a Heuristic Device

Increasing high school students' understanding of biology remains a daunting challenge – perhaps even more so in the age of molecular biology. Lewis and

Wood-Robinson (p. 190) conclude their year 2000 biology education research article entitled "Genes, Chromosomes, Cell Division and Inheritance – Do Students See any Relationship?" with the words: "It would appear that the science education which these students [age 14–16] received, as reflected in their knowledge and understanding of genetics, provided neither a firm basis for future training as a scientist nor a useful preparation for personal interactions with science in their adult lives."

Nearly a century ago, in a report on the major problems of U.S. science teaching, a committee of the Central Association of Science and Mathematics Teachers offered the following set of suggestions to improve instruction in U.S. high school biology courses (Galloway, 1910):

(1) "More emphasis on 'reasoning out' rather than memorization."
(2) "More attention to developing a 'problem-solving attitude' and a 'problem-raising attitude' on the part of students."
(3) "More applications of the subject to the everyday life of the pupil and the community. . . ."
(4) More emphasis on the incompleteness of the subject and glimpses into the great questions yet to be solved by investigators."
(5) "Less coverage of territory; the course should progress no faster than pupils can go with understanding."

Today's research findings and reform recommendations tend to echo, validate, and reaffirm these 90-year-old suggestions as eminently conducive to sound biology learning and the attaining of biological literacy (Hurd, 1970). Might the history of U.S. biology instruction be underutilized as a heuristic to guide contemporary instructional research and practice? Might this generalization be true for other fields as well?

REFERENCES

American Association for the Advancement of Science (1989). *Science for all Americans*. Washington, DC: Author.

American Association for the Advancement of Science (1993). *Benchmarks for science literacy*. New York: Cambridge University Press.

American Society of Naturalists (1999). *Evolution, science, and society*. Chicago: University of Chicago Press.

BSCS (1990). *Science for life and living*. Dubuque, IA: Kendall-Hunt.

BSCS (1993). *Developing biological literacy*. Colorado Springs, CO: Author.

Bybee, R. W. (1986). The Sisyphean question in science education: What should the scientifically and technologically literate person know, value, and do – As a citizen? In: R. W. Bybee (Ed.), *Science-Technology-Society 1985 NSTA Yearbook* (pp. 79–93). Washington, DC: National Science Teachers Association.

Carr, E., & Ogle, D. (1987). K-W-L plus: A strategy for comprehension and summarization. *Journal of Reading, 30*, 626–631.

Darwin, C. (1859). *On the origin of species by means of natural selection*. London: J. Murray.

Donovan, M. S., Bransford, J. D., & Pellegrino, J. W. (Eds) (1999). *How people learn: Bridging research and practice*. Washington, DC: National Academy Press

Duschl, R. A. (1990). *Restructuring science education: The importance of theories and their development*. New York: Teachers College Press.

Fensham, P. J., & Kass, H. (1988). Inconsistent and discrepant events in science instruction. *Studies in Science Education, 15*, 1–16.

Gabel, D. L. (Ed.) (1994). *Handbook of research on science teaching and learning*. New York: Macmillan.

Galloway, T. W. (Ed.) (1910). Report of the Committee on Fundamentals of the Central Association of Science and Mathematics Teachers. *School Science and Mathematics, 10*, 801–813.

Gardener, P. L., & Gauld, C. (1990). Labwork and students' attitudes. In: E. Hegarty-Hazel (Ed.), *The Student Laboratory and the Science Curriculum*. London: Routledge.

Gowin, D. B. (1981). *Educating*. Ithaca, NY: Cornell University Press.

Hagen, J., Allchin, D., & Singer, F. (1996). *Doing biology*. New York: HarperCollins College Publishers.

Harlen, W. (1999a). *Effective teaching of science: A review of research*. [SCRE Publication #142, Using Research Series 21]. Edinburgh, Scotland: Scottish Council for Research in Education.

Harlen, W. (1999b). Raising standards of achievement in science. *SCRE Research in Education, 64*, 1–4.

Hodson, D. (1993). Re-thinking old ways: Towards a more critical approach to practical work in school science. *Studies in Science Education, 22*, 85–142.

Hurd, P. D. (1970). Scientific enlightenment for an age of science. *The Science Teacher, 37*, 13–14.

Hyerle, D. (1996). *Visual tools for constructing knowledge*. Alexandria, VA: Association for Supervision and Curriculum Development.

Lamarck, J. (1809). *Philosophie Zoologique. Paris: H. Agasse*. (Translated by Elliott, H. [1914]. London: Macmillan and Company.)

Langer, E. J. (1997). *The power of mindful learning*. Reading, MA: Addison Wesley.

Lewis, J., & Wood-Robinson, C. (2000). Genes, chromosomes, cell division, and inheritance – Do students see any relationship? *International Journal of Science Education, 22*(2), 177–195.

Manning, P. (1999). *Islands of hope*. Winston-Salem, NC: John F. Blair Publishing.

Mayer, W. (1978). *Biology teacher's handbook* (3rd ed.). Colorado Springs, CO: BSCS.

Mayr, E. (1982). *The growth of biological thought: Diversity, evolution and inheritance*. Cambridge, MA: The Bellnap Press of Harvard University Press.

National Academy of Sciences (1998). *Teaching about evolution and the nature of science*. Washington, DC: National Academy Press.

National Research Council (Committee on High School Biology Education, 1990). *Fulfilling the promise: Biology education in the nation's schools*. Washington, DC: National Academy Press.

National Research Council (1996). *National science education standards*. Washington, DC: National Academy Press.

National Research Council (Bransford, J. D., Brown, A. L., & Cocking, R. R., Eds.) (1999). *How people learn: Brain, mind, experience, and school.* Washington, DC: National Academy Press.

National Research Council (2000). *Inquiry and the national science education standards: A guide for teaching and learning.* Washington, DC: National Academy Press.

Nelson, G. (2000, July 28). Biology textbooks do not tell the story. *Science, 289*, 639.

Paivio, A. (1971). *Imagery and verbal processes.* New York: Holt, Rinehart, & Winston.

Peterson, N., & Jungck, J. (1988, March/April). Problem-posing, problem-solving, and persuasion in biology education. *Academic Computing*, pp. 14–17, 48–50.

Spiegel, G. F. & Barufaldi, J. P. (1994). The effects of a combination of text structure and graphic postorganizers on recall and retention of science knowledge. *Journal of Research in Science Teaching, 31*(9), 913–932.

Strike, K. A., & Posner, G. J. (1985). A conceptual change view of learning and understanding. In: L. West & A. Pines (Eds), *Cognitive Structure and Conceptual Change* (pp. 211–231). Orlando, FL: Academic Press.

Treagust, D. (1988). Development and use of diagnostic tests to evaluate students' misconceptions in science. *International Journal of Science Education, 10*, 159–169.

Udovic, D. (1997). *The workshop biology project.* Eugene, OR: Department of Biology, University of Oregon.

Wallace, A. R. (1858, August). On the tendency of varieties to depart indefinitely from the original type. *Journal of the Proceedings of the Linnean Society: Zoology, 3*(9), 45–52.

Wurman, R. S. (1999). *Understanding.* Newport, RI: TED Conferences Inc.

APPENDIX A

A Sample of Innovative Biology Teaching Activities

What's Going on Out There?

From your author's interviews of some of today's high school biology teachers and his visits to their classrooms, the following activities, not listed in any special order, may be representative of the changes or shifts in biology instruction and activities away from predominantly traditional "teacher-telling and cookbook-type" instructional methods of the past – although there is a wide range in the penetrability of these innovations, mediated by differences in resourcing and by various biology educators' spheres of influence, nation-wide.

Biology teachers were found to have students (a) doing concept mapping and semantic network mapping of biology topics via Inspiration™ and SemNet™ computer software; (b) conducting collaborative learning projects about microbial biofilms; (c) doing E-mail questioning of distant biology experts; (d) using on-line botanical image bases and compatible image processing software;

(e) using peer teaching dyads to master classical genetics; (f) doing cross-age tutoring of middle school students about bacteria; (g) doing volunteer work for environmental action projects such as reforestation, motor oil recycling, and marine debris clean-up; (h) taking course-culminating, distant biology field trips (e.g. the Galapagos Islands, Costa Rica, Hawaii); (i) doing opt-out virtual dissection of 16 species via software, CD-ROMs, and videos; (j) using real-time, video-camera-enhanced demonstration dissections supplemented by photo dissection manuals; (k) using image-based assessment of biological understanding; (l) joining national environmental biology data-collection-and-sharing networks; (m) persuasion of peers via student-designed Microsoft PowerPoint™ presentations of student research; (n) doing World Wide Web biology lessons, searches, and web quests; (o) using calculator based laboratories (CBL's) for monitoring pH and doing colorimetry work; (p) video-recording students' video microscopy and Flexcam™ work; (q) doing on-line migration tracking by plotting the latest posted data daily; (r) using computer simulation and modeling of lengthy, subcellular, or dangerous biological processes; (s) doing desktop/web site publishing of a student research journal; (t) using miniature, battery-powered, in-field, electronic data-loggers; (u) participating in university biotechnology materials-and-equipment lending programs; (v) using real-time computer probeware and sensors for graphing and data analysis; (w) making student-directed biology research videos followed by in-class screening and critique; (x) using commercial research organisms (e.g. Wisconsin Fast Plants™, WowBugs™, C-ferns™) and experiment kits; (y) writing, costuming, propping, and role-playing interactive historical vignettes (history of biology) for the class; (z) using bioinformatic and genomic data bases found on the World Wide Web; (aa) experiencing the process of PCR; (bb) performing an immunoassay simulation; (cc) conducting a field study of toxicants in a freshwater stream; (dd) Bottle Biology™ apparatus construction and experimentation; (ee) doing plant tissue culture in the biology classroom; (ff) performing plant dissections; (gg) clay-modeling of cell structures and sequences in developmental biology; (hh) using preservice biology teachers as teaching laboratory interns; (ii) writing to learn biology; (jj) performing video motion analyses of animal behaviors; and (kk) solving human anatomy/physiology case-study problems.

Within the past decade, the availability of biology teaching resources and instructional alternatives has increased dramatically. The real question now is: What shall I choose that best teaches what I have already planned to teach? It's very easy to get side-tracked! Too many teachers are overwhelmed by the plethora of biology learning tools, strategies, and activities for their use – and so they teach a high school biology course that is more of a spontaneous and random walk through biology than a well-plotted sailing course. Another danger lies in using corporate-financed, attractively packaged "free" teaching materials that may be far from neutral on issues that affect their financial viability. That's where research-based learning theory, sound biology education research findings, and NRC science reform document principles can aid teachers in their instructional decision making. This may even mean changing comfortable practices that have long been accepted as productive. For example, your author suggests using no single-day laboratory lessons – any inquiry-based laboratory worth doing takes a substantial amount of time to carry out, and for students to understand mindfully and meaningfully, and these ought to involve problem posing, problem solving, and persuasion of peers – none of which should be rushed. Prelaboratory and postlaboratory preparatory and analytical sessions also help to maximize learning. Given a constant time frame and human information processing capacity, we can gain greater depth of biology learning only at the expense of restricting the breadth of biology learning.

THE TEACHING AND LEARNING OF PHYSICS

Jim Minstrell and Pamela A. Kraus

A. INTRODUCTION

The teaching and learning of physics has undergone dramatic change in the past 30 years. As was the case when Hestenes (1979) confronted the notion that teaching is solely an art and not a science, the challenge in science teaching has been on crossing the gap from research to practice. We know from research on learning that students' initial understanding and skills play a crucial role in their learning. However, moving students from their initial understandings to reaching the goals of a physics course, a deep conceptual understanding of the content with the ability to reason and think scientifically, has been challenging for students, teachers, and researchers. In this chapter we present a framework that describes how exemplary teachers, using well-engineered lessons, close the gap between research and practice. We discuss in detail four aspects of an instructional framework. As a part of this description, we elaborate on the twelve generic guidelines from chapter 1, as they apply to teaching introductory physics in the K-12 school setting. But, first a few general comments about the sources behind our exposition.

Exemplary teaching of physics has been greatly influenced by work from three distinct groups; cognitive scientists, discipline-based researchers of learning and teaching of physics, and classroom science teachers. Cognitive scientists have given us insight into how people learn. A recent publication by the National Resource Center summarizes some relevant findings on learning (Donovan, 1999). First, learners come to the learning situation with

Subject-Specific Instructional Methods and Activities, Volume 8, pages 215–238.
2001 by Elsevier Science Ltd.
ISBN: 0-7623-0615-7

preconceptions that don't change unless they are appropriately addressed specifically in instruction. Second, learners need experiences with phenomena and with organizing these experiences into meaningful conceptions. Finally, learners need opportunities to reflect on and monitor their own learning relative to the goals of learning. The report goes on to suggest that instruction should be centered on the learner, on deep understanding of important content and on on-going assessment.

Researchers in physics education help identify preconceptions and specific student difficulties, and they engineer lessons to address the difficulties. Physics teachers adapt and test the lessons under their local conditions and make recommendations to developers and to other teachers about how to tune the lessons for practical use in their own classrooms. It is from the results of all three of these efforts that we base our remarks on the teaching of physics. Although the efforts have come from apparently distinct groups (cognitive scientists, physics education researchers, and science teachers), the following represents our attempt to assemble the separate contributions into a coherent whole. It is through the integration of these perspectives that a rich pedagogical content knowledge for teaching physics is created that also relates to depth of knowledge of the content and the management of day-to-day activities in the classroom.

B. INSTRUCTIONAL FRAMEWORK

The bulk of this chapter is structured around a discussion of four aspects of a framework that appears to us to be common to all exemplary physics teaching. All four aspects are interrelated and when applied together represent, we believe, a complete instructional setting in physics. These four aspects are:

(1) Identifying and monitoring students' thinking
(2) Testing and modifying students' understanding and promoting change
(3) Teaching and learning for a conceptual understanding
(4) Applying learning to build a functional understanding

The first two aspects describe instruction that is centered on the learner's understanding. Teachers incorporating the first two aspects spend a significant portion of instructional time listening to what students are saying throughout instruction and adapting their instruction accordingly. The second two aspects describe the choices teachers make both before and during the course of study. They establish the goals of instruction and help teachers know when they have reached the intended goals.

Traditional physics teaching typically discounted the first two aspects and reduced the second pair to merely stating conclusions from physics and then having students use the stated ideas (often summarized in a formula) in solving numerical problems. What we are suggesting in our framework is that a significant portion (upwards to half the time) of a unit is spent on generating understanding (the first two aspects), moving from individual students' preconceptions to a class set of ideas that organize the phenomena. Compared to traditional instruction, this changes the focus to knowing "how we know" and "why we believe" the derived ideas. The other portion of class time, reflected in the second two aspects of our framework, is spent on organizing those generative experiences around the powerful ideas of physics, the scientific reasoning processes through which they were derived and coming to see the utility of those ideas.

Those readers who have been around science education for two or more decades, may notice that our framework has roots in "The Learning Cycle" used in early elementary science curricula such as the *Science Curriculum Improvement Study (SCIS)* (Karplus, 1974). That framework had three parts (in this case phases). It began with *Exploration*, wherein children had opportunities to handle objects, see what happens and generate questions and try things out. This initial phase is limited to children's preconceptions. Out of these first-hand experiences manipulating objects and thinking about their own ideas came an *Invention* phase in which the teacher was expected to define new concepts to help children organize experiences and the children discussed the validity of these new ideas in light of experiences. The final phase was *Discovery* of situations in which the newly invented ideas could be used or applied. This later phase was to provide for mastery and retention. Our suggested framework adds detail and perhaps some additional purpose to the phases of "The Learning Cycle".

More recent elementary hands-on inquiry science curriculum, such as *Insights* (Education Development Center, 1977) and *Science and Technology for Children (STC)* (National Science Resources Center, 1997), use a framework also consistent with the framework we take for our organization. Their four phases include the following:

(1) Getting Started/Focus
(2) Exploring and Discovering/Explore
(3) Processing for Meaning/Reflect
(4) Extending Ideas/Apply

One of the differences between this and our framework, which we will elaborate in the following discussion, is that they proceed through their phases sequentially. The four aspects of our framework need to be considered throughout all the cycles of instruction. While one aspect may be emphasized, exemplary teachers have the entire framework in place at all times. Despite the differences, we take these phases from the elementary curricula to be validation that our framework is functional from the elementary through high school for organizing physics learning and teaching activities.

In the discussion of the framework below, we describe the four aspects and lend specificity to the generic instructional guidelines. We use several examples from existing elementary, middle and high school physics curricula. These examples are from programs that have demonstrated success in improving students' understanding of physics. The four aspects appear to us to be common to all these programs, even if they are not explicitly stated as such. In addition, all the programs have based much of their development on research in the field of teaching and learning and on the expert advice of teachers. Several have also built on results from research on human cognition.

1. Identifying and Monitoring Students' Thinking

Before elaborating on this first aspect of the framework it is important to say that teachers have to know where they are going. An important part of teaching any subject is knowing the goals of the teaching and learning. It seems obvious that it is important to clarify the learning goals, but many teachers do not take this important step. Ask many physics teachers what their goals are and they respond with information about the activity of the students rather than the goal of the learning. For example, "I was covering section . . . of the text," or "I was doing the dropped ball experiment." The activities are not the goals but the means to get to the learning goals.

Clarifying the learning goals gives the expert teacher a sense of purpose to know where the class is headed. In the case of physics the general goal is deep understanding of the big, powerful ideas, the scientific inquiry processes, and the nature of the discipline of physics, such as the content of the goals described in the National Science Education Standards and the Benchmarks for Science Literac. (American Association for the Advancement of Science, 1993; National Research Council, 1996).

After clarifying the goals of learning, teachers need to know initially and during instruction, where students' thinking is with respect to the learning goal. If students have not yet reached the goal, what are they thinking. Most teachers believe they are continually monitoring students' thinking. But, too often the monitoring is only with respect to whether students "have the right idea" or not. Seldom is the ongoing assessment to find out what ideas students do have if they don't have the "right" idea. Exemplary teaching of physics includes identifying what ideas, reasoning and procedures students are using when they aren't thinking in a way consistent with the content learning goals.

To appropriately direct instruction, it is important to know the specific problematic thinking of students. The more clearly a teacher can describe the trouble, the more likely they are to be able to choose, adapt or design instruction to address the specific difficulty. This is a place where the research on students' misconceptions and the like plays an important role. The need for instruction to identify and address students' preconceptions is one of the most important conclusions from research on how people learn (Donovan, 1999).

Minstrell and colleagues have taken the research on students' conceptions and catalogued them into what they call "facets of thinking." (Minstrell, 1992) The term "facets" was used to avoid the negative connotation of "misconceptions." Many of the ideas constructed by students have positive aspects as well as problematic aspects. They loosely define facet to mean any construction of one or more pieces of knowledge, reasoning, or procedure used by a student to answer a particular question. The facets are organized into clusters. Each cluster includes the facets associated with explaining a particular event (e.g. explaining falling bodies), or the cluster describes the apparent meaning of some big idea in physics (e.g. meaning of average velocity). Facet clusters are used to assist in designing and adapting instruction (Minstrell, in press a). Teachers using the facet clusters relevant to particular learning goals keep them handy as they are teaching that portion of the curriculum. It gives them a quick reference from which they can identify which sort of thinking, which facets, are apparently being invoked by which students. In a recent project, teachers and developers working together are compiling general guidelines and particular lessons to address particular problematic facets. By adopting this emphasis on identifying students' thinking and adapting instruction to address dominant problematic facets, teachers have increased the end of year performance of their classes by an average of 15% (Minstrell, in press b).

Tools have been developed that assist students and teachers in conducting formative assessment, monitoring learning while students are developing their ideas. For example, Hunt, Minstrell and colleagues have developed a computerized DIAGNOSER that serves pairs of questions from which facet diagnoses are made. The first question is in the vein of "what would happen if . . .?" and the second is around "what reasoning best justifies the answer you chose?" While the questions are in multiple-choice format, each answer choice for each question is associated with a particular facet. Then the student is given a prescription for something to do or think about that should foster improved understanding (Hunt, 1996; Levidow, 1991; Minstrell, 2000b).

The practice of monitoring students thinking has begun to permeate the more traditional modes of instruction, such as the lecture. Mazur (1997) developed an innovative, more interactive lecture method for the large courses at the university. *Peer Instruction* breaks the lengthy lecture into several short cycles of instruction. Using well-engineered questions, or *ConcepTests*, the lecturer first checks the types of reasoning students are using. After students discuss their ideas with their peers, a poll is taken of the students in the class. The instructor then adjusts the subsequent mini-lecture to address specific ideas held by the students. They reported significant and substantial improvement in student performance in the class by monitoring and attending to students' thinking (Crouch, submitted 2000).

At some universities and high schools, polling the class is assisted by a computerized system called *Classtalk*. Students select responses on small hand-held devices that are recorded and displayed for the teacher on a central computer at the lecturer's desk. Class percentages and individual responses are displayed (Dufresne, 1996).

In addition to assessing after instruction to make adjustments to what might next be done, exemplary physics teachers may also elicit students' ideas even before beginning the unit of instruction. Linn et al. (2000a) suggest this is a common activity structure for the well-engineered science lessons in Japan. These science teachers routinely elicit students' initial ideas and opinions about natural situations. One purpose of doing this is to make connections between the new work and students' possible interests. A second purpose is to help elicit students' ideas and questions that will become the focus of a next set of investigations.

Expert physical science teachers in this country also ask elicitation questions at the beginning of a unit. This activity not only helps teachers identify students' initial ideas, but the subsequent discussions of the answers and justifications for the answers can help students become aware of the ideas of their fellow students. (DiSessa, 1998; Goldberg, 1997; Minstrell, in press b) Students report that they don't particularly like having to answer elicitation questions before they have had instruction. But they say the activity is good to do because it gives them a chance to see the sorts of issues and situations they need to be able to handle by the end of the instructional activities. It helps them focus on the goals for learning. Clarifying the goals of learning and monitoring learning with respect to the goals are other important findings in how people learn (Donovan, 1999).

These sorts of elicitation and diagnostic questions are critical to building shorter cycles of "quality control" in the education process. Expert teachers use these as a part of their formative assessment. For the most part, they are not used for grading except for giving students credit for putting in a good honest effort. The teachers are asking students to honestly express their thinking. "Give me your best answer at this time." "I want to know what sense you make of this situation." "What is there in your past experience that seems to be relevant and helps you to make sense of this situation?" "I won't grade you on this except that you seem to put in an honest effort." These are some of the comments we have heard from teachers as they administer these initial elicitation questions or the diagnostic questions embedded within instruction.

This brings up another important general guideline related to identifying and monitoring student thinking. The verbal interaction among students and between students and the teacher needs to be supportive of free and honest expression. Classroom participants need to encourage and support expression of ideas, but at the same time they need to be critical of the ideas. This involves a delicate balance. In a class discussion, expert teachers can respectfully catch one students' comment without evaluating it and reflect it back out to the class to consider and comment on the validity of the idea or the evidence used or not used in support of the idea (Van Zee, 1997). The teacher guides fellow students to be respectful of expression, but also guides students to be critical of the ideas. "That is an interesting idea. How do you know that?" and "What evidence have we seen for that?" are important questions to stimulate scientific thinking. Showing genuine interest in hearing what students are saying by asking for more detail or for clarification of meaning foster and encourage expression. Expert teachers can ask students a question and then exhibit wait-time, giving the student adequate time to respond fully. Then, rather than evaluating what the student has said, the expert teacher pauses to see if the student needs or wants to say more by way of clarification or elaboration (Rowe, 1972). Appropriate wait-time and pause-time have been shown to foster more complete responses by learners and to effect deeper learning, especially by students who were previously identified as underachievers.

2. *Testing and Modifying Students' Understanding Promoting Change*

One of the general goals of teaching physics is for students to experience how knowledge is developed in science. Addressing this goal involves a weaving together of two of the generic guidelines: Strategy Teaching and Thoughtful Discourse. In the physics class, the teacher teaches strategies of thinking and action, through modeling thinking and actions and by scaffolding students' use of strategies. The teacher may sometimes deliberately model the thinking expected of students by thinking aloud. In doing so, she helps to make the processes of science more explicit and can even model appropriate metacognitive activities.

> Hmm, I see these various arrangements of pulleys all lifting the same weight. But for some arrangements, I don't have to pull as hard as for others in order to lift the same weight. Why might that be? Let me try them all more carefully. I see that I have to pull the rope farther in some arrangements than for others. I wonder if how hard I have to pull to raise the weight is related to the different distances I have to pull the rope? But, wait, how far I pull the rope depends on how high I want to lift the weight. I'm getting confused. I'd better be more careful and maybe keep some of these variables constant. I will keep the weight the same in all the arrangements I try, and I will also keep the distance I raise the weight the same. Now, with those variables controlled, maybe I can see if the distance I pull the end of the rope is related to how hard I have to pull. I'd better pull with a spring scale so I can measure how hard I pull, and I will measure the distance I have to pull the rope to raise the weight that preset distance. And maybe I'd better write my results down in a data table so I can remember them all.

Modeling expected thought patterns has been shown to transfer to students. Teachers who model actions and who describe why they are doing what they are doing have found students mimicking the "think aloud" during their interactions with materials and with fellow students. This fosters metacognitive activity that has also been characteristic of how people learn (Donovan, 1999).

Sometimes the teacher uses questions to guide students' learning to use processes typical of scientific inquiry. The following is a brief vignette in which the teacher scaffolds the initial part of a discussion about data from an investigation with pulleys. In this example the teacher is modeling cooperative learning in groups where members need to assist each other in being clear about their observations and their inferences as they construct explanations. She also attempts to respond to each individual's needs, interests and capabilities. She ensures that each student will have an opportunity to learn.

> Teacher: So, now that you've worked with these pulleys and recorded your data, what do you notice? [An open-ended question asking for students' observations.]
> Sue: Some numbers are the same and some are different.
> Mike: Duh. That's obvious.
> Teacher: Mike, remember we need to let people speak their ideas. But, you can ask the speaker to be clear. [The teacher manages the discussion by reminding what sorts of comments are not OK (evaluative put downs) and what sorts of talk can further the understanding of the participants (questions of clarification or explanation).]
> Mike: Sorry. Do you see a pattern in the numbers?
> Sue: Well, it looks like the bigger one number gets the smaller the other gets.
> Teacher: What do you mean? What numbers? [The teacher models asking for clarity.]
> Sue: The bigger the distance the smaller the pull.
> Mike: Yeah. How come? It seems like we're getting something for nothing, cuz sometimes you don't have to pull as hard.

Teacher: Good question. [The teacher encourages spontaneous questions requesting explanation.]
Sue: But you're not getting something for nothing. You have to pull the rope farther when you pull less hard.
Sarah: Yeah. In fact when you pull twice as far, you only have to pull about half as hard.
Teacher: Does that sort of relation seem to hold for the other data? [The teacher is encouraging students to seek additional evidence.]
Sarah: Yes. When you only have to pull about a third as hard, you have to pull the rope about three times as far.
Mike: But how come? How come it is sometimes easier and sometimes harder when we pull different distances?
Sarah: I think it is because there are more ropes.
Teacher: What do you mean more ropes? [The teacher is seeking more clarity for this new idea.]
Sarah: See, when we only pull a third as hard there are sort of three ropes holding up the weight and pulley here. But, with this one where we had to pull with full strength, there is only one rope lifting the weight.
Mike: Oh yeah, and when there are two ropes holding it up, you only have to pull half as hard.
Teacher: Does the number of ropes rule hold for the rest of the data? [The teacher is again seeking more evidence.]
Mike: Seems to.
Teacher: So, Sarah, why should more ropes also relate to more distance? [The teacher moves the conversation back to a critical question asked early by Mike and for which the teacher believes Sarah is ready to construct an explanation.]
Sarah: Because it relates to . . . (Teacher interrupts)
Teacher: Excuse me. What do you mean by "it"? [The teacher seeks clarity by asking for antecedent of "it".]
Sarah: Because more distance and more ropes holding it, the weight, up relates to less hard pull. Oh. Oh. Wait! I think I see. When I pull this rope with the one third pull, the three ropes all have to get shorter by one foot to raise the weight one foot. And to do that we have to pull the end of the rope three times as far to take up the one foot for shortening each of the three ropes.
Sue: Huh?
Teacher: Mike and Sue, do you understand what Sarah just said? OK, get Sarah to explain it to you again and keep pressing her to be clear until you both understand what she is suggesting. Then, test the idea with the other situations. You all are doing a nice job of unpacking the data you got and making sense of it. I'm confident that by pushing each other just a little more you will all understand how and why you get an advantage by using a pulley.

[The teacher expresses confidence in the group being able to come to a shared understanding and gives responsibility to some individuals to guide further discussion.]

Notice that it is the students who are testing and modifying their ideas. That is, the students are responsible for constructing and reconstructing their ideas from their observations. The teacher is assisting by managing the learning environment. She guides their development of ideas not by telling the students the conclusions, but by creating the questioning and activity environment within which the students' thinking is kept on task.

The preceding will give students an opportunity to identify and test their own ideas and perhaps build a deeper understanding of relationships, between force and distance with pulleys, for example. However, students also need the opportunity to learn the skills of scientific inquiry. The preceding lesson gave students an opportunity to learn *by* inquiry, but research and extensive experience indicate that students don't learn to do inquiry simply by learning scientific principles through inquiry (Woolnough, 2000). Students need instruction that explicitly addresses the teaching and learning of the skills of scientific inquiry. The following vignette demonstrates how a teacher might guide students in learning specific skills of scientific inquiry.

[This discussion would occur after students have come to the conclusion that force of the pull is inversely related to the distance that the line must be pulled. So, the students have apparently generated the principle that was the content goal of the lesson.]

Teacher: OK, you all seemed to arrive at this principle for pulleys, but now it is time to reflect on how we came to know that. Did someone tell you that was how pulleys worked? How did you identify the relevant variables?
Sue: We just kept trying stuff to see what happened until we found the right answer.
Teacher: Is that really what you did? How did you know even what you were trying to do?
Mike: Well in the beginning we had lots of different arrangements of pulleys. And when we tried all of them we noticed that different pulley arrangements were harder or easier to pull to raise the heavy weight.
Teacher: Then what? Go on. How did you decide what you needed to do? Did I need to tell you?
Mike: No, some of us just started wondering. How come some are easier to pull than others? You just said that was a good question. You didn't answer the question.
Sarah: Yeah. We just said, "cool, lets find out." You encouraged us to find out what we could.
Teacher: So, Sue, what did happen? What did you find out? [The teacher distributes the responsibility to a student who was not very clear early in the discussion.]
Sue: Well, we noticed that we pulled more rope for some arrangements than we did others.
Teacher: So why was that important to do? How did that end up being helpful in the long run?
Sue: Well, we were noticing that the more rope we pulled seemed like less force.
Sarah: Yeah. There was a pattern there. The distance seemed to be related to the pull.
Teacher: You were looking for a relationship between the variable you were interested in (the bigger or smaller force needed) and some other variable. The force of the pull was the outcome variable. And you were looking for other variables that might relate to the outcome of more or less pulling force.
Sue: And the other variable was the distance we pulled the rope.

Teacher: So, you were able to find a relationship between the force (outcome variable) and the distance.
Mike: Except that doesn't really explain why it works. It just says there is a relationship between the numbers. Why should that relationship work?
Teacher: Great question, and an important question. You did good to find a pattern between the sets of numbers for each of those two variables. But, next comes Mike's question of why should that relationship work. In science we frequently will first identify the relevant variables. Then, determine and test a potential relationship between the variables, like you did. But in science, we also try to explain why the relationship works, why it makes sense to us. [The teacher is summarizing important points from the discussion so far.]
Sarah: And that is when I noticed the number of ropes that were attached to the lower pulley and weight. When there were three lines holding up on the weight and lower pulley, we had to pull the rope three feet to take up the slack in each of the three lines. That is why there is a relationship between the force of the pull and the distance we had to pull the rope to raise the weight.
Teacher: Nice summary Sarah. Did it matter how high you raised the weight?
Mike: Yeah. Otherwise we couldn't compare the distances. We had to make it a fair test between the systems.
Teacher: Good thinking. Keeping some variables constant through the experiment is called controlling that variable. You controlled the height variable, keeping the height you raised the weight the same, so it would be a fair test of your hypothesized relationship between the outcome variable and the variable (distance) you thought seemed related to the force.

So, what you have done is like what scientists do. They get curious about some interesting outcome variable? Then they look for other variables that might be related, and when they find one, they test to be sure that that variable is related to the outcome variable. Meanwhile, they control other variables that might make a difference, while they are testing the potentially relevant variable. Finally, they try to explain why the relationship works the way the way it does. They try to make sense of the relationship. You all were doing science. Yes, true, you didn't make a discovery that was new to the world, but you did discover something new to you. And, from what you showed me yesterday, you understand the pulleys pretty well. You can design arrangements that will require a predictable amount of force and rope pulling to raise a weight a given amount. [Yesterday they had a brief performance test.] Someday you may use the same sort of scientific approach you used here to learn some new idea that nobody has ever thought of before. That is why we will sometimes have these discussions about how science works as well as having the usual discussions about the specific content ideas of science. I want you to be able to use these processes of science.

If we want students to understand the nature and processes of science as well as the specific content principles, we need to make the processes a specific piece of the activities and discussion in class. The goal needs to have its own explicit agenda. It is sometimes useful to think of multiple agendas in the classroom. In the case of this story, there is one agenda about learning the principles related to pulleys, and there is a second agenda about learning about the nature and processes of science. Each agenda must be explicit. Each demands its own time in the foreground of the classroom experience while the other agenda moves into the background (Minstrell, 2000a).

Experiments and especially full investigations are frequently messy. To help students learn to conduct investigations on their own, high quality physics programs help guide and support students as they develop ideas about content and process skills related to investigation. Now software can help scaffold the strategies and cooperative learning. In the *Constructing Physics Understanding (CPU) Project* students, working in groups, are guided by software while conducting investigations and formulating conclusions or explanations (Goldberg, 1997). In other activities they are encouraged to design and conduct investigations to test their own hypotheses. The data are accumulated on computer and shared with other groups at the point each group finishes a particular set of activities. Students are encouraged to seek and store evidence in support of the powerful ideas resulting from consensus discussions about their experiments and other activities.

In the *Computer as Learning Partner (CLP) Project*, middle school students' strategies for testing and modifying their ideas are assisted through using the computer to organize observations and inferences (Linn, 2000b). For example, students are led by the software to state their hypotheses about insulation qualities of various materials. Then, the system poses situations for which students make predictions based on their hypotheses. The program presents tables and graphical axes on which students record their observations. Next they are asked for their conclusions and for specific evidence from their data in support of those conclusions. In the process they are asked to reflect on how these results and conclusions relate to their initial ideas. Like the CPU program, there are specific opportunities along the way for students to present their ideas individually and note individual ideas divergent from those of the cooperative working group.

In all these examples students are encouraged to make their own sense of the experiences. Expert physics teachers are clever about adapting curriculum and instruction that will build on useful student ideas and challenge problematic ideas to foster development of student thinking about the physical world.

3. Teaching and Learning for Conceptual Understanding

In the first two aspects of the instructional framework, we described the importance of teachers paying close attention to the thoughts and actions of their students. Exemplary teachers monitor, evaluate and adjust their teaching throughout instruction. In the next two sections, we discuss the role of establishing challenging learning goals, engineering instruction to help meet these goals, and knowing when these goals have been met.

Physics includes a large number of topics, many of which are considered fundamental to understanding of the discipline. This is epitomized in university level introductory physics texts, which typically exceed 1000 pages in length. While most teachers and curriculum developers at the K-12 level espouse depth over breadth, in practice this is difficult to achieve. Deciding what to "cover" in a physics course is one of the more challenging tasks for teachers. Articulating the curriculum from elementary through high school can address some of the issues of coverage. However, once the content is chosen, the more important task for teachers is ensuring that the focus is on a conceptual understanding of those topics and on organizing this understanding into a rich conceptual structure (Donovan, 1999).

The following is offered as an illustration comparing curricula that develops a conceptual understanding to that which simply "covers" the topic. Current electricity is a unit often studied in physics courses beginning as early as third grade and continuing through high school. However, only a few curricula and textbooks at the middle and high school level focus on the conceptual understanding of simple electric circuits. Instead, the mathematical expressions for Ohm's law and Kirchhoff's rules are typically presented and students are asked to practice applying these expressions to determine currents, voltages and equivalent resistances in various circuits.

In classrooms with accompanying labs, the laboratory activity often involves following a set of instructions to measure quantities in circuits set up by the instructor. Typical of many traditional physics laboratory experiments, quite often the goal is simply to verify that the rules apply to "real" circuits rather than uncover the relationships through testing and modifying their own theories. While the students have their hands on real equipment, they are rarely challenged to "discover" or explain relationships for themselves. (This contrasts with the pulley investigation described in the previous section where the students had to not only grapple with "discovering" the relationships, but needed to make sense of those relationships during their investigation.)

Instead of developing a conceptual understanding for circuits, this more traditional instructional treatment encourages students to simply learn formulas to apply without a conceptual basis for applying them. The result is often very poor performance on relatively simple tasks such as comparing the brightness of bulbs in the three simple circuits shown in Fig. 1 (McDermott, 1992). While students taught as described above have the information to correctly rank the brightness of the bulbs ($A = D = E > B = C$), they often fail to do so because of a lack of a conceptual understanding of the situation.

In contrast, curricula designed to help students develop a conceptual understanding of simple circuits begin by asking students to determine the

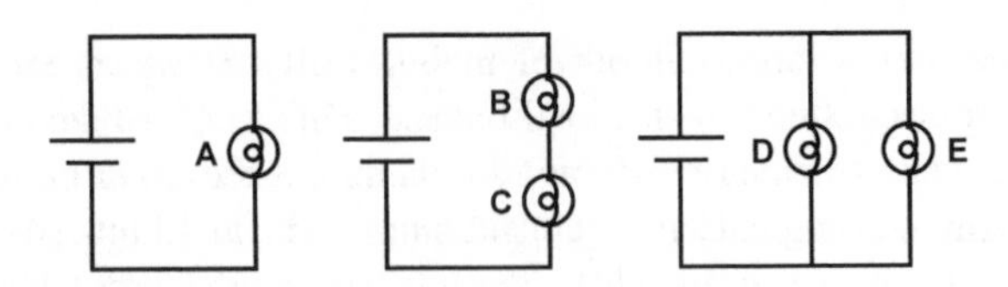

Fig. 1. Bulb ranking task: Three circuits with ideal batteries and identical bulbs.

necessary conditions of a complete circuit. Students determine these conditions after investigating lighting a light bulb using one battery, one wire and one bulb. Through teacher- and student-guided investigations and probing questions, students begin to build a model for the behavior of current in circuits. In this process, students' initial ideas, such as, the battery is a source of constant current or that current is "used up" in a circuit, are challenged. Students must sort out for themselves the observations that do and do not support their ideas. This is done through discussing their ideas with their peers, organizing the results from their investigations and finally, reflecting on their initial ideas. The process of building this model for current (and subsequently for voltage) is slow and often requires revisiting an idea several times in new contexts. However, developing models to explain phenomena is fundamental to the scientific reasoning process. In addition, without a coherent model, students often fail to consistently apply ideas to new situations.

Students with a conceptual model for the behavior of current in simple circuits would rank the brightness of the bulbs in Fig. 1 by comparing the currents through the bulbs. A correct conceptual explanation begins with recognizing that the current supplied by the battery in the first circuit will all travel through bulb A since this is a single path, series circuit (i.e. there is nowhere else for the current to go.) The current in the first circuit will be greater than the current in the second circuit, because there are two obstacles (bulbs) in series, posing more resistance and thus allowing less current to flow through the second circuit. Whatever current travels from the battery through bulb B must also go through bulb C, since there is only a single route. So, both bulbs B and C will be equally bright. However, each will be dimmer than bulb A because the current is less. The battery in the third circuit will supply a current that is twice as great as the battery in the first circuit. The third circuit provides two routes for the current. Even though each route is just as resistive as the one route in the first circuit, having two routes means the third circuit will pose less total resistance than in the first circuit. The current will divide equally between the branch with bulb D and the branch with bulb E, since the resistance of each branch is the same. Thus, bulbs D and E will be equally bright and the same brightness as bulb A.

The conceptual understanding illustrated in the above explanation is essential in order to apply Ohm's law and Kirchhoff's rules successfully to this and more complicated circuits. As a consequence, this conceptual understanding underpins students' ability to solve both conceptual and mathematical physics problems. (Dufresne, 1995; Mestre, 1993)

An effective instructional approach to enhancing students' conceptual understanding is the use of analogies (Clement, 1993). Effective analogies are those that the student has offered as an example or are systems with which the student has prior experience. For the analogy to enhance students understanding, students should identify both the parts of the system that are analogous and the parts that are not. For example, the third circuit in Fig. 1 might be compared to a road full of traffic where a second lane has opened. With a second lane, the flow of cars can double because the total resistance to the flow is less than if only one lane was open. In this analogy, the two lanes of traffic correspond to the two branches of the circuit with bulbs D and E. The increased rate at which the traffic can flow with two lanes open corresponds to the increased current from the battery in the third circuit with two pathways compared to the first circuit with only a single pathway. However, the resistance of each branch does not have an analogous part in the simple traffic system.

Another important feature of this aspect of the framework is to ensure that instruction, guided by student and teacher investigations, is not diminished to activities for the sake of activity. While designing and conducting investigations are important skills in all the sciences including physics, these skills must be learned in the context of developing an understanding of the concepts fundamental in physics in order to understand the physics. Isolated investigations not tied to the learning goals of the course lead to a fragmented view of physics and undermine the role of the investigations – to help students construct an understanding of the concepts.

Students attain a deep conceptual understanding of the concepts through analysis of their investigations, relating the outcomes to their hypothesis, and comparing their results with others in the class. Unlike physics laboratory experiments traditionally used in instruction, the investigations with the greatest impact on deepening students' conceptual understanding do not have an outcome known in advance by students. Instead, the results of these investigations are interpreted by the students in the context of the ideas they are testing out.

It is during the analysis of results that teachers need to push students' thinking, as illustrated in the pulley dialog. The individual, small group and whole class discourse is fundamental to deepening students understanding of the concepts they are testing out. Posing questions to challenge student thinking and extend their ideas is fundamental to exemplary teaching. Rarely is the quickest path to the "right" answer the one that leads to the greatest insight.

The *Modeling Method* of physics teaching exemplifies this discourse (Wells, 1995). In a course taught using the *Modeling Method*, students spend a large portion of class time designing and conducting investigations around developing a few important models in physics. As these models are fit into scientific theories, students give structure to their knowledge of phenomena. A significant and important part of class time is spent analyzing the results of student-led investigations. After each investigation, students are required to summarize their findings using both qualitative and quantitative elements. Through the use of small whiteboards, teachers using the *Modeling Method* require students to orally present their investigation to the rest of the class. During this presentation, teachers facilitate the discussion by encouraging the other groups to question the presenters' findings in comparison with their own. The teacher will also ask specific questions designed to challenge certain misconceptions held by the students. Knowing what questions to ask and when to ask them is not easy. In addition, it may take several questions to hone in on a particular student difficulty. The art of guiding these discussions typically takes many years to perfect.

In the *Modeling Method*, it is recommended that teachers carefully choose the order of presentations by the groups. The order is chosen to encourage the maximum discussion by the rest of the class. By choosing a group with a unique procedure, investigation and/or result to go first, other groups will have specific issues to discuss when comparing the presentation to their own findings. The ensuing discussion often elicits additional student difficulties not previously known to the teachers, thus illuminating the path of future investigations to both the teacher and the students. This sequence of discourse effectively engages learners in reflecting on their ideas. This reflection will either strengthen the students' models or leave the students dissatisfied with their ideas, which is necessary if they are going to want to change the way they are thinking (Posner, 1982).

By focusing on teaching and learning for a conceptual understanding, several of the guiding principals of good teaching described in the first chapter of this volume are integrated into the instruction. Three principles in particular, Curricular Alignment, Coherent Content, and Thoughtful Discourse, are explicitly embedded in this aspect of the framework. The important ideas are that: the content is focused around powerful ideas in physics; the curriculum components work together to achieve the goals of the course; deep conceptual understanding is emphasized over broad coverage; students are taught through guided-inquiry involving small, sequenced steps; the teacher acts as a facilitator of building understanding; questioning strategies and classroom dialogs play an important role in instruction. Knowing the learning targets and aligning curriculum activities and teaching strategies with the desired learning goals fosters an organization into a conceptual structure. All these components of exemplary teaching are necessary in order to address the final aspect of the framework described next.

4. Applying Learning to Build a Functional Understanding

The development of a functional understanding (i.e. the ability to chose, reason and apply relevant concepts in situations not previously studied (McDermott, 1993)) is the ultimate goal of exemplary physics teaching. Helping students build a functional understanding takes time and practice. In addition, knowing when or if students have reached this goal is an ongoing challenge.

The first three aspects of this instructional framework provide the environment to learn the knowledge, thinking skills, and conceptual understanding required to have a functional understanding of the topic. The final aspect of the framework for instruction requires students to apply, generalize and reflect upon what they have learned.

Problem solving is one arena in which physics students are often asked to demonstrate a functional understanding. Research on expert versus novice problem solving has revealed that it is more than just knowledge of the subject that is important for success in this important application. While experts do have a large body of knowledge to draw upon, their success lies not in having the knowledge but from knowing how and when to use this knowledge. To know how and when to use the knowledge, experts have given a structure to their knowledge such that it is easily accessible (Dufresne, 1987). Accessibility of knowledge is another important feature of how people learn (Donovan, 1999). Additionally, when categorizing a problem, experts typically look beyond the surface features and attempt to identify the underlying principles that might apply to the situation, while novices focus heavily on the surface features (Chi, 1981). Finally, students who have become successful at performing a qualitative analysis of a situation show improvement in their problem solving abilities (Mestre, 1993).

The implications of this research on instruction are that in order to help students become better at applying their learning to problem solving, instruction should emphasize helping students obtain a deep conceptual understanding that is also functional. This involves practice with analyzing situations qualitatively (Mestre, 1993). In addition, as new qualitative knowledge is acquired, the application of this new knowledge should be in contexts in which existing knowledge must be considered. These connections help build the foundation of understanding and helps students structure what they are learning.

Minds-On Physics is a high school curriculum that has put into practice the results of this problem-solving research (Leonard, 1999). Students' functional understanding of forces and motion is constructed through a series of guided activities. Students build a conceptual understanding of force and motion along side the physics formalism. As students progress through the activities, they are asked to apply what they are learning in new and more complex situations. In addition, they are challenged not simply to solve novel problems, but to determine what information is needed to solve the problem given a particular approach. One such activity is described below.

Near the end of the *Interactions* module, students are asked to consider six situations. In each situation, the initial position and speed of an object is specified, but the other information provided is limited, varying with the situation. Students are reminded of three generic steps in the process of solving dynamics problems; (1) identify the forces and determine the net force given the information in the problem, (2) use this net force and Newton's second law to determine the acceleration of the object given the mass of the object, (3) use kinematics relations to find the change in position and velocity of the object. The students' task is to decided if they have enough information to determine the object's displacement and velocity. If not, they are asked to explain which of the three steps in solving the problem they cannot complete. The situations vary from a cart on an inclined plane, to a pendulum, to a cart compressing a spring. Note that the situations have very different surface features. In analyzing each situation, students come to recognize the essential and non-essential information in each problem. In addition they arrive at the conclusion that the acceleration must be constant during the motion to use the kinematics expressions.

Students' sense of ownership of their learning is an important component in achieving a functional understanding of the material. The *ThinkerTools* curriculum uses two assessment strategies to help students develop ownership of their learning (White, 1995, 1998). The first is ongoing self assessment and the second is a project-based summative assessment. This program provides another example of capitalizing on the need to help students learn to monitor their own learning.

ThinkerTools is a curriculum designed for middle school students emphasizing scientific inquiry and modeling. Students spend a significant time working through activities and investigations designed to help them build a conceptual understanding of force and motion. They accumulate many experiences with how different forces affect the motion of objects through experiences with computer-simulated motion and "real life" motion. After each activity, students are asked to assess one component of their performance on that particular activity. Possible components include communicating well, reasoning carefully, using the tools of science, designing good experiments, making connections, understanding the processes of inquiry, and understanding the science. Students not only give themselves a score for one of these criteria in each activity, they are asked to justify their score. Justifying a score is a skill that takes time to learn. However, research has shown that not only does that process of self assessment help deepen students understanding (particularly with low achievers), but the better the students are at justifying their self-assessment scores throughout the course, the higher their performance on their final research projects (White, 1998).

Projects are one way for teachers to assess for a functional understanding. The final projects students' conduct in *ThinkerTools* embed the skills of scientific inquiry with applying and generalizing concepts learned throughout the course. Students are encouraged to do investigations in situations they have not studied before. Suggested situations include circular motion, collisions, fluid resistance, and air resistance. As with all the other activities in the *ThinkerTools* curriculum, students participate in the assessment of their own and their peers' final projects.

Performance assessments are another tool that can both help build and test for a functional understanding. Well-constructed items extend students thinking and help them generalize their learning. In addition, performance assessments can look at other factors that might be getting in the way of students developing a functional understanding such as inadequate process or reasoning skills.

One example of a performance assessment that meets multiple goals often involves designing an investigation to test out an idea, conducting the investigation, and making a conclusion about the initial idea on the basis of the investigation. For example, in one unit of an innovative middle school curriculum, *Model Assisted Reasoning in Science (MARS)*, students study the concepts of mass and density through building mental models (Raghavan, 1993). On a final performance assessment in this unit students are asked to consider the idea that the density of a liquid depends on how thick the liquid is, which, by the way, is a common belief among students. Students must design and conduct an experiment to test this idea. After the experiment they must explain what their results tells them about the idea that density depends on thickness of a liquid. There are many important levels to evaluate students' performance on this assessment: experimental design, having an operational definition of density, ability to measure mass and volume, skill at comparing sets of data, ability to reason from data, drawing conclusions from data, and relating conclusions to initial ideas. This well designed assessment also would help students confront possible lingering ideas not addressed earlier in the unit when developing a model for density.

The level of application and generalization required of students in demonstrating their functional understanding of the concepts in any course depends upon the developmental level of the students. For the younger students, the assessed context may not be very different than the context in which they learned the concepts, while older students might be required to apply concepts to very different situations. It should be emphasized that both formative and summative assessments throughout exemplary instruction should be of the understanding or the application of knowledge, not of the knowledge itself.

Several of the generic principles for good teaching described in the first chapter of this volume are woven throughout the final aspect of the framework. They include Practice and Application Activities, Scaffolding Students' Task Engagement, Goal Oriented Assessment, and Achievement Expectations. Specifically, the development of a functional understanding requires that: students have opportunities to practice and apply their learning; students are provided with timely feedback; students become progressively more independent at pursuing their own inquiry; assessments are used to make decisions regarding subsequent instruction; students participate in assessing their own work; achievement expectations are high in order to help students to reach a functional understanding of the important concepts in the course.

C. SUMMARY

Effective physics instruction will include all aspects of the framework described in this chapter. Teachers need to identify students' problematic ideas and continually monitor students' thinking. Teachers should focus on helping students generate, test, and modify their ideas. Teaching should promote a deep understanding of observations, inferences and relations organized in a conceptual framework. Exemplary teaching should result in a functional understanding of ideas and processes that can be applied appropriately across multiple and even novel context.

While one aspect of the framework may play more of role during certain parts of a unit, they are all present in the expert teachers' minds as they help move students from their initial ideas to reaching the learning goals of the course. This is accomplished through small cycles of instruction that are informed through a continuous monitoring of students' conceptual understanding. Each cycle bringing the understanding of the student and the learning goals for the course closer together, thus crossing the gap from research to standards-based, exemplary classroom practice.

It is important to emphasize that exemplary physics instruction requires exemplary physics teachers. The teaching skills required in the framework described are not embedded in a particular curriculum or series of activities, but reside with the individual teacher. Teachers often spend years refining these skills and adapting activities to reach all the learners under their charge.

One important component to refining these skills is the depth to which a teacher understands and can apply the concepts they are teaching. Knowing the content at a deep level is required to help students achieve a functional understanding of the concepts. The path to acquiring this depth of understanding is the same for the teacher as it is for the student. The preparation and professional development of teachers must be in environments in which all four aspects of the framework are in place.

In addition, knowledge of the nature of learning will impact the implementation of the aspects of this instructional framework. It is not enough for the teacher to have an understanding of the material; they must also have an understanding of how one comes to know what they know. Finally, the teacher needs to develop pedagogical content knowledge. They need to have available extensive knowledge of particular curricular activities, how students will respond to the activities, and what learning can be expected from the activities. Teachers need to know the learning targets, the initial thinking of students, and how to cross the gap from initial ideas to functional conceptual understanding.

This task of becoming an exemplary teacher may seem daunting, especially for those new to teaching. However, as with every profession, expertise is not something one attains; expertise is something for which we continually strive.

ACKNOWLEDGMENTS

We thank our colleagues who teach physics at all levels. It is their work we are attempting to honor and report in this chapter. The writing of this chapter has been partially supported from National Science Foundation Grants REC-9906098 and REC-9972999. The ideas expressed here are those of the authors and do not necessarily represent the thoughts of the foundation.

REFERENCES

American Association for the Advancement of Science (1993). *Benchmarks for Science Literacy.* New York: Oxford University Press, Inc.

Chi, M. T. H. et al. (1981). *Expertise in Problem Solving* (Technical report LRDC-1981/3). Pittsburgh.

Clement, J. (1993). Using bridging analogies and anchoring intuitions to deal with students' preconceptions in physics. *Journal of Research in Science Teaching*, *30*(10), 1241–1257.

Crouch, C., & Mazur, E. (Submitted 2000). Peer Instruction: Eight Years of Experience and Results. *American Journal of Physics*.

DiSessa, A. A., & Minstrell, J. (1998). Cultivating conceptual change with benchmark lessons. In: J. G. Greeno & S. Goldman (Eds), *Thinking Practices in Mathematics and Science Learning*. Mahwah: Erlbaum Associates.

Donovan, M., Bransford, J., & Pellegrino, J. (Eds) (1999). *How People Learn: Bridging Research and Practice*. Washington DC: National Research Council.

Dufresne, R., Gerace, W., Hardiman, P., & Mestre, J. (1987). Hierarchically structured problem solving in elementary mechanics: guiding novices' problem analysis. In: J. Novak (Ed.), *Proceedings of the 2nd International Seminar "Misconception and Educational Strategies in Science and Mathematics* (Vol. 3, pp. 116–130). Ithaca: Cornell University.

Dufresne, R., Leonard, W., & Gerace, W. (1995). *A qualitative model for the storage of domain-specific knowledge and its implications for problem-solving* (Technical report). Amherst: University of Massachusetts.

Dufresne, R., Gerace, W., Leonard, W., Mestre, J., & Wenk, L. (1996). Classtalk: A classroom communication system for active learning. *Journal of Computing in Higher Education*, *7*, 3–47.

Education Development Center (1977). *Insights: An elementary hands-on science curriculum.* Kendall/Hunt.

Goldberg, F., Heller, P., & Bendall, S. (1997). Constructing Physics Understanding.

Hestenes, D. (1979). Wherefore a science of teaching? *The Physics Teacher*, *17*(4), 125–242.

Hunt, E., & Minstrell, J. (1996). Effective instruction in science and mathematics: Psychological principles and social constraints. *Issues in Education*, *2*(2), 123–162.

Karplus, R., & Lawson, C. (1974). *SCIS Teacher's Handbook* (pp. 179). Berkeley: Lawrence Hall of Science, University of California, Berkeley.

Leonard, W., Dufresne, R., Gerace, W., & Mestre, J. (1999). *Minds-On Physics* (1st ed.) (Vol. Interactions, Activities and Reader). Kendall/Hunt.

Levidow, B., Hunt, E., & McKee, C. (1991). The DIAGNOSER: A HyperCard tool for building theoretically based tutorials. *Behavioral research, methods, instruments, and computers*, *23*(2), 249–252.

Linn, M., Lewis, C., Tsuchida I., & Butler Songer, N. (2000a). Beyond Fourth-Grade Science: Why do U.S. and Japanese students diverge? *Educational Researcher*, *29*(3), 4–14.

Linn, M., & Hsi, S. (2000b). *Computers, Teachers, and Peers*. Science Learning Partners: Lawrence Erlbaum Associates.

Mazur, E. (1997). *Peer Instruction: A user's manual*. Upper Saddle River: Prentice Hall.

McDermott, L. C., & Shaffer, P. S. (1992). Research as a guide for curriculum development: An example from introductory electricity. Part I: Investigation of student understanding. *American Journal of Physics*, *60*(11), 994–1003.

McDermott, L. C. (1993). How we teach and how students learn – A mismatch? *American Journal of Physics*, *61*, 295–298.

Mestre, J., Dufresne, R., Gerace, W., Hardiman, P., & Touger, J. (1993). Promoting skilled problem-solving behavior among beginning physics students. *Journal of Research in Science Teaching*, *30*(3), 303–317.

Minstrell, J. (1992). Facets of students' knowledge and relevant instruction. In: R. Duit, F. Goldberg & H. Niedderer (Eds), *Research in Physics Learning: Theoretical Issues and Empirical Studies* (pp. 110–128). Kiel: IPN.

Minstrell, J. (2000a). Implications for Teaching and Learning Inquiry: A Summary. In: J. Minstrell & E. van Zee (Eds), *Inquiring into Inquiry Learning and Teaching in Science*. Washington: American Association for the Advancement of Science.

Minstrell, J. (2000b). Student Thinking and Related Assessment: Creating a Facet Assessment-based Learning Environment. In: J. Pellegrino, L. Jones & K. Mitchell (Eds), *Grading the Nation's Report Card: Research from the Evaluation of NAEP.* Washington DC: National Academy Press.

Minstrell, J. (in press a). Facets of students' thinking: Designing to cross the gap from research to standards-based practice. In: K. Crowley, C. D. Schunn & T. Okada (Eds), *Designing for Science: Implications for Professional, Instructional, and Everyday Science*. Mahwah: Lawrence Erlbaum Associates.

Minstrell, J. (in press b). The role of the teacher in making sense of classroom experiences and effecting better learning. In: D. Klahr & S. Carver (Eds), *Cognition and Instruction: 25 Years of Progress*. Mahwah: Lawrence Erlbaum Associates.

National Research Council (1996). *National Science Education Standards*. Washington, DC: National Academy Press.

National Science Resources Center (1997). *Science and Technology for Children*. Burlington: Carolina Biological Supply Company.

Posner, G. J., K. A. Strike, P. W. Hewson, & W. A. Gertzog (1982). Accommodation of a scientific conception: Toward a theory of conceptual change. *Science Education*, *66*(2), 211–227.

Raghavan, K., Kesidou, S., & Sartoris, M. (1993). *Model-centered curriculum for model-based reasoning*. Paper presented at the Third International Seminar on Misconceptions and Educational Strategies in Science and Mathematics, Ithaca, NY.

Rowe, M. B. (1972). *Wait-Time and Rewards and Instructional Variables: Their influence on language, logic and fate control*. Paper presented at the National Association of Research in Science Teaching, Chicago, IL.

Van Zee, E., & Minstrell, J. (1997). Using Questioning to Guide Student Thinking. *Journal of the Learning Sciences*, *2*(6), 227–268.

Wells, M., Hestenes, D., & Swackhammer, G. (1995). A Modeling Method for high school physics instruction. *American Journal of Physics*, *63*(7), 606–619.

White, B. (1995). The ThinkerTools project: Computer microworlds as conceptual tools for facilitating scientific inquiry. In: S. M. Glynn & R. Duit (Eds), *Learning Science in the Schools: Research Reforming Practice* (pp. 201–227). Mahwah, New Jersey: Lawrence Erlbaum Associates.

White, B., & Frederiksen, J. (1998). Inquiry, modeling, and metacognition: Making science accessible to all students. *Cognition and Instruction*, *16*(1), 3–118.

Woolnough, B. (2000). Appropriate Practical Work for School Science – Making it Practical and Making it Science. In: J. Minstrell & E. van Zee, (Eds), *Inquiring into Inquiry Learning and Teaching in Science*. Washington: American Association for the Advancement of Science.

CHEMISTRY: A MATTER OF UNDERSTANDING REPRESENTATIONS

David F. Treagust and Gail Chittleborough

INTRODUCTION

The generic aspects of effective teaching outlined in Chapter One generally are applicable to the teaching of chemistry. This chapter describes approaches to teaching and assessment that have been shown to be effective in enhancing students' knowledge and understanding of chemistry. The teaching methods are grounded in the conceptualisation of knowledge as a tentative human construction widely known as constructivism (Bodner, 1986). This conceptualisation is reflected in "connecting with students' prior knowledge" as Brophy (2001, p. 2) outlined as good teaching. Further, the teaching approaches are described within the context of their uniqueness to chemistry as a discipline. Without defining these contexts, the instructional approaches would be less meaningful and lacking in interpretation to the reader. Shulman's (1986) emphasis on teachers' pedagogical content knowledge illustrates the importance of teachers' skills and wisdom in selecting appropriate teaching approaches for particular student needs to ensure that teaching occurs within their zone of proximal development.

The first section of this chapter is a discussion of the unique problems of teaching and learning chemistry. The important issue of the different levels of understanding involved in learning about chemical phenomena is examined as well as the difficulties in comprehending the specialised language used in chemistry and distinguishing it from the everyday meanings of identical or similar terms. The second section of this chapter looks at research on

Subject-Specific Instructional Methods and Activities, Volume 8, pages 239–267.

ISBN: 0-7623-0615-7

instruction related specifically to different representations, such as the variety of models and analogies used in chemistry to enhance the learning of entities and their related explanations. The third and largest section of this chapter examines a range of approaches to teaching and assessment and their specific use based on researched outcomes. These approaches include concept mapping, problem-solving, laboratory work – both in wet labs and in computer-mediated environments, cooperative learning, the use of varied teaching resources and diagnostic assessment techniques. This variety helps to develop an optimal program of instruction. The chapter ends with a review and reflection of the significance of educational research about teaching chemistry and how this relates to the generic features of good teaching.

THE UNIQUE PROBLEMS OF TEACHING CHEMISTRY

Chemistry is a difficult subject to teach because it has many abstract concepts that are totally unfamiliar to students whose personally constructed representations are often in conflict with scientifically accepted explanations. Although the rote learning of chemical formulae and facts are essential for long-term memory, this alone does not challenge the learners' understanding. Only a personal reconstruction of the chemical concepts by the learner helps achieve meaningful learning (Johnstone, 1997). Indeed, Boo (1998) reports that students can produce correct answers to particular questions without understanding the chemical concepts and concluded, from a study in Singapore of Grade 12 students' understanding of chemical bonding, that many students did not have an understanding of the nature of science and inferred that this compounded the effects of their understanding of the chemistry. This anomaly of achieving correct answers despite using conceptually unacceptable strategies that reflect little understanding also has been identified in the areas of stoichiometry and equilibrium (BouJaoude & Barakat, 2000; Huddle & Pillay, 1996).

Human beings have a natural curiosity and when observations are made which seem bizarre or different they try to explain what is observed. At an early age, children already have personal well-developed approaches for generating their own explanations (Metz, 1998). According to Nakhleh (1994), difficulties in learning chemistry can be attributed to the students not constructing a sound understanding of basic chemical concepts so that there is no foundation upon which to build advanced concepts. Students' conceptions which differ from the commonly accepted scientific meaning are referred to as misconceptions or alternative conceptions depending on the author's understanding of the nature of knowledge (Garnett, Garnett & Hackling, 1995b; Nakhleh, 1994). While chemistry introduces students to new ideas and a new vocabulary, Renstrom, Andersson and Marton (1990) conclude that science teaching only "changes their [students'] views of things in the world around them to a very limited extent" (p. 567). The inference is that students do not link or apply the new knowledge that they learn at school to everyday life.

Research shows that what students already know, in other words their preconceptions, has a serious impact on their learning of new material (Garnett et al., 1995b; Taber, 1996) because these are the foundations on which new knowledge is built. Consequently, students' everyday experiences and intuitive logic should be recognised by teachers in order to develop students' scientific explanations (Andersson, 1990). Teaching strategies based on constructivist principles include providing students with opportunities to restructure their conceptions through discussions and to reflect on their understandings, using demonstrations, experiencing conflict situations and taking greater responsibility for their own learning (Garnett et al., 1995b).

Three Levels of Understanding Chemical Phenomena

The study of chemistry is essentially about the nature of matter which itself is complex, having abstract concepts that are complicated by the levels at which they can be portrayed (Johnstone, 1993). Research shows that effective understanding is achieved by being able to interrelate the three levels of representation of matter which are described as being at the:

- Symbolic level – comprising a large variety of pictorial representations, algebraic and computational forms.
- Microscopic level – comprising the particulate level, which can be used to describe the movement of electrons, molecules, particles or atoms.
- Macroscopic level – comprising references to students' everyday experiences.

Research shows that many secondary and college students, and even some teachers, have difficulty transferring from one level to another (Boo, 1998; Gabel, 1998). As Kozma and Russell (1997, p. 949) point out, "understanding chemistry relies on making sense of the invisible and the untouchable." Explaining chemical reactions demands that a mental picture or model is developed to represent the microscopic particles in the substances being observed. Observations of changes in colour or volume or the evolution of a gas for example, reveal nothing about the microscopic behaviour of the chemicals involved. Yet, explanations are nearly always at the microscopic level – a level which cannot be observed – but is described and explained using symbols by which personal mental models are constructed. Unfortunately, students often transfer the macroscopic properties of a substance to its microscopic particles, believing for example that sulfur is yellow, so the atoms of sulfur are yellow also. Indeed, this is not surprising considering the graphical representation of yellow circles in textbooks to represent the atoms (Andersson, 1990). To overcome this problem, Gabel (1998) recommends that teachers provide physical examples or at least descriptions of the chemicals in the problems, so that students can establish their own links between the three major levels for portraying the chemical phenomena. In a study into students' understanding of acids and bases, Nakhleh and Krajcik (1994) reported that students' explanations made many more references to the macroscopic level than the microscopic level and more about the microscopic than the symbolic, indicating that they are more confident describing these chemicals at the macroscopic level. Given this finding, which is supported by other studies, it is somewhat surprising that so few chemistry curricula emphasise the chemistry of students' everyday experiences (Garnett et al., 1995b). Fortunately, there are exceptions. The use of familiar items in chemistry laboratory work has been used to reinforce the link between chemistry and home, resulting in improved students' perceptions of chemistry (Ramsden, 1994; Roberts, Selco & Wacks 1996). In England, the Salters Chemistry course incorporated a constructivist approach using familiar chemicals as the starting point to motivate students and create a positive classroom climate (Campbell, Laxonby, Millar, Nicolson, Ramsden & Waddington, 1994).

In a similar way, commonsense and first hand experiences support students' belief in the continuous nature of matter whereas in school chemistry lessons the discontinuous or particulate nature of matter is introduced and used to explain scientific phenomena. One outcome of this dichotomy is for students to use both beliefs simultaneously (Krnel, Watson & Glazar, 1998; Renstrom et al., 1990); this position is supported by Dierks (1989) who argues that there is no value in destroying the continuous model of matter for the particulate model because they can be used side by side. Macroscopic, continuous descriptions are used for continuous meaning and chemical symbols should be restricted to the microscopic, discontinuous meanings where molecules and atoms are involved. However, the confusion between the microscopic and macroscopic nature of matter is well documented (Andersson, 1990; Garnett et al., 1995b; Krnel et al., 1998) and gives rise to students confusing dissolving and melting, associating heat with weight and matter, having difficulty accepting the conservation of matter and mass when some substances appear to disappear, understanding the transformation of water from solid to a liquid but being unable to transfer this phenomena to other substances, accepting the "disappearance" of liquids during evaporation, and believing that if a gas is formed then it changes into air.

The chemistry teacher faces a formidable intellectual challenge because attempts to simplify chemical concepts often lead to confusion such as students only associating elements with atoms and molecules with compounds. Other confusions are when teachers and textbooks categorize reactions as chemical or physical even though these changes are not absolute categories but are more of a continuum (Palmer & Treagust, 1996). Similarly, melting and dissolving often do not occur in isolation, and the categorisation of a reaction as only one type is not always possible. Fensham (1994) provides examples of these anomalies: "heating washing soda leads not to melting as it would appear, but to sodium and sulfate ions dissolving in the solid's water of crystallisation, hydrogen chloride gas reacts with water as it dissolves to form ionic species that were not present in the original gas" (p. 19). In other situations, students apply the octet rule for molecular bonding indiscriminately – in areas where it is not appropriate – leading to the reference of "molecules" of sodium chloride based on the electronic transfer of one electron from a sodium atom to a chlorine atom (Taber, 1997). Learning chemistry is not simple, and well-informed teaching practices such as reinforcing the links between the three major levels portraying chemical phenomena – microscopic, macroscopic and symbolic – are needed to ensure that students do not develop entrenched alternative conceptions. Such an approach aligns with the effective teaching principle of coherent content described in the introductory chapter.

Many chemistry courses fail to show the relevance, usefulness and applicability of chemistry to everyday life. One way to achieve this might be to use common terms alongside chemical names so those students can relate the chemicals with everyday experiences; simple examples are sodium bicarbonate as baking soda, acetic acid as vinegar and sodium chloride as common salt. Another aspect of relevance is to refer to everyday chemical events such as heating, combustion, solids, liquids and gases changing phase and, melting and boiling in the specific chemical context that also includes microscopic and symbolic representations. Such a decision is important because it will challenge pre-conceptions or the alternative conceptions that many students have developed for these terms (Krnel et al., 1998). Although it is understandable how these conceptions have arisen, they are not scientifically acceptable and have to be unlearnt or challenged so that new conceptions can be better understood. One way that this difficult task can be addressed is through cooperation and sharing of ideas between students (Garnett et al., 1995b; Schmidt, 1997) and for the teacher to be aware of common alternative conceptions and have strategies in place to help students reconstruct their conceptual frameworks (Taber, 1998). Teacher's awareness of students' background, ideas and experiences helps create a supportive classroom climate, one of Brophy's 12 principles of effective teaching.

Language

Chemistry has its own special language, but the same words used in everyday speech also have different meanings and these frequently give rise to learning difficulties. When terms are used in a scientific context, it is assumed that the scientific meaning will be understood (Garnett et al., 1995b). However, confusion and frustration arise when teachers and textbooks use everyday vocabulary in explanations of chemical phenomena assuming that students understand the special chemical meanings of the terms being used (Nakhleh, 1994; Schmidt, 1997). In classrooms, often no distinction is made between the scientific meaning and the commonplace meanings of our vocabulary such as driving force, precipitation, energy, or bond (Boo, 1998; Fensham, 1994). Even within chemistry, there are several meanings for the same word and students confronted with the same words with different meanings become confused (Selinger, 1998). For example, 'pure' can refer to the cleanliness of a substance, not its chemical nature, 'mixture' refers to something physically combined together, not the chemical nature of, for example, glass or blood or drinking water. Students' experiences are mainly with mixtures; however, their perception is that these substances, such as brass, lemonade, wine or tap water are chemically pure. Words as different as dissolving and melting, which are obvious to teachers, are frequently confused when used by students who have insufficient background or experience with which to distinguish these terms and consequently the teacher's meaning is not communicated clearly (Fensham, 1994). These basic alternative conceptions reveal a weak foundation on which to build further chemical knowledge.

Particular words such as particle, molecule, ion, atom and substance are often misused and misinterpreted. Terms such as donated, shared and accepted, which were initially used metaphorically, now have developed new literal meanings in their chemical use (Taber, 1998). The historical development of chemical terms means that some names are misleading; for example, Schmidt (1997) reported that many students conclude that the prefix "ox" in oxidation indicates that oxygen is involved in all redox reactions; a more enhanced understanding of oxidation and reduction can only be achieved with a more universally acceptable definition of the concept using oxidation numbers (Garnett & Treagust, 1992). Similarly, students' understanding of the terms neutralisation and neutral that result from common and historical meanings conflicts with the acceptable chemical meaning because as Schmidt (1997, p. 132) explains, "in proton transfer reactions, acids and bases never consume each other." Nevertheless, this issue can be used to advantage by discussing the historical development of the phenomena and why the particular term now is considered misleading (Schmidt, 1997).

The anthropomorphic use of language in chemical explanations such as the commonly used phrases 'the atom wanted or needed to gain or lose electrons' and 'the atom was happy' are used with the intention of helping students identify with the topic. However, those explanations develop misunderstandings such as students associating forces between the nucleus and the electrons as only applying to the atom's "own" electrons (Taber, 1998; Treagust & Harrison, 1999). When teachers speak about water being made of oxygen and hydrogen, students can interpret this to mean that water is a mixture of the two gases; such an interpretation can be confusing when it contradicts students' knowledge of the properties of water (Renstrom, et al., 1990). Garnett and Treagust (1992, p. 132) report that the movement of ions in solution is often described with phrases like "ions carry a charge" which may be misinterpreted by students to mean that the ions carry the electrons. The precise and consistent use of language along with detailed particulate descriptions of the microscopic nature of matter can improve students' interpretations (Fensham, 1994; Garnett et al., 1995b).

TEACHING CHEMISTRY USING DIFFERENT TYPES OF REPRESENTATIONS

A chemical model is a representation of an object or a concept which can exist in numerous forms, such as ball and stick models, space-filling models, structural formula, chemical equations and computer models (Gilbert & Boulter, 1998; Hardwicke, 1995). Types of representation can be a static model, an analogy, a metaphor, a dynamic model, a picture, a diagram, a graph, an idea, a simulation or anything that can be used to develop a mental model in the learner. Each type of model adds in a unique way to the construction of the students' understanding of the chemistry under investigation.

Models and Modelling

The use of models and modelling in chemistry is arguably constructivist and it is likely that students' visualisation of models fosters conceptual development and conceptual changes by inducing gestalt shifts in learners' mental models (Norman, 1983). Model-based teaching and learning is consistent with personal and social constructivist theories of learning where the focus is on the learner, with all learning being dependent on language and communication (Cosgrove & Schaverien, 1997; Gergen 1995; Yager 1991). Indeed, the introduction of model-based reasoning is a highly desirable skill, but it does require extensive instruction and practice within the culture of the classroom (Stephens, McRobbie & Lucas, 1999). The use of models can encourage discussion and the articulation of explanations that encourages students to evaluate and assess the logic of their thinking (Raghavan & Glaser, 1995). However, more often, secondary students perceive a different representation as a new thing to learn rather than a means to explain what is to be learned.

The use of models and modeling in chemistry teaching is a common practice that engages students by helping them to develop their own mental models of chemical compounds. It is practically impossible for chemical phenomena to be explained without the use of models. However, despite this common use of models, studies have shown that students misunderstand the reasons for using models and modeling (Renstrom et al., 1990; Treagust, Chittleborough & Mamiala, 2001). Many secondary students view models only as copies of the scientific phenomena (Grosslight, Unger, Jay & Smith, 1991) and their understanding of the role of models frequently is seen as being simplistic. Even university students have limited experience with models and only a small percentage of these students have an abstract understanding of model use in chemistry (Ingham & Gilbert, 1991). Teachers' level of understanding of models also has been described as limited because they have a simplified understanding of models and modeling in science (Van Driel & Verloop, 1999). Nevertheless, modeling is a common, intrinsic behavior used in everyday life and also in the chemistry laboratory. Understanding models and their role in the development of scientific ideas is part of the chemistry teachers' personal philosophy of science and is central to his or her pedagogy (Selley, 1981). Gilbert, Boulter and Elmer (2000) and Harrison and Treagust (1996) recommend that teachers be educated to use models in a more scientific manner consistent with Brophy's effective principle of strategy teaching.

The use of concrete models and pictorial representations has been shown to be beneficial to students' understanding of chemical concepts (Gabel & Sherwood, 1980; Harrison & Treagust, 1996). However, the extensive and accepted process of using models has made the model appear as "fact" to many teachers and students (Boo, 1998; Harrison & Treagust, 1996). Frequently, students do not differentiate between models and they do not regard models differently from the observed characteristic that the model is trying to explain (Harrison & Treagust, 1996). Indeed, teachers and textbooks often represent atoms and molecules as real and factual, forgetting the origins of their evolution from theoretical models of matter. The strengths and limitations of each model need to be discussed so that students can assess its accuracy and merit (Hardwicke, 1995). A more accurate approach may be to represent the acceptable explanations as making use of particular chemical models which when used consistently explain the observation (Justi & Gilbert, 1999).

In comparing the perceptions of experts and novices on a variety of chemical representations, Kozma and Russell (1997) concluded that novices used only one form of representation and rarely could transform to other forms, whereas the experts transformed easily. Novices relied on the surface features, for example lines, numbers and color, to classify the representations, whereas experts used an underlying and meaningful basis for their categorization. The study highlighted the need for representational competence including an understanding of the features, merits and differences of each form and showed the significance of computer animations in linking the various representations.

Chemical models generated by computer graphics provide excellent detail and dynamics to illustrate molecular size, shape, bonding and electronic structure of chemical compounds, as is shown in the electron density distributions by Shusterman and Shusterman (1997), and may help overcome the difficulties that students have in visualising the particulate nature of matter (Balaban, 1999; Williamson & Abraham, 1995). In comparing a computer package DTMM (Desktop Molecular Modeler) with traditional instructional methods, Barnea and Dori (1999) reported that the 15 year old chemistry students using the computer molecular modeling program had improved visualisation of chemical substances, better understanding of the model concept, as well as a better understanding of the bonding structure of molecules. Similarly, Copolo and Hounshell (1995) reported improved retention on a test of isomeric identification by students using both computer and ball and stick models, indicating the effectiveness of concrete instructional aids in teaching abstract concepts. However, these students performed poorly on items using two dimensional representations, illustrating the difficult task of mental transference that needs to be given consideration by teachers because it is pivotal in developing mental models (Copolo & Hounshell, 1995; Kozma & Russell, 1997). This difficulty may be enhanced through the use of computer programs that are able to transfer between the three different levels of representation from macroscopic, microscopic and symbolic so students can compare representations (Ealy, 1999; Herron & Nurrenbern, 1999). In this way, learning theories guide the use of technology and the pedagogical value of these new methods can be established. Computer modeling for teaching and learning chemistry is expanding into secondary schools as resources become available (Beckwith & Nelson, 1998).

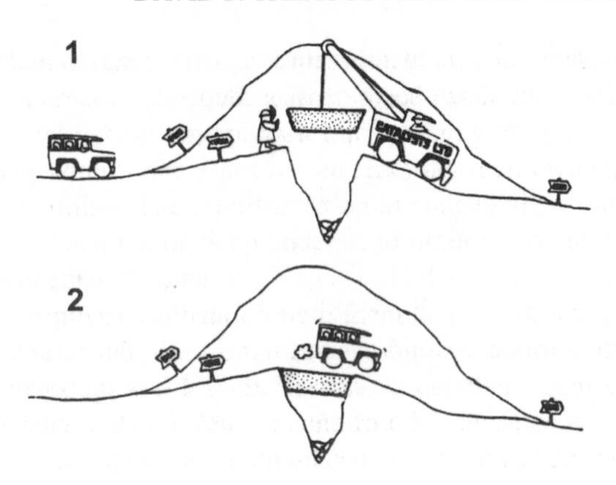

Fig. 1. An example of a pictorial analogy. From Hunter, Simpson & Stranks (1976, p. 257). Copyright. Reprinted with permission.

Analogies

Analogies make new abstract information more concrete and easier to imagine by using what the student already knows and is familiar with, and linking it to new, unfamiliar ideas (Collins & Gentner, 1987). Analogies provide new perspectives for viewing and an opportunity for reorganizing and linking knowledge which results in relational understanding as opposed to rote learning (Duit, 1991; Glynn, 1994). For example, the analogy of 12 in a dozen for the 6.02×10^{23} particles in a mole, is found in nearly all high school textbooks (Staver & Lumpe, 1993) as is the analogy to show the magnitude of Avogadro's number using oranges, in which a mole of oranges would form a sphere the size of the earth. A frequently used analogy for the use of a catalyst in a chemical reaction is shown in Fig. 1.

Research has shown that analogies are more effective with students of lower reasoning abilities and that higher order thinkers enhance their learning through constructing their own analogies (Duit, 1991). Analogies are motivationally stimulating, but the choice of analogies and the teaching of their limitations are critical to their success (Gabel, 1998; Staver & Lumpe, 1993). For example, using the dancing partners analogy when teaching the concept of limiting reagents is valuable as long as the source and target are clearly identified by all students. Even so, research has shown that teachers do not always use analogies effectively and that alternative conceptions often arise from their use (Gabel, 1998). The *Teaching-with-Analogies* (Glynn, 1994) approach and the *Focus-Action-Reflection* (Treagust, Harrison & Venville, 1998) method of teaching with analogies outline effective teaching strategies to avoid these difficulties.

SPECIFIC APPROACHES TO TEACHING AND ASSESSMENT

Effective teaching principles can be manifested through specific teaching and assessment tasks. Given opportunities, activities, practice and feedback, the tasks of concept mapping, problem solving, laboratory work, using new resources and diagnostic assessment techniques can complement each other and can cater for individual differences.

Concept Mapping

The most important single factor influencing learning is what the learner already knows (Ausubel, 1968) and this is the foundation of a constructivist epistemology which is an accepted process of knowledge construction (Novak, 1991). The abstract and difficult nature of chemistry often means that students fail to achieve meaningful learning or form relationships between various chemical concepts. Instead of representing science in the traditional format, as a large body of knowledge to be mastered, teachers should represent science as an evolving framework of concepts and conceptual relationships, which are constructed, not discovered, by the learner. The strategy of having students actively construct their own conceptual links throughout the course of study has proven to be successful in improving students' understanding (Novak, 1991). Concept maps are diagrammatic representations of key concepts, usually structured hierarchically and the relationships between concepts are indicated by linking words or phrases which help to visualize knowledge (Nakhleh, 1994). Ideally, concept maps are constructed by the students and convey their current level of understanding. In addition, concept maps may be used in instruction to identify the organisation of the conceptual domain to be taught as in the topic of covalent bonding directed towards Grade 11 and 12 students (Tan & Treagust, 1999), (see Fig. 2).

Concept maps are proven instructional tools used to create meaningful learning because they require learners to reveal their conceptual understandings and misunderstandings of the interrelationships of various concepts. The reconceptualisation of ideas that challenge incorrect conceptions can be assisted with concept maps (Novak, 1990). Diagnostic testing instruments have been formulated from concept maps and propositional statements for particular chemistry topics, making it easier for teachers and students to identify these alternative conceptions (Treagust, 1995).

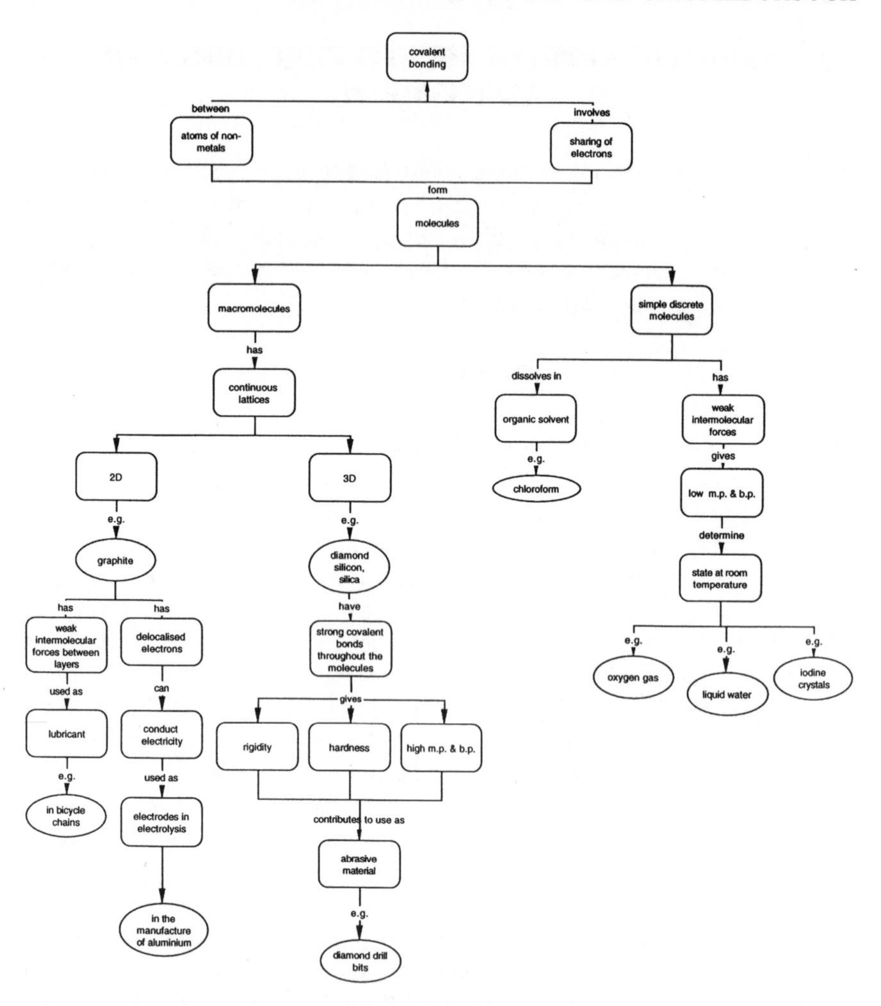

Fig. 2. Example of a concept map for the topic of covalent bonding in grades 11 and 12. Adapted from Tan and Treagust (1999) with permission.

Concept Maps as Instructional Tools

Research into the use of concept mapping in chemistry classes is positive and on-going and shows that the value of concept mapping in helping students make connections or links between concepts is undisputed (Cullen, 1990; Novak, 1990; Regis & Giorgio, 1996). Initially, students need to be taught the skill of constructing concept maps and given practice and feedback so that this skill is developed (Novak, 1991; Regis & Giorgio, 1996). Although concept mapping is very time consuming, its use can be justified by the achievement of meaningful learning. Having students construct concept maps before and after specific instruction allows them to reflect on their own conceptual organization and the changes that have occurred through their new learning. The changes in the structure of conceptual maps are considered to correspond to changes in cognitive structure due to learning. For example, Wilson (1996) found significant relationships between students' concept maps and their ability to recall and apply knowledge in tests on chemical equilibrium.

Students often fail to recognize the concepts being demonstrated in laboratory activities. The use of concept maps describing pre- and post-laboratory work have proven to be successful in identifying and focussing on the relationships between the concepts involved in the particular experiment as well as increasing students' confidence (Markow & Lonning, 1998). Nahkleh (1994) reported that concept maps revealed much information about the students' understanding and about how they have constructed their knowledge. The number of acceptable and unacceptable links, the hierarchical distribution and the cross-links gives some insight into students' understanding. In brief, concept mapping facilitates meaningful learning by building "networks of connected knowledge" (Brophy, 2000 p. 11) identified as an effective teaching principle.

Concept maps sometimes are used in science textbooks as summaries at the end of chapters to provide the knowledge structure of the content as well as linking interrelationships between concepts in various chapters. These can be used by teachers and students as learning aids; however, because the students have not constructed the concept map personally, these maps do not ensure or indicate student understanding.

Concept Maps as an Assessment Tool

The personal and idiosyncratic nature of concept maps creates difficulties for their use in common assessment. Nevertheless, numerous studies and evaluation techniques have been proposed and they are worthy of investigation though their validity and reliability have proven to be inconsistent (Ruiz-Primo & Shavelson, 1996). Assessment of concept maps in terms of the degree of differentiation, number of nodes, the level of connectivity and integration indicate the level of students' understanding, with all these aspects increasing with higher achievers (Wilson, 1994; 1998). Concept maps can give insight into the ontological categories that reflect the organization of the students' knowledge. As it is time-consuming for students to draw and for teachers to analyze individual concept maps, electronic versions such as *Inspiration* and *Semnet* can generate stimulating concept maps quickly (Fisher 1991; Inspiration Software Inc, 2000). Considering that experts have elaborate and differentiated knowledge structures, Wilson (1998) proposes that the construction of concept maps should be a primary objective in chemistry teaching because strategies to promote the construction of more elaborate and differentiated knowledge structures representing different ontological categories is more likely to provide a base for understanding than memorizing knowledge items.

Concept map building through group work provides a non-threatening way for students to express their understanding verbally and diagrammatically, with discourse among students helping to clarify their understanding. Reflective and positive discussion with teacher and peers helps students recognize and modify knowledge structures in which scientifically unacceptable links are reworked into scientifically acceptable links.

Problem Solving

Problem solving is a commonly used teaching method in chemistry because it challenges students' understanding of the subject matter and requires them to apply the concepts that they have learned (Gabel & Bunce, 1994). The characteristics of good problem solvers, according to Herron and Greenbowe (1986), include a good command of the basic facts and principles, the ability to construct appropriate representations of the problem, the ability to use general reasoning strategies that allow logical connections among the elements of the problem and the ability to apply several verification strategies, in which the problem representation is checked against the given facts, the solution is deemed logically sound, the calculations are checked for errors and the problem solved is the problem presented. Similarly, the methodology for problem solving outlined by Hanson and Wolfskill (2000) includes identifying the problem and the important issues, evaluating information, planning a solution, executing a plan, validating the solution and assessing an understanding of the solution.

Despite the existence of these strategies, students have difficulty solving chemistry problems because their understanding of the relevant concepts is often not sufficient, they use memorised algorithms without thinking, and have difficulty transferring between macroscopic and microscopic levels of understanding (Staver & Lumpe, 1995). Chemistry teaching comprises both algorithmic problems, which can be completed on a formula basis and often do not indicate understanding, as well as conceptual problems, which require students to display their understanding. When faced with conceptual problems, research shows that students do not rely on their reasoning skills or conceptual understanding, but often resort to using algorithms without understanding the problem (Niaz & Robinson, 1992). Mason, Shell and Crawley (1997) report that students performed better at algorithmic problems compared to conceptual problems even though they take longer to complete the algorithmic problems; however, successful completion of algorithmic problems did not guarantee understanding of the chemical concepts. The large variety of problem solving strategies generated by students reveals competing and conflicting frameworks which may be attributed to their alternative conceptions, reasoning strategies and prior knowledge (Astudillo & Niaz, 1996; BouJaoude & Barakat, 2000). Solving problems conceptually requires students to transfer between the three levels of representation and reconstruct the problem in terms of their own understanding.

Strategies to help improve problem solving ability include using analogies, models, diagrams and verbal and visual descriptions. Noh and Sharmann (1997) reported that pictorial materials at the molecular level do not improve students' ability to solve problems, but do improve their ability to construct correct scientific concepts. For solving problems on the mole concept, Staver and Lumpe (1993) described students' understanding of the mole concept as being vague and their inability to transfer between the macro and atomic levels affected their ability to do the problems. They recommended that students use analogous representations for the mole to provide a more concrete experience such as using different sized shot gun pellets to represent different atoms and arbitrarily assigning one size as the standard mass and then measuring the other sized pellets relative to the assigned mass, so that an analogous Avogadro's constant can be determined. Gabel, Briner and Haines (1992) demonstrated the advantages of integrating the three levels of understanding of chemical phenomena in solving chemical problems by ensuring students have seen or are aware of the macroscopic properties of the chemicals, by using three dimensional models to represent the microscopic level, and by using diagrams to represent the symbolic level of understanding.

A computer program called *The Mole Environment* contains graduated problem solving study-ware which specifically incorporates the concepts of the three levels of representation, the idea that matter is particulate, and environmental awareness to promote a global understanding rather than algorithmic learning of the mole concept (Dori & Hameiri, 1998). This program resulted in improved student understanding of the mole concept and an appreciation of its application to environmental aspects. Alternative computer interactive problem sets in chemistry have been shown to be effective because they provide the students with immediate feedback, they can be personalised to suit individual students, and the results can be stored easily (Bodner & Domin, 2000; Spain, 1996). The introduction of problem-solving teaching into the laboratory was shown by Gallet (1998) to improve students' interpretive capacities, motivation and communication skills.

Laboratory Work

Laboratory work is an essential and usually a compulsory component of any chemistry course with experiments carefully chosen to provide macroscopic examples of the concepts being taught. Because conceptual understanding is usually explained at the microscopic and symbolic level, laboratory activities are an essential and often the only component of the macroscopic representation level. Nevertheless, laboratory work is often criticised for not being relevant to the coursework, and being a recipe task in which the students simply follow the instructions without understanding what they are doing (Gallet, 1998). Consequently, along with factors of cost, safety and time, the importance of laboratory work in chemistry has diminished in recent years, although computer simulations have increased especially for dangerous, expensive and time-consuming experiments (Lunetta, 1998).

Importance of Laboratory Work

In response to this criticism, teachers have tried many innovative techniques to motivate their students. Adopting a constructivist approach and identifying students' preconceptions have proven to be successful by adapting experimental methods to promote inquiry through challenges, predictions and experimental design (Clough & Clark, 1994). These decisions also have implications for promoting teamwork, interest and confidence such as adapting experiments that require students to include a publicly expressed prediction which promote enthusiasm and rivalry (Plumsky, 1996). For example, after weighing and heating a sample of potassium chlorate until it completely decomposes, students are required to calculate the mass of the final product in the test tube. The teacher weighs the tube and the grade for the group is determined by the level of accuracy of their prediction within balance error. In a similar manner, the Predict-Observe-Explain (P.O.E.) approach, commonly used with demonstrations, requires students to make predictions about the outcome of an experiment, then observe the demonstration of the experiment and finally explain his or her observations and prediction (Liew & Treagust, 1995). This active learning approach is aimed at promoting critical thinking, and improving self-confidence and communication skills; the teacher observes and listens, gives students more thinking time and accepts students' ideas instead of judging them.

Clough and Clark (1994) maintain that teachers are the critical component in this process through their effective articulation and communication to the students. The importance of pre-laboratory preparation is crucial considering that what students already know determines what they will learn. The laboratory can be used to serve important links between the macroscopic observations and the microscopic representations and any techniques that facilitate these links are valuable. The teacher plays a key role in bridging the chasms between the three levels of representations because the variety of methods, technology and style of laboratory work all contain inherent problems that can result in student misunderstandings. Research has shown that students' attitude towards chemistry is enhanced when the laboratory activities are related to the theory being studied and when the rules of behaviour expected in the laboratory are clearly outlined (Wong & Fraser, 1996).

Unfortunately, laboratory work is not always done in a scientifically correct or appropriate manner and this is exacerbated when students are more interested in obtaining predetermined or expected results than in understanding the significance of the results. Gallet (1998) refers to "student osmosis" (p. 72) in report writing, and Ritchie and Rigano (1996) to "fudging the results" (p. 13) when describing the scientific practices of the students. This is indeed disappointing, but is a result of using recipe-driven practicals designed to achieve near perfect results and rewarding these results (Ritchie & Rigano, 1996). This state of affairs can be rectified by teachers promoting intellectual honesty by practising authentic scientific processes in the laboratory. There is greater value in doing open-ended experiments, which do not have pre-determined or expected results but do require students to explain their results in a scientific manner. Improvements have been seen in students' skills of identifying variables, hypothesising, planning, carrying out experiments and interpreting data through these methods (Garnett, Garnett, & Hackling, 1995a).

In a response to cost and safety, many schools have adopted a micro-scale approach to chemistry. Plastic, simple, often improvised equipment is claimed to achieve the learning objectives more effectively or as well as large-scale glassware and metalware (Beasley, 1993; Ritchie & Rigano, 1996). Similarly, Simms (1994) reports that micro-scale chemistry promotes independent work and Cooper, Conway and Guseman (1995) observed an increase in dialogue between students, an improved attitude of the students towards chemistry and an improved score in academic tests (Comprehensive Test of Basic Skills,

American Chemical Society, and Advanced Placement Chemistry) when compared to previous year results.

Execution, Recording and Assessment of Laboratory Activities

The working space in the brain is limited and Johnstone (1997) is critical of the overload that instruction in laboratory manuals can demand, forcing students to adopt a recipe-like procedure. The technical, unfamiliar language often used in laboratory manuals puts additional demands on students' short term memory reserves (Gabel, 1998). Robinson (1998) believes that visual images in laboratory manuals can assist students' understanding. This view is supported by Dechsri, Jones and Heikkinen (1997) who reported improved student performance in the cognitive, affective and manipulative domains of the laboratory as a result of including pictures and diagrams with the text. Pre- and post-laboratory discussions can be used to identify any alternative conceptions (Nakhleh & Krajcik, 1994).

Laboratory activities should develop students' skills in experimental techniques such as observing, classifying, using laboratory equipment, as well as applying conceptual knowledge, developing procedural knowledge and applying inquiry tactics such as identifying variables and interpreting data (Garnett et al., 1995a). Various strategies can be used to record laboratory activities. A typical practical report should include sections on preliminary trials, planning, performing, interpreting, communicating and feedback. Reports can be presented in a variety of formats such as posters, web pages (Rigeman, 1998), power point presentations, or concept maps (Markow & Lonning, 1998). Writing and talking about chemistry improves the understanding of the concepts (Kozma & Russell, 1997) and using a critical peer review system can improve both the writing style and analysis of practical reports (Newell, 1998). The video and animation powers of the computer and the use of chemical models can be utilised in chemistry tests so that the test items include the three levels of chemical representation (Bowen, 1998). A holistic assessment of laboratory investigations is recommended by Garnett et al. (1995a), with caution to over-valuing specific scientific skills at the cost of assessing the students' understanding of the whole investigation.

Relevance of Laboratory Activities to Real Life

Motivation and interest can be achieved by giving a real-life perspective to laboratory work such as simulating a forensic chemistry problem (Long, 1995), having a mock trial using role playing (Kimbrough, Dyckes & Mlady, 1995) or managing the chemistry of a swimming pool (Bieron, McCarthy & Kermis, 1996). Experiments exposing clear scientific method using everyday items such as baking soda, vinegar, shampoo and sugar in practical assessment tasks have been used to "reinforce the connection between science and the students out of school experience" (Doran, Chan & Tamir, 1998, p. 131). Experiments to investigate environmental issues such as the presence of chloro-fluorides in household chemicals, in a chemistry program in Germany, helped students to "learn something that is useful to manage everyday situations" (Klemmer, Hutter-Klemmer & Howard, 1996, p. 59). These examples support the significance of curricular alignment with relevance to students' lives inside and outside school as discussed as one of Brophy's principles of effective teaching.

The Use of Technology in the Laboratory

New technology allows for reliable and accurate data to be accessible in the laboratory. Nakhleh and Krajcik (1994) compared the use of different technologies namely chemical indicators, pH meters and micro-computer-based laboratories (MBL) to perform titrations. Overall, students performing titrations using the MBL method made greater improvement in their conceptual understanding and appeared to be more engaged in the experiment compared to other methods. The use of data-loggers or microcomputers is ideal for those experiments that are too quick or too slow to record easily. The computer's interactivity, visual imaging and the possibility of using the visual display as an additional memory source expand its application to laboratory sessions so that a continual reference is available (Dechsri, Jones & Heikkinen, 1997).

Interactive software allows the student to perform virtual experiments along with providing macroscopic, microscopic and symbolic representations. For example, the *Electrochemical Cells Workbench* reported by Greenbowe (1994) allows students to manipulate the experimental apparatus, chemicals, and instruments in order to design and build an experiment. Greenbowe (1994) claims that the "interactive multimedia program becomes a problem-solving tool, a conceptualizer, and a tutorial for students" (p. 557).

Co-operative Learning

For students to construct meaningful knowledge networks, teaching needs to provide opportunities for being engaged in motivating, interactive activities. Ensuring that dialogue between students focuses on their understanding of chemical concepts can be beneficial in the construction of knowledge networks (Nakhleh, Lowrey & Mitchell, 1996) and improvement of their attitude towards chemistry (Gabel, 1998; Garnett et al., 1995b; Huddle & Pillay, 1996).

Small group learning, which can be used for a problem-solving activity, laboratory work, concept mapping or research task, requires positive interdependence of the members of the group in order to develop teamwork skills. Approaches such as the jigsaw structure involves the formation of a transient group made up of one person from each base group. People in the transient group then become experts at solving one problem and return to their base group to explain to their group-mates how to solve this problem (Towns, Kreke & Fields, 2000). Quality discussion can challenge alternative conceptions and provide a non-threatening forum for the learners to express their own ideas and gain feedback from others, instead of from the teacher in the form of marks or comments (Myers, Lim, Maschak & Stahl, 1996). Strategies to promote purposeful talk include providing a positive classroom climate, ensuring that students have sufficient time for discussion, and developing structured tasks that promote student input. Process workshops in which students work in teams doing active learning tasks such as critical thinking, problem solving, guided discovery and reflection have been shown to improve attitude, interest and results among students (Hanson & Wolfskill, 2000). Bowen's (2000) review of numerous studies into cooperative learning effects on both high school students' and college students' chemistry achievements concluded that "the medium students' performance in a cooperative learning environment is 14 percentile points higher than in a traditional course" (p. 118). The collaboration generated from cooperative learning illustrates the value of thoughtful discourse that can improve understanding, confidence and develop teamwork skills.

The Uses Of Varied Resources

The textbook is still an essential resource for chemistry teaching but increasingly is being augmented by computer software. A user-friendly textbook should provide viable explanations that can supplement classroom work taking into consideration students' reading level, the appropriate depth of information required and including photographs, diagrams, graphs, problems, worked solutions to problems and summaries.

With increasing levels of computer literacy among students, the growing range of computer software targeting specific needs in chemistry can serve as useful aids when integrated into chemistry teaching programs. However simply having access to computer software and multimedia is not enough to affect students learning and without teacher facilitation the resources may not be used to their full potential. Computer programs and videos that provide simulations, animations and pictures of experiments help students to explore new concepts being introduced. Examples of using these media are playing sections of a video with the sound off, asking students to discuss and create their own dialogue, having students make a list of the analogies used in the video and making predictions of the outcomes of virtual laboratory sessions (Rodrigues & Corrigan, 1998). Chemistry CD-ROMs, for example the *Periodic Table CD*, can provide a variety of information such as chemical data, pictures, animation, video movies, laboratory techniques and molecular models but they can be very costly (Banks & Holmes, 1995). Importantly, these multimedia tools often provide representations of chemistry at the macroscopic, microscopic and symbolic levels. However, CD-ROMs provide a rich source of visual aids but are a supplement, not a substitute for, laboratories (Brooks & Brooks, 1996). The use and development of multimedia software which contain text, sound, images, animations and interactivity is growing with claims that the new technology will empower the student and change the emphasis from teacher-centred to more student-centred learning where the teacher is not the primary source of information (Tan & Tan, 1997).

The use of networked multimedia chemistry computer programs at the college and university level is now common (Smith & Stovall, 1996). Inevitably, some of these methods will filter into high school chemistry teaching where the integration of molecular modeling computer software has been shown to improve high school students' understanding of bonding and three dimensional structures (Barnea & Dori, 1999). The accessibility of some molecular modeling programs through the Internet introduces new possibilities. Although these new approaches are still in their infancy, they are not seen as alternative teaching methods but rather adjuncts to foster more effective instruction. The motivational and fun aspects of the technology enhance learning (Powers, 1998) and the ability to customise the software to meet the learner's individual learning style is a valuable asset; however, caution must be heeded that the educational tool adds nothing more than "bells and whistles" (Powers, 1998, p. 318).

The importance of developing students' skills in locating information also is essential. The rapid evolution of on-line information, computer programs and databases in addition to traditional references has greatly extended the accessibility of students to information and provides more current information. While this availability empowers the student, practice is needed to refine search strategies so that time is not wasted along with augmentation of the students' computer skills (Penhale & Stratton, 1994).

Diagnostic Assessment Techniques

Research suggests that current assessment procedures distort and narrow instruction, misrepresent the nature of the subject, and underscore inequities in access to education. Alternative forms of assessment, such as performance tasks and portfolios are needed that might permit the assessment of thinking rather than the possession of information (Wolf, Bixby, Glenn & Gardner, 1991). However, the use of multiple-choice items to examine students' understanding in specified and limited content areas of science has been ignored largely because of the negative association with behaviourism. Indeed, such a concern is justified when authors do not specify the origin of the distracters and when items do not investigate conceptual understanding.

Treagust (1995) described an approach using two tiers of multiple-choice distracters per item to diagnose students' conceptual understanding of specified content areas in science. In these items, the first tier of multiple-choice distracters involves a content response while the second tier involves a reasoning response. The source of the items is specified clearly; use of concept maps and propositional knowledge statements needed to understand the concepts document the conceptual area. The items are developed based on the known literature and responses from students to free-response items and interviews. An essential aspect of this method is the necessity to field-test items so that, as far as possible, they are representative of the range of students' responses to the concept being investigated. Using this method, diagnostic test items have been developed in covalent bonding and structure (Peterson, Treagust & Garnett, 1989), bonding (Tan & Treagust, 1999) and chemical equilibrium (Tyson, Treagust & Bucat, 1999). These diagnostic tests provide an opportunity for teachers to diagnose student learning without the need for interviews.

CONCLUSION

According to research on learning and instruction discussed in this chapter, the generic guidelines to good teaching described by Brophy are generally supported within the context of the chemistry classroom. This chapter began by attending to specific issues about the unique nature of chemistry as a discipline, particularly what may be termed theoretical issues. The three issues initially examined were the need to identify what students already know about a topic and build on that knowledge – this is the notion of constructivism; the need for instruction to be attentive to the three levels of representation involving understanding of phenomena at the macroscopic, microscopic and symbolic levels; and the need for generating appropriate and effective mental models for interpreting abstract chemical concepts. Without these essential instructional considerations, it is our contention, and that of others whom we cite, that good teaching in the chemistry classroom is not likely to eventuate.

The chapter subsequently described research findings on different instructional approaches that have been shown to enhance student learning outcomes. Significant and proven effective teaching approaches and strategies applicable to chemistry include the use of concept maps, cooperative learning, problem-solving, and laboratory work. Research also has shown that instruction can be enhanced by the integration of a wide variety of print-based and technologically-based teaching resources. In addition, specially designed diagnostic tests as an integral part of instruction can be used to assess students' meaningful understanding and identify their misunderstandings. In describing chemistry education research using each of these strategies or approaches, we also have shown how they are examples of particular generic features of good teaching identified by Brophy.

REFERENCES

Andersson, B. (1990). Pupils' conceptions of matter and its transformations (age 12–16). *Studies in Science Education, 18*, 53–85.

Astudillo, L. R., & Niaz, M. (1996). Reasoning strategies used by students to solve stoichiometry problems and its relationship to alternative conceptions, prior knowledge and cognitive variables. *Journal of Science Education and Technology, 5*(2), 131–139.

Ausubel, D. P. (1968). *Educational psychology: A cognitive view.* New York: Holt, Rinehart and Winston.

Balaban, A. T. (1999). Visual chemistry: Three-dimensional perception of chemical structures. *Journal of Science Education and Technology, 8*(4), 251–255.

Banks, A. J., & Holmes, J. L. (1995). The periodic table CD. *Journal of Chemical Education, 72*(5), 409.

Barnea, N., & Dori, Y. J. (1999). High-school chemistry students' performance and gender differences in a computerised molecular modeling learning environment. *Journal of Science Education and Technology, 8*(4), 257–271.

Beasley, W. F. (1993). Down-scaling chemistry: The quiet revolution. *Australian Science Teachers Journal, 39*(3), 50–52.

Beckwith, E. K., & Nelson, C. (1998). The chemviz project. *Learning and Leading With Technology, 25*(6), 17–19.

Bieron, J. F., McCarthy, P. J., & Kermis, T. W. (1996). A new approach to the general chemistry laboratory. *Journal of Chemical Education, 73*(11), 1021–1023.

Bodner, G. M. (1986). Constructivism: A theory of knowledge. *Journal of Chemical Education, 63*, 873–878.

Bodner, G. M., & Domin, D. S. (2000). Mental models: The role of representations in problem solving in chemistry. *University Chemistry Education, 4*, 22–28.

Boo, H. K. (1998). Students' understandings of chemical bonds and the energetics of chemical reactions. *Journal of Research in Science Teaching*, *35*(5), 569–581.

BouJaoude, S., & Barakat, H. (2000). Secondary school students' difficulties with stoichiometry. *School Science Review*, *81*(296), 91–98.

Bowen, C. W. (1998). Item design considerations for computer-based testing of student learning in chemistry. *Journal of Chemical Education*, *75*(9), 1172–1175.

Bowen, C. W. (2000). A quantitative literature review of cooperative learning effects on high school and college chemistry achievement. *Journal of Chemical Education*, *77*(1), 116–119.

Brooks, H. B., & Brooks, D. W. (1996). The emerging role of CD-ROMs in teaching chemistry. *Journal of Science Education and Technology*, *5*(3), 203–215.

Brophy, J. (2001). Introduction. In: J. Brophy (Ed.), *Subject-specific instructional methods and activities*. (Vol. 8, pp. 1–23). Oxford, UK: Elsevier Science Ltd.

Campbell, B., Laxonby, J., Millar, R., Nicolson, P., Ramsden, J., & Waddington, D. (1994). Science: The Salters' approach – a case study of the process of large scale curriculum development. *Science Education*, *78*(5), 15–47.

Clough, M. P., & Clark, R. (1994). Cookbooks and constructivism. *Science-Teacher*, *61*(2), 34–37.

Collins, A., & Gentner, D. (1987). How people construct mental models. In: D. Q. Holland & N. Quinn (Eds), *Cultural models in language and thought* (Vol. 1, pp. 243–265). New York: University of Cambridge.

Cooper, S., Conway, K., & Guseman, P. (1995). Making the most of microscale. *The Science Teacher*, *62*(1), 46–49.

Copolo, C. F., & Hounshell, P. B. (1995). Using three dimensional models to teach molecular structures in high school chemistry. *Journal of Science Education and Technology*, *4*(4), 295–305.

Cosgrove, M., & Schaverien, L. (1997). Models of science education. In: J. Gilbert (Ed.), *Exploring models and modelling in science and technology education* (Vol. 1, pp. 20–34). Bulmershe, Reading, U.K.: The University of Reading,

Cullen, J. (1990). Using concept maps in chemistry: An alternative view. *Journal of Research in Science Teaching*, *27*(10), 1067–1068.

Dechsri, P., Jones, L. L., & Heikkinen, H. W. (1997). Effect of a laboratory manual design incorporating visual information-processing aids on student learning and attitudes. *Journal of Research in Science Teaching*, *34*(9), 891–904.

Dierks, W. (1989). An approach to the educational problem of introducing the discontinuum concept in secondary chemistry teaching and an attempted solution. In: P. L. Linjse, P. Licht, W. deVos & A. J. Waarlo (Eds), *Relating macroscopic phenomena to microscopic particles* (pp. 177–82). Proceedings of a seminar at the Centre for Science and Mathematics Education, University of Utrecht.

Doran, P. R., Chan, F., & Tamir, P. (1998). *Science educators guide to assessment*. Arlington: National Science Teachers Association.

Dori, Y. J., & Hameiri, M. (1998). The "mole environment" studyware: Applying multi-dimensional analysis to quantitative chemistry problems. *International Journal of Science Education*, *20*(3), 317–333.

Duit, R. (1991). On the role of analogies and metaphors in learning science. *Science Education*, *75*(6), 649–672.

Ealy, J. B. (1999). A student evaluation of molecular modeling in first year college chemistry. *Journal of Science Education and Technology*, *8*(4), 309–321.

Fisher, K. M. (1991). Semnet: A tool for personal knowledge construction. In: P. A. M. Kommers, D. H. Jonassen & J. T. Mayes (Eds), *Cognitive Tools for Learning*. Berlin: Springer-Verlag.

Fensham, P. J. (1994). Beginning to teach chemistry. In: P. J. Fensham, R. F. Gunstone & R. T. White (Eds), *The Content of Science: A Constructivist Approach to its Teaching and Learning*. London: The Falmer Press.

Gabel, D. (1998). The complexity of chemistry and implications for teaching. In: B. J. Fraser & K. G. Tobin (Eds), *International Handbook of Science Education* (Vol. 1, pp. 233–248). Great Britain: Kluwer Academic Publishers.

Gabel, D., Briner, D., & Haines, D. (1992). Modelling with magnets. *The Science Teacher, 59*(3), 58–63.

Gabel, D. L. & Bunce, D. M. (1994). Research on problem solving: Chemistry. In: D. L. Gabel (Ed.), *Handbook of Research on Science Teaching and Learning* (pp. 301–326). New York, NY: McMillan Publishing Co.

Gabel, D., & Sherwood, R. (1980). The effect of student manipulation of molecular models on chemistry achievement according to Piagetian level. *Journal of Research in Science Teaching, 17*(1), 75–81.

Gallet, C. (1998). Problem-solving teaching in the chemistry laboratory: Leaving the cooks . . . *Journal of Chemical Education, 75*(1), 72–77.

Garnett, P. J., Garnett, P. J., & Hackling, M. W. (1995a). Refocusing the chemistry lab: A case for laboratory-based investigations. *Australian Science Teachers Journal, 41*(2), 26–32.

Garnett, P. J., Garnett, P. J., & Hackling, M. W. (1995b). Students' alternative conceptions in chemistry: A review of research and implications for teaching and learning. *Studies in Science Education, 25*, 69–95.

Garnett, P. J., & Treagust, D. F. (1992). Conceptual difficulties experienced by senior high school students of electrochemistry: Electric circuits and oxidation-reduction equations. *Journal of Research in Science Teaching, 29*, 121–142.

Gergen K. J. (1995). Social construction and the educational process. In: L. P. Steffe & J. Gale (Eds), *Constructivism in education* (pp. 17–39). Hillsdale, New Jersey: Lawrence Erlbaum Associates.

Gilbert, J. K., & Boulter, C. J. (1998). Learning science through models and modelling. In: B. J. Fraser & K. J. Tobin (Ed.), *International handbook of science education* (pp. 53–66). Dordrecht, The Netherlands: Kluwer Academic Publishers.

Gilbert, J. K., Boulter, C. J., & Elmer, R. (2000). Positioning models in science education and in design and technology education. In: J. K. Gilbert & C. J. Boulter (Eds), *Developing models in science education* (pp. 3–18). Dordrecht: Kluwer.

Glynn, S. M. (1994). *Teaching science with analogies: A strategy for teachers and textbook authors*. Athens, GA: National Reading Research Center.

Greenbowe, T. (1994). An interactive multimedia software program for exploring electrochemical cells. *Journal of Chemical Education, 71*(7), 555–557.

Grosslight, L., Unger, C., Jay, E., & Smith, C. (1991). Understanding models and their use in science: Conceptions of middle and high school students and experts. *Journal of Research in Science Teaching, 28*(9), 799–822.

Hanson, D., & Wolfskill, T. (2000). Process workshop – a new model for instruction. *Journal of Chemical Education, 77*(1), 120–130.

Hardwicke, A. J. (1995). Using molecular models to teach chemistry: Part 1 using models. *School Science Review, 77*(278), 59–64.

Harrison, A. G., & Treagust, D. F. (1996). Secondary students' mental models of atoms and molecules: Implications for teaching chemistry. *Science Education*, *80*(5), 509–534.

Herron, J. D., & Greenbowe, T. J. (1986). What can we do about Sue: A case study of competence. *Journal of Chemical Education*, *63*(6), 528–531.

Herron, J. D., & Nurrenbern, S. C. (1999). Chemical education research: Improving chemistry learning. *Journal of Chemical Education*, *76*(10), 1354–1361.

Huddle P. A., & Pillay, A. E. (1996). An in-depth study of misconceptions in stoichiometry and chemical equilibrium at a South African university. *Journal of Research in Science Teaching*, *33*(1), 65–77.

Hunter, R. J., Simpson P. G., & Stranks, D. R. (1976). *Chemical science*. Sydney: Science Press.

Ingham, A. I., & Gilbert, J. K. (1991). The use of analogue models by students of chemistry at higher education level. *International Journal of Science Education*, *13*(2), 203–215.

Inspiration Software Inc [online accessed 12/8/2000] URL:http:www.inspiration.com

Johnstone, A. H. (1993.) The development of chemistry teaching: A changing response to changing demand. *Journal of Chemical Education*, *70*(9), 701–705.

Johnstone, A. H. (1997). Chemistry teaching – science or alchemy. *Journal of Chemical Education*, *74*(3), 262–268.

Justi, R., & Gilbert, J. K. (1999). History and philosophy of science through models: The case of chemical kinetics. *Science and Education*, *8*, 287–307.

Kimbrough, D. R., Dyckes, D. F., & Mlady, G. W. (1995). Teaching science and public policy through role playing. *Journal of Chemical Education*, *72*(4), 295–296.

Klemmer, G., Hutter-Klemmer, L., & Howard, E. (1996). Chemistry education and environmental awareness. *School Science Review*, *77*(280), 55–61.

Kozma, R. B., & Russell, J. (1997). Multimedia and understanding: Expert and novice responses to different representations of chemical phenomena. *Journal of Research in Science Teaching*, *34*(9), 949–968.

Krnel, D., Watson, R., & Glazar, S. A. (1998). Survey of research related to the development of the concept of 'matter'. *International Journal of Science Education*, *20*(3), 257–289.

Liew, C. W., & Treagust, D. F. (1995). A predict-observe-explain teaching sequence for learning about students' understanding of heat and expansion of liquids. *Australian Science Teachers Journal*, *41*(1), 68–71.

Long, G. A. (1995). Simulation of a forensic chemistry problem: A multidisciplinary project for secondary school chemistry students. *Journal of Chemical Education*, *72*(9), 803–804.

Lunetta, V. N. (1998). The school science laboratory. In: B. J. Fraser & K. G. Tobin (Eds), *International handbook of science education* (pp. 81–96). Dordrecht, The Netherlands: Kluwer.

Markow, P. G., & Lonning, R. A. (1998). Usefulness of concept maps in college chemistry laboratories: Students' perceptions and effects on achievement. *Journal of Research in Science Teaching*, *35*(9), 1015–1029.

Mason, D. S., Shell, D. F., & Crawley, F. E. (1997). Differences in problem-solving by non-science majors in introductory chemistry on paired algorithmic-conceptual problems. *Journal of Research in Science Teaching*, *34*(9), 905–923.

Metz, K. E. (1998). Scientific inquiry within reach of young children. In: B. J. Fraser & K. G. Tobin (Eds), *International handbook of science education* (pp. 81–96). Dordrecht, The Netherlands: Kluwer.

Myers, J. J., Lim, W., Maschak, J., & Stahl, R. J. (1996). Promoting purposeful talk in the cooperative science classroom. In: R. J. Stahl (Ed.), *Cooperative learning in science* (Vol. 1, pp. 433). New York: Addison Wesley Publishing Company.

Nakhleh, M. B. (1994). Chemical education research in the laboratory environment. How can research uncover what students are learning? *Journal of Chemical Education*, *71*(3), 201–205.

Nakhleh, M. B., & Krajcik, J. S. (1994). Influence on levels of information as presented by different technologies on students' understanding of acid, base, and pH concepts. *Journal of Research in Science Teaching*, *31*(10), 1077–1096.

Nakhleh, M. B., Lowrey, K. A., & Mitchell, R. C. (1996). Narrowing the gap between concepts and algorithms in freshman chemistry. *Journal of Chemical Education*, *73*(8), 758–762.

Newell, J., A. (1998). Using peer review in the undergraduate laboratory. *Chemical Engineering Education*, *32*(3), 194–196.

Niaz, M., & Robinson, W. R. (1992). Manipulation of logical structure of chemistry problems and its effect on student peformance. *Journal of Research in Science Teaching*, *29*(3), 211–226.

Noh, T., & Sharmann, L. C. (1997). Instructional Influence of a molecular-level: Pictorial presentation of matter on students' conceptions and problem-solving ability. *Journal of Research in Science Teaching*, *34*(2), 199–217.

Norman, D. A. (1983). Some observations on mental models. In: D. Gentner & A. L. Stevens (Eds), *Mental models* (pp. 7–14). Hillsdale, NJ: Erlbaum.

Novak, J. D. (1990). Concept mapping: A useful tool for science education. *Journal of Research in Science Teaching*, *27*(10), 937–949.

Novak, J. D. (1991). Clarify with concept maps. *The Science Teacher*, *58*(7), 44–49.

Palmer, W. P., & Treagust, D. F. (1996). Physical and chemical change in textbooks: An initial view. *Research in Science Education*, *26*, 129–140.

Penhale, S. J., & Stratton, W. (1994). Online searching assignments in a chemistry course for nonscience majors. *Journal of Chemical Education*, *71*(3), 227–229.

Peterson, R. F., Treagust, D. F., & Garnett, P. J. (1989). Development and application of a diagnostic instrument to evaluate grade 11 and 12 students' concepts of covalent bonding and structure following a course of instruction. *Journal of Research in Science Teaching*, *26*, 301–314.

Plumsky, R. (1996). Transmuted labs. *Journal of Chemical Education*, *73*(5), 451–454.

Powers, P. (1998). One path to using multimedia in chemistry courses. *Journal of College Science Teaching*, *27*(5), 317–318.

Raghavan, K., & Glaser, R. (1995). Model-based analysis and reasoning in science: The MARS curriculum. *Science Education*, *79*(1), 37–61.

Ramsden, J. (1994). Context and activity based science in action. *School Science Review*, *75*(272), 7–14.

Regis, A., & Giorgio, A. P. (1996). Concept maps in chemistry education. *Journal of Chemical Education*, *73*(11), 1084–1088.

Renstrom, L., Andersson, B., & Marton, F. (1990). Students' conceptions of matter. *Journal of Educational Psychology*, *82*(3), 555–569.

Rigeman, S. (1998). The convergent evolution of a chemistry project: Using laboratory posters as a platform for web page construction. *Journal of Chemical Education*, *75*(6), 727–730.

Ritchie, S. M., & Rigano, D. L. (1996). School labs as sites for fraudulent practice: How can students be dissuaded from fudging? *Australian Science Teachers Journal*, *42*(2), 13–16.

Roberts, J. L., Selco, J. I., & Wacks D. B. (1996). Mother earth chemistry: A laboratory course for non-majors. *Journal of Chemical Education*, *73*(8), 779–781.

Robinson, W. R. (1998). A view of the science education research literature: Visual aids in laboratory manuals improve comprehension. *Journal of Chemical Education*, *75*(3), 282–283.

Rodrigues, S., & Corrigan, D. (1998). *Including IT: Managing information technology in the science classroom.* Annandale NSW: User Friendly Resource Enterprises Ltd.

Ruiz-Primo, M. A., & Shavelson, R. J. (1996). Problems and issues in the use of concept maps in science assessment. *Journal of Research in Science Teaching*, *33*(6), 569–600.

Schmidt, H. J. (1997). Students' misconceptions – looking for a pattern. *Science Education*, *81*(2), 123–135.

Selinger, B. (1998). *Chemistry in the market place*. Sydney: Harcourt Brace & Company,

Selley, N. J. (1981). The place of alternative models in school science. *School Science Review*, *63*, 252–259.

Shulman, L. S. (1986). Those who understand : Knowledge growth in teaching. *Educational Researcher*, *15*(2), 4–14.

Shusterman, G. P., & Shusterman, A. J. (1997). Teaching chemistry with electron density models. *Journal of Chemical Education*, *74*(7), 771–776.

Simms, J. (1994). Maximize with microscale. *The Science Teacher*, *61*(2), 30–33.

Smith, S., & Stovall, I. (1996). Networked instructional chemistry: using technology to teach chemistry. *Journal of Chemical Education*, *73*(10), 911–915.

Spain, J. D. (1996). Electronic homework: Computer-interactive problem sets for general chemistry. *Journal of Chemical Education*, *73*(3), 222–225.

Staver, J. R., & Lumpe, A. T. (1993). A content analysis of the presentation of the mole concept in chemistry textbooks. *Journal of Research in Science Teaching*, *30*(4), 321–337.

Staver, J. R., & Lumpe, A. T. (1995). Two investigations of students' understanding of the mole concept and its use in problem solving. *Journal of Research in Science Teaching*, *32*(2), 177–193.

Stephens, S., McRobbie, C. J., & Lucas, K. B. (1999). Model-based reasoning in a year 10 classroom. *Research in Science Education*, *29*(2), 189–208.

Taber, K. S. (1996). Chlorine is an oxide, heat causes molecules to melt, and sodium reacts badly in chlorine: a survey of the background knowledge of one A-level chemistry class. *School Science Review*, *78*(282), 39–48.

Taber, K. S. (1997). Student understanding of ionic bonding: molecular versus electrostatic framework? *School Science Review*, *78*(285), 85–95.

Taber, K. S. (1998). An alternative conceptual framework from the chemistry education. *International Journal of Science Education*, *20*(5), 597–608.

Tan, D. K.-W., & Treagust, D. F. (1999). Evaluating students' understanding of chemical bonding. *School Science Review*, *81*, 75–83.

Tan, S. T., & Tan, L. K. (1997). CHEMMAT: Adaptive multimedia courseware for chemistry. *Journal of Science Education and Technology*, *6*(1), 71–79.

Towns, M. H., Kreke, K., & Fields, A. (2000). An action research project: Student perspectives on small-group learning in chemistry. *Journal of Chemical Education*, *77*(1), 111–115.

Treagust, D. F. (1995). Diagnostic assessment of students' science knowledge. In: S. M. Glynn & R. Duit (Eds), *Learning science in the schools: research reforming practice* (Vol. 1, p. 379). Mahwah, New Jersey: Lawrence Erlbaum.

Treagust, D. F., Chittleborough, G., & Mamalia, L. T. (2001, April) *Learning introductory organic chemistry: Secondary students' understanding of the role of models and the development of scientific ideas*. Paper presented at the annual meeting of the American Educational Research Association, Seattle, WA.

Treagust, D. F., & Harrison, A. G. (1999). The genesis of effective scientific explanations for the classroom. In: J. Loughran (Ed.), *Researching teaching: methodologies and practices for understanding pedagogy.* London: Falmer Press

Treagust, D. F., Harrison, A. G., & Venville, G. J. (1998). Teaching science effectively with analogies: An approach for preservice and inservice teacher education. *Journal of Science Teacher Education*, *9*(2), 85–101.

Tyson, L., Treagust, D. F. & Bucat, R. B. (1999). The complexity of teaching and learning chemical equilibrium. *Journal of Chemical Education*, *76*(4), 554–558.

Van Driel, J. H., & Verloop, N. (1999). Teachers' knowledge of models and modelling in science. *International Journal of Science Education*, *21*(11), 1141–1153.

Williamson, V. M., & Abraham, M. R. (1995). The effects of computer animation on the particulate mental models of college chemistry students. *Journal of Research in Science Teaching*, *32*(5), 521–534.

Wilson, J. M. (1994). Network representations of knowledge about chemical equilibrium: variations and achievement. *Journal of Research in Science Teaching*, *31*(10), 1133–1147.

Wilson, J. M. (1996). Concept maps about chemical equilibrium and students' achievement scores. *Research in Science Education*, *26*(2), 169–185.

Wilson, J. M. (1998). Differences in knowledge networks about acids and bases of year-12, undergraduate and postgraduate chemistry students. *Research in Science Education*, *28*(4), 429–446.

Wolf, D., Bixby, J., Glenn III, J., & Gardner, H. (1991). To use their minds well: Investigating new forms of student assessment. In: G. Grant (Ed.), *Review of research in education* (pp. 31–74). Washington, DC: American Educational Research Association.

Wong, A. F., & Fraser, B. J. (1996). Environment-attitude associations in the chemistry laboratory classroom. *Research in Science and Technological Education*, *14*(1), 91–102.

Yager, R. E. (1991). The constructivist learning model: towards real reform in science education. *The Science Teacher*, *58*(6), 52–57.

EARTH SCIENCE

Richard A. Duschl and Michael J. Smith

INTRODUCTION

The second half of the 20th century has been a time of rapid changes in the character of the earth sciences. The transition is one moving away from a focus on the surface geology of mapping and mining and toward a focus on global change and earth systems (Mayer & Armstrong, 1990). A science once dominated by historical types of explanations revealing the 'story in the rocks' is slowly evolving into a causal modeling science for making predictions (e.g. climate changes, earthquakes, volcanic eruptions, flooding, hurricanes) and understanding how human activity effects global change. Thus, not surprisingly the information or subject matter and the cognitive and the material tools needed to engage in earth science inquiry have shifted as well.

John McPhee elegantly captures the character of this change in his book *Rising From the Plain*. The story is about the geology of Wyoming told through the experiences of a high country, Rocky Mountain regional field geologist – Dave Love. For Love, geology was a kind of story telling. Through experiences of touching different geologic structures, you piece together the implied tectonics; i.e. the story in the rocks. He, like many other geologists relate to the Hindu fable of blind men and the elephant (each individual feeling a different part of the elephant comes to different opinions of what it is). But field geologists like Dave Love are a diminishing breed.

> In recent years, the number of ways to feel the elephant has importantly increased. While science has assimilated such instruments as the scanning transmission electron microscope, the inductively coupled plasma spectrophotometer, and the $^{39}Ar/^{40}Ar$ laser microbe . . . the percentage of geologists has steadily diminished who go out in the summer and deal with rock, and the number of people has commensurately risen who work the year around in fluorescent light with their noses on printouts (McPhee, 1986, p 146).

Subject-Specific Instructional Methods and Activities, Volume 8, pages 269–290.
2001 by Elsevier Science Ltd.
ISBN: 0-7623-0615-7

Feeding facts and fragments of the earth into laboratory machines is referred to as "black-box geology" carried out by "analog geologists" (McPhee, 1986, p. 146). At the beginning of the 1900s and well into the 20th century geology and the other earth sciences were essentially atheoretical disciplines. The theories and paradigms that influenced geological inquiry came from the physical sciences – i.e. physics and chemistry. In the words of Thomas Kuhn (1970), geology and the other earth sciences (e.g. oceanography, climatology, planetary geology, meteorology) were immature sciences since they lacked an organizing paradigm.

The rapid changes in technology during the 20th century have significantly impacted all the sciences. But the impact on the earth and geological sciences were nothing short of revolutionary. For during this past century the geological sciences made the transition from an immature science to a mature paradigm driven science. The complete story of the gathering and compilation of the various forms of evidence and reasoning that led to this and other revolutions in the earth sciences is clearly beyond the scope and intent of this chapter. What is within the purview for this chapter though is a more general discussion of how observational techniques, tools and guiding conceptions (e.g. theoretical frameworks) have evolved and along the way shifted what comes to count as doing earth science research and inquiry. In brief, descriptive and reductionist inquiry approaches to the earth sciences have given way to model-based inquiry approaches. The focus on local mapping and mining, and human-observational techniques have yielded to global mapping and modelling, and instrument-observational techniques. New tools and new techniques have literally made it possible for students of the earth sciences to have direct access to the raw data and models being used to study planet earth.

In this chapter, we set out to describe the changes in earth science inquiry and the implications these changes have for K-12 science instruction. As a way of introduction, we first turn to two brief descriptions of information explosions that changed the earth sciences. The approach will be one of describing how technological advances forged and developed new observational tools (e.g. satellites) and theoretical frameworks (e.g. plate tectonics) that have moved us further and further away from sense-perception data sources to theory-driven data sources. Next, we review the arguments and positions for the earth systems approach to teaching Earth Science. A section that develops the position that earth science contexts are well suited for problem-based learning frameworks and project-based inquiry frameworks follows this. The final section discusses how both the structure of knowledge in the earth sciences and the epistemological goals of the earth sciences: (1) shape generic pedagogical practices, and (2) inform curriculum formats and goals.

THE TALE OF TWO INFORMATION EXPLOSIONS

New material sciences, new modes of transportation, and communication innovations have enabled earth scientists to explore farther and probe deeper. Where once geologists were restricted to the data presented at the surface of the Earth in mountain exposures, road cut exposures and mining exposures, technological developments have literally opened up the Earth to inspection. Technologies in oil exploration, for example, made possible deep drilling techniques. These techniques enabled the discovery of the Mohorovicic discontinuity – the Earth's interior boundary between lighter/less dense continental rock (lithosphere) and heavier/denser oceanic rock (asthenosphere). The advancement of weather balloon technologies has contributed to our understanding of the atmosphere's structure and composition. The turning point was the year 1957, which was dubbed the International Geophysical Year or IGY. International teams of scientists set up earth observation systems around the globe but most notably on Antarctica. To get a sense of how technology has impacted our thinking about earth science education, let us turn to two events, one historical and one contemporary, which reveal how inquiry in the earth sciences has changed.

Two information explosions, both aided by computer technology and both related to military activities, have fundamentally altered how we think of planet earth. The first information explosion contributed to the development of the Theory of Plate Tectonics. The second information explosion is ongoing and contributing to theories of global warming.

The information explosion that contributed to the development of plate tectonics has two components. Part one was computer modelling the first map of ocean-floors. Naval vessels traversing the oceans during World War II would run their sonar (sound waves) to keep an eye out for enemy submarines. What was being generated, however, were profiles of the ocean-floor. At the end of the war, there were literally thousands of miles of this data. A suggestion was made to use the newly developed computers to compile this data to provide a picture/map of the Atlantic ocean-floor. Maps of the other ocean-floors followed. The compilation of this data into a single map, an enormous feat in 1950 computer science, revealed that the ocean floor wasn't flat as was believed but rather was composed of vast mountain chains called ridges and deep trenches. For the first time we have a physical model of the ocean floor.

Part two of the first information explosion involves earthquakes. During the Cold War sensitive seismographs were placed around the world to monitor underground testing of nuclear weapons. The seismographs did indeed signal when atomic tests were occurring. But these earthquake-detecting instruments also recorded and revealed the location of hitherto unknown low intensity (<5.0 on the Richter scale) earthquakes. A decade of data showed shallow focus earthquakes on the ocean-floor aligning with the mid-ocean ridge system. The data also showed that deep focus earthquakes (>70 km) correlated with the position of deep-sea trenches. A consideration of the two data sets by J. Tuzo Wilson led to his proposal that the Earth was comprised of moving plates. At the edges of these plates, interaction with other plates caused earthquakes and the formation of volcanoes. The Pacific 'Ring of Fire' was explained.

The second information explosion was the release of classified satellite imagery data. For approximately 30 years, Landsat satellites have been snapping pictures of the surface of the Earth. Again, the Cold War was a major influence of the deployment of this technology. Every few weeks a satellite would pass over the same region. Employing 5 bands of electromagnetic radiation (3 visible light bands and 2 infrared light bands) details about the surface of the earth are revealed. The amount of information is enormous and computer technologies have now made this data available for public use.

Through the efforts of NASA and NOAA administrators and scientists, this database is now available to citizens, scientists, universities and governments. Schools with online capabilities can access the data. Such databases make it possible for local, regional, and global constituencies to conduct inquiries about environmental and climatic changes.

Consider the Terra mission that monitors the vital signs of the planet (King & Herring, 2000). Terra is a complex, $1.3 billion Earth satellite launched in December 1999 that is designed to measure 16 of the 24 characteristics scientists have identified as factors that play a role in determining climate. Terra is part of the Earth Observing System (EOS), a NASA program designed to gather data for developing models of climate and climate changes. An advanced computer network (EOS Data and Information System (EOSDIS), receives and processes the information. Four centers in the U.S. archive the measurements from Terra, which are then made available to scientists and civilians alike. The data produced from Terra coupled with the archived Landsat data provide inquiry opportunities to monitor global climate and engage in land and resource management.

The 16 vital signs being monitored by Terra are aerosols, air temperature, clouds, fires, glaciers, land temperature, land use, natural disasters, ocean productivity, ocean temperature, pollution, radiation, sea ice, snow cover, vegetation, water vapor. Table 1 provides a listing of the specific Terra instruments and the vital signs each monitors (King & Herring, 2000).

Table 1. List of the Research Instruments on the Terra Satellite and Vital Signs Each Will Monitor.

Terra Instrument	Vital Signs being Monitored
ASTER – Advanced Spaceborne Thermal Emisssion and reflection Radiometer	clouds, glaciers, land temp., land use, natural disasters, sea ice, snow cover, vegetation
CERES – Clouds and the Earth's Radiant Energy System	radiation, clouds
MOPITT – Measurements Of Pollution In The Troposphere	pollution
MISR – Multi-angle Imaging SpectroRadiometer	aerosols, clouds, land use, natural disasters, vegetation
MODIS – MODerate-resolution Imaging Spectroradiometer	aerosols, air temp., clouds, fires, land temp., land use, natural disasters, ocean productivity, ocean temperature, radiation, sea ice, snow cover, vegetation, water vapor

EARTH SYSTEM SCIENCE

The two information technology stories above are but two examples of how the focus of Earth Science has shifted to a global scale. But this global perspective can be used to monitor and study regional and local processes. Examining the list of vital signs being monitored by the five Terra instruments, ones gains an appreciation for the earth systems students should be learning about in schools. The high technology and supercomputer capabilities for data processing and data modeling are forcing scientists to reconsider the relationship between science sub-disciplines and their mode of inquiry. The changes taking place in the earth sciences are reported in the "Bretherton Report" (Earth System Sciences Committee, 1988). Chaired by meteorologist Francis Bretherton and comprised of scientists from the three government funded agencies charged with monitoring and studying the earth (e.g. National Aeronautic and Space Administration; U.S. Geological Survey; and National Oceanographic and Atmospheric Administration), the committee deliberations led to reconceptualizing the processes and goals for the study of the earth. The approach is called earth system science.

> The Earth Systems Science Committee (ESSC) set forth the following goal: "To obtain a scientific understanding of the entire Earth system on a global scale by describing how its component parts and their interactions have evolved, how they function, and how they may continue to evolve on all timescales" (ESSC, 1988, p. 11).

The driving force behind the ESSC goal is the fact that scientists do not yet understand the cause-and-effect relationships among Earth's lands, oceans, and atmosphere well enough to predict what, if any, impacts these rapid changes will have on future climate conditions. Scientists need to make many measurements all over the world, over a long period of time, in order to assemble the information needed to construct accurate computer models that will enable them to forecast the causes and effects of climate change. The only feasible way to collect this information is through the use of space-based Earth "remote sensors". Consequently, NASA's Earth Observing System has begun an international study of planet Earth that is comprised of three main components: (1) a series of satellites specially designed to study the complexities of global change; (2) an advanced computer network for processing, storing, and distributing data (called EOSDIS); and (3) teams of scientists all over the world who will study the data.

Earth systems were organized on two time scales – one on millions of years and one on decades and centuries. Plate tectonics and organic evolution are studied on the time scale of millions of years. Global warming and acid rain are studied on the shorter time scale. Today, there is compelling scientific evidence that human activities have attained the magnitude of a geological force and are speeding up the rates of global changes. For example, carbon dioxide levels have risen 25% since the industrial revolution and humans have transformed about 40% of the world's land surface. Given that fact that shorter time scale changes are increasingly being influenced by the world's human population, "[a]n understanding of these short-term global changes is essential for the health of future generations of humans and of the planet as a whole" (Mayer, 1995). There is a strong implication that the K-12 earth science curriculum should be organized around the earth systems science guidelines.

In Earth system science, the planet Earth is viewed as evolving as a synergistic physical system of interrelated phenomena, processes and cycles. Given the concerns that humankind is impacting Earth's physical climate system, a broader concept of Earth as a system is emerging. Within this concept, knowledge from the traditional Earth science disciplines of geology, meteorology and oceanography along with biology is being gleaned and integrated to form a physical basis for Earth system science. The broader concept of Earth system science has also come to include societal dimensions and the recognition that humanity plays an ever-increasing role in global change. Fundamental to the Earth system science approach is the need to emphasize relevant interactions of physical, biological and dynamical processes that extend over spatial scales from microns to astronomical units (AUs), and over time scales of milliseconds to billions of years.

This concept of the Earth as a complex and dynamic entity of interrelated subsystems implies that there is no process or phenomenon within the Earth system that occurs in complete isolation from other elements of the system. While this system view is elegant and satisfying philosophically, it presents an enormous challenge to researchers and educators attempting to quantify the breadth of the system's elements, states and processes within the classroom. No individual, academic department or university is able to develop and offer the enormous depth and breadth of knowledge such a paradigm demands. Only by joining partners and experts from different disciplines can the diversity and complexity of Earth system science be fully appreciated (Johnson et al., 1997, 1998a, b).

The Earth system science concept fosters synthesis and the development of a holistic model in which disciplinary process and action lead to synergistic interdisciplinary relevance. However, the development both conceptually and physically of the Earth system model and its quantitative assessment in the classroom and laboratory is a continuing, formative processes which requires nurturing and commitment to eclectic learning beyond any individual discipline. The intersection of disciplinary specialties often provides the most fertile and interesting fields for study, and opportunity for students. The challenge for educators to develop and offer courses in the classroom that provide this deeper understanding of the Earth system is awesome. In seeking to construct an overarching interdisciplinary framework of process and state of the system, Earth system science must also retain the strength of traditional disciplines for understanding fundamentals and complex interactions (Kalb et al., 1997).

The initial views of the Earth Systems (ES) approach to earth science education emerged from a conference that brought science educators and Bretherton Report geoscientists together in Washington DC in April 1988. The core issue was to abandon "the reductionist approach of focusing on the specific contributions of certain scientific disciplines in understanding concepts and process with their defined domain" and replace it with a curriculum framework that relates the concepts and processes "to the earth system in which they operate and interact with other processes and concepts" (Mayer & Armstrong, 1990; p. 155).

The problem is that whereas scientists function in interdisciplinary teams to pursue scientific inquiries (for example, more than 800 scientists from many disciplines are working on the numerous datasets collected by the Terra mission), our science education curriculum frameworks still honor the separate discipline model of teaching science. For example, many earth science texts subdivide the study of earth into discrete subjects (geology, meteorology, oceanography, space science, etc.) and topics (rock cycle, volcanoes, ocean currents, moon phases, seasons, etc.). Unfortunately, disconnecting the topics from each other, and from any specific place and time undermines the more holistic character of Earth science. Mayer (1995) maintains that the 'hard' science approach has been unable to provide adequate insight into the complex processes of the earth systems, illustrating the severe limitations of reductionist science for studying processes at they occur in the real world. (p. 379) The move to earth systems, and thus to an integrated science approach, according to Mayer and Armstrong (1990) recognizes; (1) the need for curriculum to reflect our social and economic systems and the basic understanding of the nature of scientific investigation; (2) how science disciplines are now intimately intertwined; (3) mathematics as an essential tool of modern science; (4) application of science in industry and global developments; and (5) the need for citizens to understand how technology is used in our society, to see the role evidence has as the real authority in science, and to see the power of theories in the investigation of nature.

The Mayer and Armstrong report of the conference outlines the goals and concepts that are a prerequisite for an evolving 21st century view of Planet Earth. The goals are presented in Table 2. The addition of the Stewardship and Appreciation goals reflect significant changes from the last major effort in earth science curriculum reform; namely the 1960s NSF funded Earth Science Curriculum Project. Essentially, earth science is now adopting a human face.

Earth system science concepts are central to the Earth and Space Science theme of the National Science Education Standards (NRC, 1996). The system

Table 2. Goals Set Forth for Earth Systems Education (Mayer & Armstrong, 1991).

Goal	Description of Goal
Scientific Thought	Each citizen will be able to understand the nature of scientific inquiry using the historical, descriptive and experimental processes of the earth sciences.
Knowledge	Each citizen will be able to describe and explain earth processes and features and anticipate changes in them.
Stewardship	Each citizen will be able to respond in an informed way to environmental and resource issues.
Appreciation	Each citizen will be able to develop an aesthetic appreciation of the earth.

approach to the study of the Earth has become widely accepted as a framework from which to pose disciplinary and interdisciplinary questions in relationship to humankind. Earth system science forms the foundation of NASA's Earth science vision (Bretherton, 1988) and is now the basis of the NSF GEO long range planning effort as part of the nation's global change research efforts (NSF, 1999). Earth system science is also the approach recommended in *Shaping the Future* (AGU, 1997), a report on the future of undergraduate Earth science education as well as the theme of the ongoing Digital Library for Earth System Education (DLESE, 1999; Manduca & Mogk, 1999).

EARTH SYSTEM CURRICULUM FRAMEWORKS

In the decade following the 1988 conferences, there have been several significant developments in the design of earth system science curriculum. A non-technological focused approach is found in Mayer's "Earth Systems Education" (Mayer, 1991; Fortner, 1991), which became manifest as PLESE (Program for Leadership in Earth Systems Education) (Mayer et al., 1992). Technological approaches to earth systems science can be found in the K-12 GLOBE Program (Global Learning and Observation to Benefit the Environment) (Rock et al., 1997); Kids as Global Scientists (Songer, 1996) and the Learning through Collaborative Visualization (CoVis) Project (Gordin, Polman & Pea, 1994; Edelson, Gordin & Pea, 1999).

PLESE emerged from a teacher enhancement initiative at Ohio State University funded by the National Science Foundation in 1990. Initial efforts within PLESE led to the formation of a set of core concepts considered prerequisite for a 21st-century view of planet Earth, and the set of seven understandings that constitute a framework for Earth systems education listed in Table 3.

The PLESE Project yielded *Science as a Study of Earth: A Resource Guide for Science Curriculum Restructure* (Mayer & Fortner, 1995) that educators could use to restructure their curriculum toward an earth systems approach. The guide contained background material on the philosophy and recommended pedagogical approaches and sample activities that K-12 teachers could incorporate into their curriculum. For example, in a PLESE activity at the high school level, students collect gas (carbon dioxide) released when a sample of limestone (calcium carbonate) is placed in concentrated hydrochloric acid. Observing the way that limestone reacts to hydrochloric acid is a common earth science activity. Where PLESE departs from traditional approach is the way in which the experiment is used to prompt students to consider the interactions between the geosphere and atmosphere as carbon dioxide moves between

Table 3. Framework for Earth Systems Education consisting of Seven Understandings Developed by Program for Leadership in Earth Systems Education (PLESE) Planning Committee (Mayer et al., 1992).

Understanding 1.	Earth is unique, a planet of rare beauty and great value.
Understanding 2.	Human activities, collective and individual, conscious and inadvertent, are seriously impacting planet Earth.
Understanding 3.	The development of scientific thinking and technology increases our ability to utilize Earth and space.
Understanding 4.	The Earth system is composed of the interacting subsystems of water, land, ice, air, and life.
Understanding 5.	Planet Earth is more than 4 billion years old and its subsystems are continually evolving.
Understanding 6.	Earth is a small subsystem of a solar system within the vast and ancient universe.
Understanding 7.	There are many people with careers that involved study of Earth's origin, processes, and evolution.

storage areas within a global geochemical cycle. Students develop the concept of "carbon sinks" (places where carbon is stored) such as carbonate rocks, oceans, atmosphere, and organic matter, and they consider how a fluctuation in the quantity of matter in one storage area inevitably impacts other Earth systems.

GLOBE is a worldwide initiative to recruit school-aged children in the gathering of data to help scientists develop models that will test the information contained in satellite images. Students in 9500 schools in 90 countries are currently part of the GLOBE effort. Students data-gathering and reporting protocols have been established for the areas of study outlined in Table 4. Participating schools collect environmental observations at or near their schools and report their data through the Internet. These data are added to NASA and NOAA global databases. Scientists use GLOBE data in their research and provide feedback to the students to enrich their science education. Global images based on GLOBE student data are displayed on the World Wide Web, enabling students and other visitors to visualize the student environmental observations.

The Kids as Global Scientists project restricts itself to meteorological investigations but also takes a global perspective and uses the Internet. Student investigations of global weather patterns are coordinated across continents

Table 4. Project GLOBE Areas of Study and Measurements Made in Each Area of Study.

Area	Measurements
Atmosphere	Precipitation pH, cloud type, cloud cover, rainfall, snowfall, and min. max, and current temperature
Hydrology	Transparency, water temperature, dissolved oxygen, pH, electrical conductivity, salinity, alkalinity, nitrate
Soil	Structure, color, consistence, texture, bulk density, particle size distribution, pH, and fertility (N, P, K) of samples taken from each horizon, soil infiltration, surface slope (in degrees), soil moisture, soil temperature, diurnal variation of soil temperature
Landcover/Biology	Qualitative land cover, quantitative land cover, dominant and co-dominant vegetation species, tree height and circumference, biomass of the herbaceous ground cover
Phenology	Lilac phenology, budburst phenology

among participation classrooms. Daily weather information and satellite images are downloaded to classrooms for students to study. New tools to support students' scientific inquiries and communication of results are a focus of the research.

The CoVis project also focuses on the support of student inquiries. Employing powerful visualization tools similar to those used by scientists, students develop and test earth system models. The CoVis researchers have so far developed the Climate Visualizer, the Radiation-Budget Visualizer, the Greenhouse Effect Visualizer and the Worldwatcher, which combines all the Visualizers (Edelson, Gordin & Pea, 1999).

At the middle school level, Worldwatcher Project students learn about the scientific factors that contribute to the controversial global warming debate. The project places students as advisors to the heads of state of several different nations, prompting students to learn about the issue as they respond to the various questions and concerns of these leaders. As expert scientists on the issue, the class is challenged to understand and explain to the heads of state what forces affect climate and what global warming actually means. Once they do this, they are asked to help the different nations of the world understand how global warming will affect them and what they can do about it. Each student team advises one country and presents a proposal that offers a set of solutions that address the concerns of their country. Stages of the project include an introduction to the basic issues of global warming, understanding the factors that contribute to temperature change, investigating the factors that determine global temperature and energy use, understanding potential consequences of atmospheric pollution on global climate, and finding solutions to these types of problems.

The WorldWatcher High School Curriculum Project is building a year-long, inquiry-based, visually intensive environmental science curriculum centered on three key issues: Populations, Resources, and Sustainability; Meeting the Demand for Energy in Southern Wisconsin; Managing Water Resources in California and Local Environmental Issues. In the first unit, students are introduced to the investigation techniques that they will use throughout the curriculum. They begin to wrestle with the problems of sustainability as they investigate the growth in human population and resource usage.

The second component of the curriculum centers on the issue of a how a community will meet the increasing demand for electricity. Students explore alternatives to power generation through HTML virtual tours, QuickTime movies, and modeling activities. Through the integration of scientific visualization, wet lab, and modeling activities, students are asked to weigh impacts on air quality, water quality, and habitat preservation in making their decision about how to meet the electricity demand. Natural variation and human induced global change are emphasized in this unit.

The third portion of the curriculum focuses in on organism interaction, plant adaption strategies, and impacts on the ecosystem due to resource allocation in the Mojave Desert and Great Central Valley in California. Students explore the use of fertilizer, pesticides, water allocation, and soil management in order to design a more sustainable environment with minimal compromise to water resources and biogeography. Ecosystem interactions, water budgets and modeling bioaccumulation are stressed in this unit. A fourth component of the curriculum allows students to apply what they have learned to their own local area. The activities that make up this unit are interwoven with the other three units over the entire year. Students study local ecosystems, collecting and analyzing data on basic physical properties, and the environmental impacts of human activities. Students also investigate a local environmental issue, using local data and media resources to analyze the problem, identify the potential threats to the environment, and make recommendations based on an evaluation of alternatives.

A program that merges a non-technological approach with a technological approach is the newly released secondary level curriculum materials *Earth-Comm* (Earth System Science in the Community – Understanding Our Environment) produced by the American Geological Institute (AGI, 2000).

EarthComm emphasizes important concepts, understandings, and abilities that all students can use to make wise decisions, think critically, and understand and appreciate the earth system. The goals of the *EarthComm* program, drawn from leading earth systems education documents, are:

(1) To teach students the principles and practices of Earth system science and to demonstrate the relevance of Earth science to their life and environment.
(2) To approach Earth science through the problem solving, community-based model in which the teacher plays the role of facilitator.
(3) To establish an expanded learning environment that incorporates fieldwork, technological access to data, and traditional classroom and laboratory activities.
(4) To support the development of communities of learners by establishing student teams and by building a greater regional and national community through telecommunication access.
(5) To utilize local and regional issues and concerns to stimulate problem-solving activities and to foster a sense of Earth stewardship by students in their communities.

EarthComm focuses on four key concepts that are elements of its overall instructional design: *relevance*, *community*, *systems*, and *inquiry*. In EarthComm the concept of relevance permeates the curriculum, but becomes particularly explicit as each chapter is introduced, and is maintained through the attention given to the Chapter Challenge. *EarthComm* makes community relevance the focus of instruction by making it the focus of the inquiry that the students undertake in each chapter. A "Chapter Challenge" calls on the students to engage with the topics, to investigate how those topics affect their lives and communities, and to express their conclusions to others through some concrete product. In EarthComm, with its earth systems science focus, a chain of connections can be drawn from virtually any earth science phenomena that leads to many other ideas, and eventually back to the students' immediate world. EarthComm uses such chains of connections to highlight the relevance of earth science to the learner.

EarthComm focuses Earth science instruction in communities. That means that EarthComm is designed to relate directly to the student's neighborhood, town, state, region, and so on. The community can be seen in EarthComm in many activities that make use of resource material, such as local maps, to describe how earth processes affect the students' community. In the first activity in "Volcanoes and Your Community," for example, students use a map of volcanoes to begin to get a sense of where there are volcanoes relative to their community.

EarthComm uses a systems metaphor to develop Earth science understandings. In a systems metaphor, the idea that "everything is connected to everything else" is important. A systems approach is more holistic, considering interactions between subsystems. Separating the subsystems – studying them in isolation – gives an inaccurate picture of the whole because the actions of one subsystem actually change the other subsystems as the activity occurs. While subsystems can be defined, their ongoing interactions with other subsystems must be constantly considered. This is the kind of thinking that EarthComm promotes. Toward that end, Earth science phenomena are considering as operating within five major subsystems, or "spheres," which interacting with and affecting each other.

An activity from an EarthComm unit called "Water Resources And Your Community" provides a very good illustration of the components of a system. In this activity students set up an apparatus in which hoses allow water to flow from a coffee can to a soda bottle (filled with sand and soil) and from the bottle into a pan. The conditions of one part of the system, such as the level of water in the coffee can, effect the flow rate (output) that goes to another part of the system (input), which has other conditions (sand and soil) that effect the flow rate to the pan.

EarthComm encourages authentic inquiry because the answers to the questions asked of students are not already entirely known by the students, teachers, or publishers before the learning begins. EarthComm incorporates inquiry through the "Chapter Challenge" component of each chapter. The challenge functions throughout the chapter as a motivation to ask how the Earth system science ideas that are being learned related to the specific communities the students are considering. Inquiry is central to the advancement of both personal and collective scientific knowledge. EarthComm supports this inquiry approach with a variety of activities in each chapter. Some are open-ended, some place the students in the position of interpreting data, some help to illustrate phenomena so that students can assess the impact the phenomena might have on their communities – but all support the emphasis on inquiry-based learning. While hands-on inquiry is part of many activities in *EarthComm,* it stands out particularly well as ideas are applied to the students' community. Following each activity is a section called "Preparing the Chapter Report" that encourages students to consider the local implications of what they have learned. In that the implications will differ for each location where *EarthComm* is taught, there can be no entirely pre-existing answers, which adds to the authenticity of the inquiry.

EARTH SCIENCE PEDAGOGICAL PRACTICES

The structure of knowledge, the epistemic goals and data-gathering processes within the Earth sciences affords certain opportunities for learning science, learning about science, and learning to do science. Contrary to the claim by Sosniak (1999) that potential combinations of subject matter, types of students and other situation factors are seemingly infinite, the Earth systems science approach employs and recognizes specific theories, methods, and aims that serve to constrain and define the field of relevant pedagogical content knowledge (PDK). I think this claim about constraining and defining PDK can be made for all the science subjects taught in K-12 schools.

By carving out the relevant domains of inquiry, Earth system science does attend to Sosniak's notion that big subject-specific ideas and concepts when well articulated might help teachers and teacher educators. But there is an assumption about the daily work of teaching in the lead chapter's 12 principles that the Earth system science curriculum challenges – namely teaching to the class as a whole, with individualization around the margins. The basis of the challenge is one that resides in the dual agenda of science education – learning what we know vs. learning how we come to know and why we believe it. The challenge represents the balance that is sought in the curriculum between learning, on the one hand, the canonical knowledge of scientific disciplines (what we know) and, on the other hand, the epistemic structures and social and economic application issues that arise during engagement in science-in-the-making activities and inquiries. The new Earth system database and technological resources coupled with the new Earth science systems framework make possible project-based science and extend inquiry-based opportunities not otherwise as easily attainable in other disciplines. Furthermore, the global, national, regional and local environmental perspectives associated with Earth system science have the potential to situate the subject matter in meaningful and relevant contexts of study. The implication is that inquiry instructional methods employing model-based learning and reasoning activities establish the grounds for best practices in the teaching and learning of the earth sciences.

The recent focus on science as inquiry in the USA also suggests previous efforts to infuse inquiry teaching approaches into science programs have not made inroads on science education practices. The National Research Council's *National Standards in Science Education* and addendum on *Teaching Science as Inquiry* (NRC, 2000), along with the soon to be released American Association for the Advancement of Science Project 2061 edited book *Teaching in the Inquiry-based Science Classroom* (Minstrell & Van Zee, 2000), clearly suggests dissatisfaction and confusion with the teaching of inquiry in school science programs. Empirical evidence of this dissatisfaction can be found in the recent Project 2061 review of U.S. middle grade science textbooks (AAAS, 1999). *All* of the major texts presently being used were judged inadequate. A general and shared perspective is that hands-on science activities and kit-based science programs do not mean either effective teaching or scientific inquiry are being addressed. Missing, is the inquiry conversation that examines the delicate dialectic between evidence and explanation or observation and theory; where one informs the other and at the same time it is being guided and constrained by the other. As Bruner (1961) suggests, you must engage in the inquiry conversation to improve the art and techniques of the inquiry conversation.

The way forward is to consider frameworks that describe and analyze the conversations of inquiry. Contemporary perspectives from:

(1) the history and philosophy of science (e.g. Ackerman, 1985; Giere, 1988; Hull, 1988; Shapin, 1994),
(2) the nature of learning and reasoning in science (Carey, 1985; Chinn & Brewer, 1993; Metz, 1995; Schauble et al., 1997; Klahr & Simon, in press)
(3) the design of science education programmes (Gitomer & Duschl, 1995, 1998; White & Frederisckson, 1998; Lehrer & Schauble, in press a & b; Izquierdo at al., in review; Polman, 2000; Krajick, Czerniak & Berger, 1998; Roth & Bowen, 1995)

have begun to shift the perspective away from final form science to a perspective of 'science-in-the-making'. Driven by a consideration of the revisionary nature of the growth of scientific knowledge and coupled with analyses of the cognitive and social practices of scientists, it is not surprising that the focus has been on engaging learners in the conversations and languages of science. During science-in-the-making episodes, the dialectical exchange or conversations between observations and theory plays out. Debates and arguments about representations, models, evidence, theories, methods, aims, are played out. A consequence of this conversation focus to learning is a conceptualization of classrooms as learning communities (Brown & Campione, 1994) and epistemic communities (Grandy, 1997; Duschl, 2000).

The Earth system science frameworks outlined above can provide rich contexts for inquiry learning. Earth system science is well suited to promote inquiry conversations. Edelson et al. (1999) identify three learning opportunities typically associated with the adoption of inquiry methods. The first is the development of general inquiry abilities like posing and pursuing researchable questions, planning and managing an investigation and analyzing and communicating results. The second opportunity provided by inquiry learning is the acquisition of specific investigation skills. The third is the opportunity to improve understanding of science concepts. Such concept improvement, according to Edelson et al. (1999), can be attained in the following ways:

- Inquiry causes students to confront the limits of their knowledge. Unexpected outcomes will lead to curiosity, which in turn leads to a motivation to learn.
- Designing inquiry activities requires science content knowledge. Thus, engaging in inquiry places a demand on the learner to gain the knowledge in order to complete the inquiry.
- Opportunities to apply scientific understanding to answering scientific questions will frequently lead learners to discover new scientific principles and refine established knowledge claims.
- Inquiry activities afford learners the chances to apply knowledge.

There are significant challenges to sustaining inquiry learning and teaching in classrooms though. Research shows that children have difficulty sustaining systematic scientific investigations (Kraijcik et al., 1998; Schauble et al., 1995). Research also shows that teachers have similar difficulties (Duschl & Gitomer, 1997). Some of the challenges identified with the successful implementation of inquiry-based learning include motivation, accessibility of investigation techniques (e.g. data collection and analysis), background knowledge, management of extended activities, and practical constraints of the learning context (e.g. resources and schedules) (Edelson et al., 1999).

Working with Earth system science frameworks Edelson et al. (1999), have developed a set of curriculum design strategies for planning and delivering inquiry-based instruction and learning. The strategies reported by Edelson et al. (1999) include: (1) using *meaningful problems* to establish motivating contexts for inquiry, (2) using *staging activities* to introduce learners to background knowledge and investigative techniques, (3) using *bridging activities* to bridge the gap between the practices of students and scientists, (4) using scaffolds that embed tacit knowledge of experts in *supportive user-interfaces*; (5) providing a library of resources as an *embedded information source*; and (6) providing *record-keeping tools* which allow students to record procedures and store products generated.

White and Frederickson (1998) report success with coupling a metacognitive model of research, the *Inquiry Cycle* with a metacognitive or reflection on thinking and inquiry process called *Reflective Assessment*. The *Inquiry Cycle* is a general model of inquiry processes and research goals with five elements – pose a *question*, generate a set of competing *predictions and* hypotheses, and plan and carry out *experiments*, analyze data to produce *models*, and then *apply* the models to various situations. While engaged in the *Inquiry Cycle's* 5 parts, learners are also guided to reflect on limitations of what is being learned and to reflect on the shortcomings of the inquiry process itself. This is done by asking students to individually self assess their performance against a set of criteria for conducting good scientific research. There are goal-oriented criteria such as "Understanding the Science", process-oriented criteria like "Being Systematic" and "Using Tools of Science" and social-oriented criteria such as "Communicating Well" and "Teamwork".

Duschl and Gitomer (1997), in research portfolio assessment learning environments, have developed a whole class instructional strategy called 'assessment conversations' for coordinating and assessing the diverse student outcomes associated with open-ended tasks. This is but one of five essential features for the design of learning environments that support full-inquiry opportunities. The five key features are:

(1) situating the curriculum unit in a *meaningful problem* solving context,
(2) encouraging students to use *multiple ways of showing understanding* (e.g. drawings, labeled diagrams, writing, etc.),
(3) engaging learners in the *public consideration of ideas* (e.g. assessment conversations),
(4) employing *formative assessment practices* in three domains – conceptual learning, representational learning, and epistemic learning;
(5) using *scientific criteria* as organizing principles to evaluate knowledge claims.

IMPLICATIONS FOR INSTRUCTION

The Earth system science framework and the inquiry models proposed by Edelson et al. (1999) White and Frederickson (1998) and Duschl and Gitomer (1997) extend and challenge the idea of coherent content presented in general teaching principle 5. One challenge is the representation of science content in a reductionist framework, as mention above, Earth system science seeks to abandon the reductionist schemes and principles for organizing content. Knowing the conceptual parts and conceptual wholes of a single domain by themselves are not powerful ideas. The implication, and the second challenge, is that the structure of the content can not be divorced from neither the criteria that selects the elements of the scientific content, the data that gives meaning to the scientific content, nor the individual and group reasoning that stitches the content elements into explanatory models. The principles that guide teachers in making the content coherent need to include as intended learning outcomes a balance between conceptual, epistemic, and reasoning goals. Polman and Pea (in press) argue that adoption of this broader view of curriculum has implications for the discourse that occurs between teachers and students. Specifically, they argue for a process called 'transformative communication' which is an interactive process of guided participation between student as active inquirer and teacher as active guide. This idea is similar to David Hammer's (1997) concept of discovery teaching wherein the diversity of students' ideas and range of thinking can enable teachers to discover the next step of instruction or guidance.

Principle 9, strategy teaching, gets closer to the issue of what constitutes coherent content by recognizing the need for comprehensive instruction to pay attention to propositional knowledge (what to do), procedural knowledge (how to do it), and conditional knowledge (when and why to do it). But again, the focus on criteria in the research programs cited above shifts the cognitive modeling processes from the construction of scientific knowledge claims to the construction *and* evaluation and revision of scientific knowledge claims. Thus, the role of investigations, demonstrations, laboratory or practical work shifts from being independent activities that verifying conceptual relationships to being sequences of activities that build and test conceptual models and inform decisions about subsequent inquiries and designs of investigations.

The availability of global databases and investigative and communication tools coupled with the relevance of environmental issues and the problems of sustainable development enable the K-12 Earth science curriculum to occupy a privileged position in science education. The Earth system science approach is well suited to support the kind of inquiry-based, model-based and metacognitive-based approaches that underpin knowing science, knowing about science, and knowing how to do science. Drawing on the cognitive theories of learning and the theoretical, methodological and goal commitments of the Earth system science community focuses the inquiry-based learning in ways that avoid Sosniak's (1999) concern that the combinations of subject matter, with types of students, with situational factors will lead to seemingly infinite options. Establish the goals and work to the goals by designing learning environments and curriculum that embrace the goals.

REFERENCES

Ackerman, R. J. (1986). *Data, Instruments and Theory: A Dialectical Approach to Understanding Science*. Princeton, NJ: Princeton University Press.

AAAS (American Association for the Advancement of Science) (1999). Heavy Texts Light on Learning. *2061 today*, *9*(2), 1–4.

AGI (American Geological Institute) (2000). *EarthComm – Earth System Science in the Community – Understanding Our Environment*. Alexandria, VA: American Geological Institute (http://www.agiweb.org/)

American Geophysical Union, Ireton, M. F. W., Manduca, C. A., & Mogk, D. W. (Eds) (1997). *Shaping the Future of Undergraduate Earth Science Education-Innovation and Change Using an Earth System Approach*. Washington, D.C.

Bretherton, F. P. (1988). Earth System Science: A Closer View. Washington, D.C.: NASA.

Brown, A., & Campione, J. (1994). Guided discovery in a community of learners. In: K. McGilly (Ed.), *Classroom Lessons: Integrating Cognitive Theory and Classroom Practice*. London: MIT Press.

Bruner, J. (1961). The Act of Discovery. *Harvard Educational Review*, *31*, 21–32.

Carey, S. (1985). *Conceptual Change in Childhood*. Cambridge, MA: Bradford Books/MIT Press.

Chinn, C., & Brewer, W. (1993). The role of anomalous data in knowledge acquisition: A theoretical framework and implications for science instruction, *Review of Educational Research*, *63*(1), 1–50.

DLESE (1999). Portal to the Future: A Digital Library for Earth System Education. http://www.dlese.org

Duschl, R. (2000). Making the Nature of Science Explicit. In: R. Millar, J. Osborne & J. Leach (Eds), *Improving Science Education Contributions from Research* (pp. 187–206). London: Open University Press.

Duschl, R., & Gitomer, D. (1997). Strategies and challenes to changing the focus of assessment and instruction in science classrooms. *Educational Assessment*, *4*(1), 337–373.

Edelson, D., Gordin, D., & Pea, R. (1999). Addressing the challenges of inquiry-based learning through technology and curriculum design. *The Journal of the Learning Sciences*, *8*(3&4), 391–450.

Fortner, R. (Ed.) (1991). Earth systems education [Special Issue]. *Science Activities*, *28*, 1.

Giere, R. (1988). *Explaining Science: A Cognitive Approach*. Chicago, IL: University of Chicago Press.

Gitomer, D., & Duschl, R. (1998). Emerging issues and practices in science assessment. In: B. Fraser & K. Tobin (Eds), *International Handbook of Science Education* (pp. 791–810). Dordrecht: Kluwer Academic Publishers.

Gitomer, D., & Duschl, R. (1995). Moving Toward a Portfolio Culture in Science Education. In: S. Glynn & R. Duit (Eds), *Learning Science in Schools: Research Reforming Practice*. Mahwah, NJ: Erlbaum.

Gordin, D., Polman, J. & Pea, R. (1994). The Climate Visualizer: Sense-making through scientific visualization. *Journal of Science Education and Technology*, *3*, 203–226.

Grandy, R. (1997). Constructivisms and objectivity: Disentangling metaphysics from pedagogy. *Science & Education*, *6*(1&2), 43–53.

Hammer, D. (1997). Discovery learning and discovery teaching. *Cognition and Instruction*, *15*(4), 485–530.

Hull, D. (1988). *Science as Process*. Chicago: University of Chicago Press.

Izquierdo, M., Sanmarti, N., Espinet, M., Garcia, M. P., & Pujol, R. M. (in review). Characterization and Foundation of School Science, *Science Education*.

Johnson, D. R., Ruzek, M., & Kalb, M. (1997). What is Earth System Science? In: *Proceedings of the 1997 International Geoscience and Remote Sensing Symposium* (pp. 688–691). Singapore.

Johnson, D. R., Ruzek, M., & Kalb, M. (1998a). Earth System Science Education: National and International Partnerships. *Eos Trans. AGU, 79*(17). Spring Meet. Suppl., S10.

Johnson, D. R., Ruzek, M., & Kalb, M. (1998b). National and International Earth System Science Education Partnerships: Approaching the New Millennium. In: *Proceedings of the 1998 International Geoscience and Remote Sensing Symposium.* Seattle.

Johnson, D. R , Ruzek, M., & Kalb, M. (1998c). Earth System Science Education and the Need for Diversity. *Eos Trans. AGU, 79*(45), Fall Meet. Suppl., F71.

Johnson, D.R , Ruzek, M., & Kalb, M. (2000). Earth System Science and the Internet. *Computers and Geosciences* Special Issue: The Year 2000 Challenges, v.26, no.6, July, pp 669–676

Kalb, M., Johnson, D., & Ruzek, M. (1997). Evolution of the Cooperative University-based Program in Earth System Science Education. *The American Meteorological Society Sixth Symposium on Education.* Long Beach, CA, pp J1–4, February 2–7, 1997.

King, M. D., & Herring, D. D. (2000). Monitoring Earth's Vital Signs. *Scientific American, 282*(4), 92–97.

Klahr, D., & Simon, H. A. (in press). Studies in scientific discovery: Complementary approaches and convergent findings. *Psychological Bulletin.*

Krajick, J., Czerniak, C. M., & Berger, C. F. (1998). *Teaching children science: A project-based approach.* Boston, MA: McGraw-Hill.

Kuhn, T. (1962/1970) *The structure of scientific revolutions* (2nd ed.). Chicago: University of Chicago Press.

Lehrer, R., & Schauble, L. (2000). Model-based reasoning in Mathematics and Science. In: R. Glaser (Ed.), *Advances in Instructional Psychology, V5,* (pp. 101–159). Mahwah, NJ:Erlbaum.

Lehrer, R., & Schauble, L. (in press b). *Children's Work with Data:Modeling in Mathematics and Science.* New York: Teachers College Press.

Manduca, C., & Mogk, D. (1999). Portal to the future: A digital library for Earth system education. Preliminary report, Coolfont Resort, Berkeley Springs, West Virginia, August 8–11. http://geo_digital_library.ou.edu/report/reportfin.html

Mayer, V., & Fortner, R. W. (Eds) (1995). *Science as a study of Earth: A resource guide for science curriculum restructure.* The Ohio State University. 246 pp.

Mayer, V. (1995). Using the earth system for integrating the science curriculum. *Science Education, 79*(4), 375–391.

Mayer, V., Armstrong, R. E., Barrow, L. H., Brown, S. M., Crowder, J. N., Fortner, R. W., Graham, M., Hoyt, W. H., Humphris, S. E., Jax, D. W., Shay, E. L., & Shropshire, K. L. (1992). The role of planet Earth in the new curriculum. *Journal of Geological Education, 40* 66–73.

Mayer, V. (1991). Earth-system science – a planetary prespective. *The Science Teacher, 58,* 31–36.

Mayer, V., & Armstrong, R. (1990). What every 17-year old should know about planet earth: The report of a conference of educators and geoscientists. *Science Education, 74*(2), 155–165.

McPhee, J. (1986). *Rising from the plains.* New York: Farrar, Straus, Giroux.

Metz, K. (1995). Reassessment of developmental constraints on children's science instruction. *Educational Researcher, 65*(2), 93–127.

Minstrell, J., & Van Zee, E. (Eds) (2000). *Teaching in the Inquiry-based Science Classroom.* Reston, VA: American Association for the Advancement of Science.

National Research Council (2000). *Inquiry and the National Science Education Standards: A Guide for Teaching and Learning.* Washington, D.C.: National Academy Press.

NSF Geosciences Beyond 2000: A Window on the First Decade (1999). http://www.geo.nsf.gov/adgeo/geo2000/start.htm

Polman, J. (2000). *Designing project-based science: Connecting learners through guided inquiry*. New York: Teachers Collge Press.

Polman, J. & Pea, R. (in press). Transformative communication as a cultural tool for guiding inquiry science. *Science Education*.

Rock, B., Blackwell, T., Miller, D., & Hardison, A. (1997). The GLOBE program: A model for international environmental education. In: K. C. Cohen (Ed.), *Internet Links for Science Education: Students-Scientist Partnerships*. New York: Plenum.

Roth, W.-M., & Bowen, G. M. (1995). Knowing and interacting: A study of culture, practices, and resources in a grade 8 open-inquiry science classroom guided by a cognitive apprenticeship metaphor. *Cognition and Instruction, 13*(1), 51–72.

Schauble, L., Leinhardt, G., & Martin, L. (1997). A framework for organizing a cumulative research agenda in informal learning contexts. *Journal of Museum Education, 22*(2&3), 3–11.

Songer, N. (1996). Exploring learning opportunities in coordinated network-enhanced classrooms: A case of kids as global scientists. *The Journal of the Learning Sciences, 5*(4), 297–327.

Sosniak, L. (1999). Professional and subject matter knowledge for teacher education. In: G. Griffin (Ed.), *The Education of Teachers (98th Yearbook of the National Society for the Study of Education)* (pp 185–204). Chicago: University of Chicago Press.

White, B., & Frederiksen, J. (1998). Inquiry, modeling, and metacognition: Making science accessible to all students. *Cognition and Instruction, 16*(1), 3–118.

SUBJECT SPECIFIC TEACHING METHODS: HISTORY

Stephen J. Thornton

INTRODUCTION

History instruction has been a staple of United States schooling for more than a century, but many students find it dull and irrelevant to their lives. The problem does not appear to be that youngsters find history itself intrinsically uninteresting. Rather, as John Goodlad (1984) put it, something strange seems to have happened to it on the way to the classroom (p. 212). In other words, it is a problem of method, the effective direction of subject matter to desired results.

American educators have long disagreed, however, about whether methods should emphasize the logical or the psychological organization of subject matter. Logical organization equates subject matter with the historian's finished product of thought. Psychological organization refers to the connection of subject matter to student experience in order to make it learnable. Of course, thoughtful educators have long recognized that the logical versus psychological are not at odds in any comprehensive view of the educational process.

Hilda Taba (1962) noted, for instance, that psychological organization should not obscure or falsify the basic ideas and relationships of the discipline (pp. 301–302). Similarly John Dewey (1938) went out of his way to admonish his self-appointed disciples who championed child-centered methods without due regard for the progressive organization of subject matter. The psychologizing of subject matter, Dewey pointedly remarked, should result in the immature learner's understanding "gradually approximat[ing] that in which subject-

Subject-Specific Instructional Methods and Activities, Volume 8, pages 291–314.

ISBN: 0-7623-0615-7

matter is presented to the skilled, mature person" (p. 74). Accordingly, the position taken in this chapter is that educators can be justified in emphasizing either the logical or the psychological end of the organizational spectrum. In any given circumstance, this emphasis will be a matter of judgment about the educational needs of a particular group of students.

The main purpose of this chapter is to evaluate some alternatives to conventional methods in history teaching. By "conventional," I mean teacher-led question-and-answer activities, student seatwork based on textbooks, watching videos, taking short-answer tests, and so on that are the mainstay methods in United States social studies classrooms (Thornton, 1994; Brophy, Alleman & O'Mahony, 2000). Although it is clear that some teachers inspire students with seemingly ordinary methods (Eisner, 1986; Wineburg & Wilson, 1991), more often these methods are an uninspiring and shallow coverage of a logical organization of subject matter. Of course, deeply interested and achievement-oriented students cope well enough with this coverage (even if their educational growth may be less than optimal).

For a far larger number of students, however, method in the service of coverage has more serious consequences. These students experience history as a collection of disconnected facts whose significance seldom becomes clear. As the progressive era historian James Harvey Robinson recognized, this method has a "specious logic and orderliness which appeals to the academic mind" but unhappily does "not inspire the learner" (cited in Taba, 1962, pp. 387–388). These learning experiences may be miseducative because over time classroom experiences shape students' ideas about routes to learning, interests, and attitudes (Stodolsky, 1988, p. 121).

More effective method can improve this situation. "Method" as I am using it here, however, is more than a carrot to grab students' attention or a stick to coerce teachers into motivating students. Rather, method entails both technique and educational purpose. Unless guided by a conception of what counts as educationally significant, method is reduced to the kind of fix-all fad that has understandably bred cynicism and greatly harmed American education. Method will not reformed by fiat, it requires rethinking teaching and teacher education (Thornton, in press, a).

PRINCIPLES OF EFFECTIVE TEACHING

Jere Brophy (pp. 1–24, this volume) has developed 12 principles of "effective teaching" that may be one place where the rethinking process can begin. As with any broad synthesis, the principles are general in nature. Their application to history methods necessarily entails detailing the subject-specific elements of instructional arrangements. As Dewey (1916) said of method: "We can distinguish a way of acting, and discuss it by itself; but the way exists only as way-of-dealing-with-material" (p. 165). In other words, the meaning of the principles must be interpreted in terms of the pedagogical demands of particular subject matters.

The 12 principles should also, in my view, be regarded as "principles." That is, as general truths from which more specific truths may be derived. In this sense, the principles may be useful for teacher decision-making. But it is not only the breadth of the principles, however, that warrants caution. As Philip Jackson (1968) warned, the inevitable immediacy and opportunism of teacher decision-making during instruction "places severe limits on the usefulness of a highly rationalistic model" of teaching (p. 167).

It seems that educators will profit most from Brophy's principles outside the classroom rather than in it. For example, when teachers plan and evaluate their instruction, they may reflect on the consonance between the principles and their methods. Similarly, the principles could well be utilized in teacher education programs. The principles seem well suited to fostering reflective dialogue about teaching. It will be important to avoid, however, the misuse of the principles by turning them into prescriptions for curriculum and instruction. Lauren Sosniak (1997) has shown how just such a transformation occurred with Bloom's taxonomy. Bloom's principles were not intended as a coercive instrument to confer greater value on "higher-order" tasks than "lower-order" tasks although that is precisely what happened to them in prescriptive policies on curriculum, instruction, and testing. However well intended, highly prescriptive policies often do more to undermine than improve teaching (Zumwalt, 1988).

APPROACH OF THIS CHAPTER

Comprehensive classifications of history methods are hard to come by. Indeed it is even difficult to identify the proper unit of analysis in method. Some speak of method expansively as the rationale and organization of the entire curricular-instructional process while others reduce it to a micro-level, technical procedure to deliver instruction (Hertzberg, 1988). The latter view is surely too narrow because it omits the normative element of method. I adopt an intermediate position that assigns the overall goals and structure of history curricula to curriculum policy-makers and instructional materials writers beyond the classroom, but considers method as primarily concerned with the teacher's direction of courses, units, lessons, and learning activities to effective results. In this scheme, the teacher is a "curricular-instructional gatekeeper," the primary decision-maker where it ultimately counts: the classroom (Thornton, 1991).

In the social studies, method customarily is directed at three elements of subject matter: knowledge, skills, and values. In the daily grind of classrooms, all three elements may be present in one lesson. Moreover, the boundaries between the elements are porous. Although I do not dwell on the distinctions among these elements in the discussion of methods to come, it is important to stipulate how I am defining each of them in this chapter.

Knowledge is subdivided into three sub-categories: facts, concepts, and generalizations. Facts are the most basic element of knowledge although the selection of facts to be taught always rests, implicitly or explicitly, on some higher-order element of knowledge that confers significance upon them (Carr, 1961). Concepts, which will be discussed at some length in the section on conceptual teaching below, are abstractions that give order to factual knowledge (e.g. manifest destiny, emancipation, revolution, New Deal, progressive era, yeoman, serf, absolutism, and ancient civilization). Unlike facts, concepts can readily be further developed and applied in new contexts.

Generalizations are considered the most complex element of knowledge. They are powerful combinations of ideas because they transfer from one situation to another and, more so than concepts, allow prediction. Because of this intellectual power, Paul Hanna and his associates developed a list of thousands of generalizations upon which to base social studies instructional programs (Hanna & Lee, 1962). Edwin Fenton (1967) warned, however, that these lists of generalizations are "inert"; such a deductive method of teaching fails to help students learn the process by which generalizations are developed (p. 13). As discussed below, one way around this dilemma is for students to build inductively their own generalizations.

Skills concern the ability to perform particular tasks. Again, the unit of analysis varies. Thus skills may be relatively specific such as the ability to place events in chronological order or read latitude and longitude. Skills are also spoken of more broadly, such as "critical thinking." Nevertheless, there is scholarly consensus that skills are most effectively learned in context rather than isolation.

Values are an inevitable part of history instruction. The inculcation of values such as patriotism has long been a common feature of history classrooms. Perhaps even more significant is the unexamined transmission of values that occurs in history teaching and instructional materials. Consequently many educators insist that any balanced conception of instruction must also extend to analysis of values rather than their mere passive reception. Value disputes sometimes arouse controversy, but it is clear that history curriculum and instruction can never be value-neutral. Rather, what are at stake is whose values will be represented and how those values are examined.

As noted, the scheme that follows adheres to a middle-range view of method. Sosniak (1999) seems correct that overspecification of subject specific methods leads, as Lee Cronbach said of aptitude treatment interaction, to a hall of mirrors that extends to infinity. As Sosniak concludes:

> the research can proceed forever, of course, as scholars continue to identify slices of content knowledge to investigate. Its meaning for practice, however, will become less significant as the studies become overly specific and reach levels of detail that are not responsive to the daily work of teaching (p. 197).

On the other hand, experience suggests that generic methods must be modified with different subject matters.

By a "middle-range" view of method I mean models specific enough to apply to actual subject matter but not so microscopic that their infinite varieties swamp rather than buoy method. Such methods should be broad enough to be adapted to structurally comparable topics (e.g. European colonization in North America and in South America) within (and sometimes, perhaps, across) school subjects. Such models should also be specific enough that what they look like in practice can be grasped.

The models of method examined below utilize learning theories, however, I maintain the ramifications of learning theory for method must always be extrapolated. Jean Piaget's theory of cognitive development, for instance, would appear to be an invaluable guide for practice, but there has been more bewilderment than clarity on how it should inform method (Noddings, 1983). In other words, understanding how students think is, by itself, an inadequate basis for method.

Accordingly classroom studies of teaching, analyses of instructional materials, and curriculum evaluations also inform this chapter. These sources can be found in particular abundance from the structure-of-the-disciplines movement of the 1960s that spawned the "new social studies." Now largely forgotten or overlooked, these sources can still be profitably mined for insights on method. Caution must be exercised, however, to avoid the "straight-jackets" of disciplinary structure and lock-step thinking models (Parsons & Shaftel, 1967, pp. 141–147) whose rigid application did so much to undermine the promise of the new social studies.

The models of history method below all share Brophy's concern for active learning. But the models, as noted, represent one of many legitimate organizational schemas. Further, many of the methods discussed in one section could be placed equally well in a different section. My intention is neither to provide a watertight classification of methods nor construct a graded hierarchy of methods. Rather I contend that several methods are useful and determination of usefulness depends on both educational purposes and practicality for a particular classroom. Moreover, since I believe depth rather than breadth will best illustrate my purposes in this chapter, some methods (e.g. museum and community study) are ignored. No disparagement of their educational significance is implied. Rather I wish to avoid a key shortcoming of many methods treatments: vagueness and lack of specificity about subject matter (Thornton, 1997). In the circumstances, it seemed wiser to select just a few methods and treat them as thoroughly as possible in the confines of one chapter.

CONCEPTUAL TEACHING

"The mainstay of the current curriculum, specific facts," Taba (1962) observed, is the "least fundamental" level of knowledge. Rather, Taba argued that the main core of school subjects ought to be "the basic ideas, concepts, and modes of thought which organize the kaleidoscope of concrete facts and events." In the case of history curriculum, she lamented that it "may be so studded with historic facts that no basic ideas about historic causation or of the essential nature of such movements as immigration in American history can emerge" (p. 269).

Forty years later, Taba's concerns still resonate. Much history instruction is scarcely more than the transmission of facts. The presence of concepts and generalizations in instructional programs may indicate no more than students memorize them (rather than develop them). As a solution to these problems, Taba and like-minded educators have proposed a variety of overlapping models that I am labeling "conceptual teaching." Unless systematic instruction is directed at it, students' conceptual learning is likely to be incomplete and, even, faulty. For example, Brophy and Janet Alleman (2000) found that primary-grades students often know that different forms of shelter existed across time and cultures. These children, however, lacked a conceptual framework to explain why these forms of shelter existed (p. 108).

Conceptually oriented educators forsake a content-coverage approach. Instead they emphasize a limited number of powerful ideas and concepts (Brophy, 1990). As Taba (1962) put it:

> a thorough study of a few examples . . . may eventuate in a better understanding . . . than does a superficial coverage of many details. . . . If one teaches for transfer and for application, a concept once mastered for one area, especially if this mastery is achieved by active learning processes, becomes applicable to any other area (p. 187).

Scholars have failed to define "concept" definitively and history does not lend itself as readily to conceptual organization as geography and the social sciences. These factors, however, qualify rather than discount its usefulness for method. "Imperialism," it was noted in the Illinois Curriculum Program (1971), "like many concepts in history, has an element of vagueness that makes it difficult to 'pin down'" (p. 67). Nonetheless, the Program designed and field-tested an imaginative method of developing the concept of imperialism.

The heart of conceptual teaching is getting students to think and value. There is an insistence that content not become an end in itself. If the content is, for example, the Spanish-American War, the conceptual aim may be the role of public opinion in the formulation of foreign policy. If the content is the reforms of the Jacksonian era, the teacher may try to develop an attitude of respect for those who devote a lifetime to concern for the welfare of others. If the subject is political parties under Washington, the conceptual aim may focus on the generalization that in a democracy a system of political parties is essential for the resolution of conflict (Davis, 1971, p. 47).

Such conceptual aims are intended to orient methods toward active student inquiry. The teacher becomes more a facilitator of student activity than knowledge dispenser. When students are presented with someone else's version of a concept, Barry Beyer (1971) argued, such presentations do not result in students internalizing the concept. As an alternative, he suggested students could be engaged in inventing their own conceptual images.

Specifically, Beyer recommended a four-step process. (He uses the example of the geographic concept of "landscape" but it could be used as well with historical concepts such as "the westward movement" or "Egypt of the pharaohs"). First, students brainstorm to list all the various implications of a word or phrase, its synonyms, or its associated terms. Second, all the terms with similar features are placed in a single group and the group labeled with a term describing the common element. Third, the groups are examined to determine any relationships that might exist among them. Fourth, the groups are arranged so as to make these relationships readily apparent. The result is a concept, a mental image of what the class has been working with (pp. 119–120).

Moving beyond individual lessons, Beyer also provides methods for lengthier periods of instruction such as courses and units that allow "learning to lead somewhere" (p. 135). For instance, he describes a study of revolutions. Beginning with some already familiar civil disturbance, students hypothesize about the nature of revolutions: What are they like? What causes them? What kind of results do they have? Next a simple conceptual model is constructed and applied to the American Revolution in order to test, broaden, and revise the hypothetical model. Different groups of students then test the model against other revolutions such as the Glorious Revolution, the Russian Revolution, and the Chinese Revolution. Accordingly, the model is revised. Each student then individually tests the hypothesis against another revolutionary event such as the Reformation, the industrial revolution, and the civil rights revolution. Finally, the students compare and contrast their model to existing theories of revolution such as Crane Brinton's *Anatomy of Revolution* (1965).

Interest in conceptual teaching among educational theorists was higher a generation ago than now. Nevertheless, occasional studies still appear that document its benefits for method and student learning (e.g. VanSledright, 1997). It may be that practically all students can benefit from conceptual teaching because it incorporates what Brophy calls (introduction, this volume) "cognitive modeling." Many students have only the slightest notion of how to distinguish what matters from what doesn't in history (Wineburg, 1991; Seixas, 1997). A steady diet of the transmission of historical facts does almost nothing to remedy this unfortunate situation. For students with a deep interest in history (and who may want to pursue its study in college), I believe conceptual teaching is an especially beneficial introduction to the structure of knowledge. Among other things, they learn, as Jerome Bruner (1960) discerned, that hypothetical and intuitive thinking is an inestimable intellectual tool (pp. 64–67; see also Noddings & Shore, 1984). These students will thrive on conceptual teaching and, even, find joy in it.

THE PRIMARY SOURCE METHOD

The use of primary sources may be the most subject specific of any method employed in the teaching of history. Primary sources are artifacts or first-hand accounts surviving from the past. Most commonly these sources are written documents but they could as well be, for instance, photographs, artworks, articles of clothing, cooking implements, tombstones, floppy disks, or a Roman aqueduct.

In contrast to the instructional norm of textbooks and other secondary sources, the primary source method in some way engages students directly with these artifacts. As "in some way" is meant to connote, this method may be interpreted as instruction based entirely on primary sources or as just one of several components of instruction. In either case, few educators would disagree that students can profit from grappling with the materials on which all historical accounts are ultimately based. As with conceptual teaching, strong proponents of the primary source method are inclined to place greater emphasis on the process of historical inquiry than a particular body of knowledge.

The intellectual lineage of the primary source method extends back to Europe in the wake of the Enlightenment. "Positivist" or "scientific" historians believed an objectively determined "ultimate" history could be constructed based on exhaustive analysis of sources. Speaking of the first *Cambridge Modern History* in 1896, Lord Acton wrote that the writing of ultimate history was not far off because "now that all information is within reach, and every problem has become capable of solution" (cited in Carr, 1961, p. 3). Left unconsidered in this scheme, of course, was that different historians and changing times would lead to changing interpretations of the meaning of the past. But it provided a powerful rationale for the study of primary sources. Inspired also by the seminar model of the German university, by the late nineteenth century the source method was introduced to American schools (Hertzberg, 1971).

Of course, school textbooks have hardly ever been without some primary source material. Current textbooks, for instance, seldom fail to include at least some of Jefferson's words in the Declaration of Independence or King's words in the "I have a dream" speech. In the nineteenth century schoolchildren were often required to recite (or even memorize) seminal statements of the national creed such as the Gettysburg Address. That primary sources should be taught remains uncontroversial, but by what method?

Designed largely for high school, the Amherst Project was one the most ambitious attempts to build instructional programs centered on the primary source method. The project evaluated both student and teacher responses to it and, consequently how the project fared in practice can be estimated more precisely than some other innovations (Amherst Project Records, c. 1961–1972).

The project produced dozens of instructional units on various topics in American history. Each unit was a compilation of primary sources and each centered on the historical method. The early units produced dealt with then staple topics in the academic study of American history such as the American Revolution and the Monroe Doctrine. Perhaps because of the social upheavals of the 1960s, topics for the later units more often included parallels with contemporary issues such as the Watts riots and poverty.

Scope and sequence was left up to the user, although the project made some suggestions about how to utilize the individual units, in whole or part, to address broader themes (e.g. combining several units to address "American values in conflict") also applicable to social studies courses other than history. The heart of the project, however, is the individual units. The "What Happened on Lexington Green?" (Bennett, 1967) unit, which is discussed below, was deemed particularly suitable for "younger students and less skilled readers" as well as a model for teacher education (Brown & Traverso, n.d., pp. 14–15).

Judged by the standards prevailing in schools, however, most units were lengthy and complex. Only selected topics in American history were dealt with. Thus, even if all the units were taught, which was clearly impossible in the time available in one school year, a great deal of the standard content of American history courses would have been omitted.

The units met their warmest reception in highly academic high schools and independent schools. Teachers reported that so much time spent on relatively specialized topics, however, sometimes became tedious for students. The process of working with the documents was often repetitive and this, too, seems to have reduced the units' appeal to students. Although the project stated that it is unnecessary to study the entirety of any unit, teachers seem to have felt some obligation to do so. Suggestions to "posthole" – that is, continue the usual chronological survey of American history but pausing periodically to study some topic in depth – also did not appear to allay teacher concerns. The project was perceived as too single-mindedly academic, too fixated on one part of what teachers saw as the proper scope of American history courses, and too little concerned with the continuity of subject matter (Thornton, in press, b).

Whatever the devotees of history who developed the Amherst project may have wished, its educational focus was narrow (Kline, 1974). In the United States, both educators and the public – across the political spectrum – expect that history will serve purposes broader than replicating what historians do (Levstik & Barton, 2000; Zimmerman, 1999). The purposes and methods of historians are not isomorphic with the purposes and methods of school educators as the tepid reception of the Amherst project testifies. One survey of teachers across several states revealed that, even though American history was the most widely taught subject among the respondents, the Amherst units were the least used of nine new social studies projects (Turner & Haley, 1975, p. 17).

Even its early advocates recognized the dangers of an exclusive reliance on the source method. History-methods pioneers Lucy Salmon (1902) and Henry Johnson (1940), for instance, enthusiastically recommended source study, but neither thought that primary sources could stand alone as the basis of instruction. Sources needed to be placed in context and this required additional methods such as textbook usage. They also worried about the high levels of analytic and synthetic thinking that formulating coherent historical accounts demanded of students (and teachers).

Relatively little research as been done on the primary-source method in actual classroom settings. My interpretation of the available theory and research suggests the method's effectiveness depends on such factors as its purposes, level of difficulty, and integration with other learning activities. Significantly, these are educational concerns rather than disciplinary concerns.

The *History 13–16* (see Shemilt, 1980) project developed in the United Kingdom is instructive on the educational possibilities of the primary source method. This curriculum was designed to introduce adolescents to historical method and is one of the few large-scale experiments on the classroom effectiveness of the primary-source method. *History 13–16*, like the Amherst Project, centers on historical method.

Disciplinary concerns – that is, topics of current interest to historians and the documents and secondary sources historians deem relevant to those topics – are the core of the Amherst Project. On the other hand, *What is History?*, the first of four segments of the *History 13–16* three-year course, features more overt attention to psychologizing subject matter. Rather than units that repeat the same documentary exercises with different topics, the units are sequenced.

Contrasting one unit from each project illustrates this different emphasis. The American project's "What Happened at Lexington Green?" (Bennett, 1967) and the British project's "The Case of Richard III and the Missing Princes" (Schools Council, 1976) were both directed at the same academically heterogeneous age group. Both units were essentially case studies aimed at answering a question: Which side fired the first shot at Lexington Green? Was Richard a wicked king who murdered his nephews to secure the throne for himself?, respectively. In both cases also, the evidence is conflicting and (ultimately) inconclusive. Both are cases that have long intrigued the popular imagination in their respective nations.

"Lexington Green" contains written eyewitness primary sources, accounts by historians, and from a range of textbooks over time. The sources are mainly lengthy and in eighteenth-century English. There are no visual illustrations. The suggested further reading is from academic history and is presumably directed at teachers. The overall impression created is scholarly and colorless.

In contrast, "Richard III" contains much shorter extracts from primary sources, fiction writers (including Shakespeare), and historians. While the documents are primary, the selection suggests they were chosen for their accessibility and appeal to adolescents. There is lavish use of pictures and photographs as well as a filmstrip. Parenthetically, as Johnson (1940) noted, with proper integration in a curriculum, the interpretation of pictures is a rigorous and invaluable exercise (pp. 182–193). Overall, through its design, use of color, illustrations, employment of words, and so forth, this unit conveys an inviting message. This message is not trivial because it may influence how and with what enthusiasm students interact with the materials (Eisner, 1979,

p. 222). Further, in the teacher's guide (Schools Council, 1976), supplementary resources are separated into books for teachers and books for students. Additional audio-visual materials are also identified.

The evaluation of *History 13–16* suggests some of the keys to its effectiveness (Shemilt, 1980). Although ultimately directed toward the logical arrangement of subject matter, this curriculum began with the relevance of history to the general education of adolescents. In this sense, "Whilst not intending the mass production of pocket historians, *History 13–16* aimed to transform classroom History into a problem-solving pursuit" (p. 87). "History," the evaluation declared, "must prove of value for pupils who will cease to pursue it beyond the age of sixteen, and justification must be sought in the mesh of adolescent needs with what the subject has to offer" (p. 2). Although the Amherst Project also insisted they were interested in general education (rather than specialized history instruction) the materials they produced send a contradictory message.

It seems clear that the educational effectiveness of primary sources is crucially dependent on how and to what purpose they are marshaled. Just handing the documents to students will leave many youngsters (and adults) lost as to their significance or how to make sense of them. This partly reflects students' socialization to uncritical acceptance of textbook narratives as the authoritative story of what happened, leaving them unequipped to confront the ambiguity and conflicting interpretations found in a range of primary sources. But it also reflects that competence in interpreting documents requires explicit instruction.

Experience with primary sources and the skill of interpreting them can be among the most rewarding and personally fulfilling approaches to historical study (Holt, 1990). Generally attempts to install primary source study as the centerpiece of history instruction, however, have failed (Thornton, in press, b). Nevertheless, we should resist the bandwagon mentality that demands either all instruction or none should be based on primary sources. Johnson (1940) probably had it right:

> School history must, in the main, be presented as ready-made information. But there can be, and should be, illustrations of the historical method sufficient to indicate the general nature of the problems behind organized history, and sufficient to give some definite training in the solution of such problems (pp. 304–305).

SIMULATION

Simulation (and closely related methods such as role play) can be a powerful method. It seldom appears, however, to have been a common method in history teaching. Nor is there much research on simulations in history or social studies education more generally (Kleg, 1991). Most available documentation on simulation is self-report and descriptive accounts. Perhaps most troubling of all is the paucity of theoretical analyses of simulation's educational purposes. This neglect is unfortunate because simulation can engage students in visceral ways.

In spite of these limitations, there are sound reasons to extol the educational potential of simulation. Anecdotal evidence and years of teaching history convince me that simulation engages most children and adolescents. Simulation actively involves the learner, both cognitively and emotionally. Whereas children and adolescents tend to regard how people in the past behaved as deficient by today's standards (Barton, 1996), simulation can help them see the world as historical figures and groups did. The motives and actions of historical figures and movements become tangible rather than merely peculiar.

Simulation is essentially a simplified model of some historical situation. Given that time sometimes forbids lengthy simulations, often the most effective simulations are relatively simple exercises developed by or adapted by the teacher. Students may simulate, for instance, the decision-making process by which President Lincoln issued the Emancipation Proclamation or the options confronted by President Kennedy during the Cuban Missile Crisis. In the former case, the simulation may follow prior instruction on how, at the outbreak of the war, Lincoln feared emancipation as an obstacle to the restoration of the union. A Cuban Missile Crisis simulation, on the other hand, might be introduced as an abstracted version of events prior to the study of the actual crisis. A further possible variant is a role-play of a well-documented event such as life at Jane Addams' Hull House or the Paris Peace Conference following World War I.

More elaborate simulations require careful preparation and debriefing. Although even complex simulations can be modified to fit available instructional time, usually they are only educationally justifiable if they form an integral component of an instructional sequence. James Banks (1970), for instance, presented an elaborate simulation game on Reconstruction. It concerns the problems that the South faced immediately after the Civil War. Students are assigned to interest groups (e.g. freedmen, aristocrats, poor whites, scalawags, carpetbaggers). Each group tries to advance its own interests in the legislative process. An important element of "reality" is that points weight the relative influence assigned to a group. For example, poor whites have more points than freedmen and aristocrats have more points than poor whites (pp. 59–63). A teacher, "Mr. Bauer," who assigned students occupational roles (e.g. clergyman, dentist, farm laborer, and teacher), developed another elaborate simulation from the economic boom of the 1920s to the Great Depression of the 1930s. The teacher intended "to recreate some of the psychological atmosphere" of that era (see Thornton, 1988, pp. 316–319).

As the preceding examples suggest, simulation games include overlapping cognitive and affective goals. Since necessarily simulations are a simplified version of actual world situations, the normative world order they encapsulate should be selected carefully. Students may learn to identify with certain value positions and legitimate some value choices. Students' decisions in the Reconstruction game (Banks, 1970), for instance, influenced not only their own lives, but also the lives of others in their simulated society (p. 59).

"Empathy" is often advanced as a goal in simulation as well as in history instruction more generally. Curiously educators representing otherwise widely divergent value positions laud this goal (Verducci, 2000). In other words, empathy as a goal has little meaning without asking with what and to what end?

Responsible educators, of course, want children to appreciate (although not always endorse) the world through eyes other than their own. In this sense, empathy can be a step beyond mere inclusion of alternative world views – which can boil down to tokenism (Noddings, in press) – to perspective taking: an honest attempt to see the world as people of other times and cultures may have experienced it. "Mr. Bauer's" simulation (Thornton, 1988), for example, succeeded with many of his students in demonstrating the feeling of helplessness and despair that the unemployed experienced during the depression. For further examinations of empathetic goals in history teaching, see Linda Levstik (1990, 1993) and Stephen Thornton (1993).

Simulation has a mixed reputation in history method because it is so easy to identify poor examples of it. Its successful use requires careful planning and pedagogy. The most significant shortcomings seem to be failure to fully integrate simulations into an ongoing instructional program and inadequate provision for debriefing where the classroom simulation is thoroughly evaluated against actual events. (Here the test is less that students acted as had people in the past than appraisal of how, why, and to what effects students had acted and how this meshed with the historical record.) Other common criticisms of simulation-as-practiced include: it can be too obvious for some students, many simulations (including some commercially produced software packages) are too simplistic, the game becoming an end in itself, and that games may foster competition rather than cooperation.

Simulation can also raise special ethical questions. For example, many educators wish for students to develop empathy with the victims of some of the most barbarous events in history such as the Middle Passage and the Holocaust.

Simulations and role plays on such events should be approached with great care, if at all. Although the development of empathy with such victims is clearly laudable and can be powerful, youngsters may be put at risk, for example, "playing" Nazi concentration-camp guards. Inhuman behavior should not be reduced to a legitimated "occupation." Moreover, as Samuel Totten (2000) argues, such simulations may diminish the complexity and horror of the worst of human conduct, in effect, trivializing it.

PROBLEM-SOLVING

This method aims at the development of reflective thought. Dewey (1971) is probably its most celebrated exponent. He argued that the development of reflective thought begins with a "a perplexed, troubled, or confused situation . . . and a cleared-up unified resolved situation at the close" (p. 106). The process of problem-solving overtly rejects premature closure. Rather the intention is to launch an investigation, develop hypotheses to solve the problem, and test these hypotheses. The student must recognize the situation in history as a problem to be solved.

Experimentation with problem-solving was an important activity when Dewey directed the Laboratory School at the University of Chicago a century ago. In a study of Chicago's early development, for instance, sixth-grade children encountered the problem of water supply. The lesson was a relatively informal use of problem-solving. As the city grew in the 1830s, water sources such as the Chicago River and wells became unsafe. A private company took over and laid pipes. These pipes turned out to be too near the surface, however, and froze in cold weather. A water shortage resulted. Students were then invited to suggest solutions to the problem. Some youngsters suggested that the company lay the pipes deeper; others recommended that the city buy out the company and manage the supply. Discussion followed of private versus municipal ownership. The children were then told that the city assumed management of water supply and the various changes that had been made to the present (cited in Tanner, 1997, pp. 70–71).

Other methods have adhered to a more formal model of problem-solving. For example, the Los Angeles schools recommended that, in the sixth-grade study of Latin America, children should learn: "The people of Mexico and the United States have many common elements in their historical backgrounds. The people of both countries have contributed to our cultural heritage" (cited in Black, Christenson, Robinson & Whitehouse, 1962, p. 171). The lesson begins with building an awareness of the problem: what cultural elements the southwestern United States shares with Mexico and the sources of these commonalties. Specific questions arise: for example, Why did the Aztecs not move into the present United States and establish themselves as rulers? Why did the explorers and settlers of the southwestern United States not move southward into Mexico?

The second step is comparison of the present problem with what is already known in order to determine what additional information must be obtained. Tentative solutions are now proposed. A solution based on the difficulty of transportation and on geographic features may be advanced. Another child may have reached the conclusion that because the Aztecs had all they required there was no necessity to expand. Discussion and directions for further research now occur.

Testing of the tentative solutions now proceeds with the teacher providing guidance concerning relevant information sources (e.g. anthropological and geographical data, knowledge of European history from the fifteenth to seventeenth centuries). The culmination of the problem-solving activity is the class evaluation of the assembled data (Black, Christenson, Robinson & Whitehouse, 1962, pp. 172–174).

The foregoing activity on Mexico would likely extend over several weeks; other briefer formal approaches to problem-solving are also possible. For example, Peter Martorella (1985) describes an exemplary problem-solving activity based on two contradictory statements by Lincoln on racial equality (pp. 109–111).

In the 1920s and 1930s Harold Rugg developed what is probably the most comprehensive problem-solving curriculum in history, geography, and the social sciences. He used these subject matters to address the current problems of the United States and the world. Rugg believed that then-existing social studies programs had failed to educate students for the world in which they lived. The logical arrangement of history subject matter championed by historians such as Johnson (1940), Rugg (1996) forcefully declared, ignored the need to base subject matter on an adequate psychology of school subjects and its accordance with human needs and human activities which satisfy them (p. 55).

Drawing from what he called "frontier thinkers" (significantly including thinkers not only from the academic disciplines of the social studies but also from the fine arts, journalism, and the humanities), Rugg based his program on the key problems confronting humankind (Rugg, 1939). Instructional materials, eventually published as textbooks, were developed and field tested. Rugg appreciated that building a comprehensive problem-solving curriculum was beyond the time, expertise, and energy of individual or small groups of teachers. The excellent materials he and his colleagues developed incorporated consonant methods such as planned recurrence of concepts, variety in learning activities, and an active role for students in problem-solving. Rugg maintained that instruction should not be shackled by the boundaries academics lay out for their disciplines. Thus, while Rugg's program is heavily oriented to history as the basis of current problems, he regularly crossed disciplinary boundaries.

B. R. Buckingham (c. 1935) presented the most comprehensive contemporary evaluation of the Rugg program. Buckingham succinctly illustrated Rugg's method with a foreign problem: the Austro-German "Anschluss" or union of the two states. "Of course," Buckingham noted, "one cannot even think of this question without resort to history" (p. 20). He went on to detail factors such as the European balance of power, geographic proximity, shared cultural and language characteristics, and so forth. Next, the then current, geopolitical consequences of the problem (which, in fact, soon came to pass) were considered. As Buckingham concluded:

> The point is that the doctrine of selection of interrelated materials from many fields – call it fusion or what you will – provides not only a way of studying the problems of the past but also of approaching and appraising problems of the present (Buckingham, c. 1935, p. 21).

A generation after Rugg, a related problems method was developed by the Harvard Social Studies Project (Oliver & Shaver, 1966). Aimed at the analysis of current societal problems, this method could be integrated into existing courses such as American history. More recently the Harvard Project has been adapted (and the problems been brought up to date) in a series of units published by the Social Science Education Consortium.

The unit, *Immigration: Pluralism and National Identity* (Glade & Giese, 1989), illustrates this method. Primary sources from the early and late twentieth century are presented. A general public issue, "Should the United States restrict immigration?" is raised. Three component issues are considered:

Ethical or value issue:	"Should all people have the right to settle where they wish?"
Definitional issue:	"Is someone who cannot find work in his/her homeland oppressed?"
Fact-exploration issue:	"Will increased immigration lower the U.S. standard of living?"

The sources seem accessible for most secondary-school students and are rich in dilemmas to spur classroom discussion. The sources are a balanced selection. Historical sources are used to draw parallels between past and present problems in order to invite judgments from students. The problems chosen are ones on which citizens in a democracy should make informed (rather than merely reactive) decisions.

Problem-solving has, with the probable exception of the 1930s when the Rugg materials were highly popular, seldom been a typical method in history instruction. This unpopularity may be partly explained by the widespread (but faulty) assumption that the sequence of steps in problem-solving must be rigidly followed. Dewey (1971) himself emphasized the phases of problem-solving need not proceed in a set order (p. 115). But as noted with disciplinary structure, insistence on rigid application undermines pedagogical inventiveness and likelihood of use. As the example on Chicago water supply demonstrates, informal and brief exercises in problem-solving are possible. The already noted unfortunate tendency in American education to demand all or nothing has tended to result in nothing for problem-solving.

Some form of problem-solving, however, seems inherent in both history and education. With some regularity, for instance, historians propose the relevance of history to problem-solving in the present (e.g. Neustadt & May, 1986). Nor is this matter confined to the United States. As one European historian observed:

> The problem of "teleology" is well known to historians. There is a tendency to notice particularly features pointing towards the present, explaining developments partly in terms of their consequences (whether or not participants were aware of their "contributions" to historical "progress"), and to ignore turnings that led nowhere. While there has been a healthy reaction against this in recent historical writing, it is still the case that certain developments appear more important from the point of view of current concerns than do others (Fulbrook, 1992, pp. xv-xvi).

Levstik (1996) noted that claims for history for "its own sake" in education have always been "specious." History must be directed to some educational purpose. In this respect, the experience of geographers may be instructive for problem-solving in school history. While modernist geographers abandoned the notion of "region," they eventually found ill-informed conceptions of it flourished among the general public (Riebsame, 2000). It seems we are stuck with people – regardless of what academic historians may wish – seeking solutions to present problems in the past. History in general education can only attempt to guide this process so it will be done as well as possible.

FUTURE DIRECTIONS

In this final section, I address three interrelated questions on methods: What kind of research has been useful? What further research is needed? What are the implications for teacher education and writers of instructional materials?

These questions are not merely interrelated; they are significantly interrelated. History method is the poorer because they have so often been addressed in a fragmentary manner.

The most useful research on method incorporates a rationale for what counts as educationally significant, actual classroom settings, and actual subject matter. It should be sufficiently specific so the reader gets a sense of what the method actually looks like. Generally focus should be on widely taught subject matters or, if in-depth treatment of special topics constituting a tiny proportion of the de facto national curriculum (e.g. the Armenian genocide, the Vietnam War) is attempted, its broader significance and applicability to other topics should be explicated. Lee Cronbach's (1966) search for "limited generalizations" – with subject matter of this nature, learning experience of this type, in this amount, produces this pattern of responses, in pupils at this level of development – still seems a useful rule of thumb for methods research (p. 77).

Even if we had Cronbach's limited generalizations, of course, research findings always need to be interpreted before they can inform practice. Research, in this sense, serves more as a heuristic to stimulate educational imagination than an algorithm that prescribes practice (Eisner, 1985, pp. 255–268). Nevertheless, ideally history method and teacher education require far more images of the desirable for educators to ponder than have been available to date.

More productive research on method also needs to begin with frank acknowledgment that the pedagogical demands and purposes of history as a school subject are not always the same as academic history. When a sixth-grade teacher plans a unit on Latin America, for instance, seldom will she be concerned with history alone. The curriculum (and student curiosity) will likely extend to geography, early cultures, and present-day life. Even in secondary schools, it appears that instructional materials are too often written with insufficient balance between the logical and psychological organization of subject matter.

A great deal of time, energy, money, and frustration have been expended in the last century over what should be taught in history. In broad terms, however, there has been long-standing and widespread support for a curriculum containing United States history and the history of other parts of the world (admittedly, with *which* other parts often generating controversy). In contrast, the effective direction of more specific subject matters to desired results – that is, method – has been mired in chronic controversy (Thornton, in press, b). To put it another way, there is agreement that the Civil War, the Aztecs, and the Renaissance should be taught: But what about them? Directed to what results? National standards in U.S. and world history may be viewed as one result of this ongoing controversy. Yet the standards are so specific and so driven by the current interests of academic historians and standardized testing that they may stifle responsiveness to student interests and pedagogical flexibility more generally.

As Nel Noddings (1998) has suggested, there may be subject specific concepts, principles, and skills that are so basic that we may rightly regard them as essential for all students. (These certainly would be far briefer than the encyclopedic national standards that appear to be chafing instruction). But if such basics could be identified we may avoid continuous distractions over whether topic x or topic y should receive more attention. More important, we could then rest assured that as long as this limited number of basics is taught somewhere – for example, the concept of imperialism or the role of women in public life could be taught in any number of eras and world regions – teachers and students would be freed up to focus on topics that interest them.

As stated, systematic research and experimentation is also needed on topics that are universally taught, but are largely unexplored in terms of method. The current work of Brophy and his colleagues (e.g. Brophy & Alleman, 2000) in the primary grades is a step in this direction, but there has been little comparable systematic work in the last 30 years. As Sosniak (1999) pointed out, such research would be of scant help to practitioners if it becomes microscopic. (Of course, there may be exceptions that are of limited scope but illustrative of how to handle topics of a particular type, for instance, complex multiple causation such as the sequence of events leading from the assassination of Archduke Franz Ferdinand to the outbreak of hostilities in World War I.) Rather major topics should be explored (e.g. the American Revolution, Meiji Japan, the Cold War). This exploration must be sufficiently broad in scope to include major topics in U.S. and world history but limited time and energy dictate that it cannot be exhaustive. Rather like teaching a unit to sixth-graders on South America, this research program would provide a brief overview of the continent and then devolve into case studies of countries representing important regional variations (e.g. Bolivia as a "Andean" country, Argentina as an "industrialized" country).

This kind of research should provide the cornerstone of method. At present the liberal arts component of teacher education is loosely coupled with the subject matter demands placed on teachers. Much of what they are expected to teach, teachers are not themselves taught in their undergraduate programs. Although space limitations forbid its full consideration here, Noddings (1999) suggests that the subject matter "in a program designed for teachers can be wonderful and sophisticated." It can provide a deep understanding of elementary subject matter "from a higher standpoint" (p. 214).

Methods courses also require constant rethinking. Over the years these courses have become increasingly divorced from the very subject matters they are supposedly directing to effective results (Thornton, 1997). Of course, multicultural education, special needs students, and "infusing" global perspectives, which *are* treated in methods textbooks, must be part of teacher education and integrated with the Civil War, World War I, and Hindu-Muslim relations in South Asia. But it should not be at the price of excluding the latter (Thornton, in press, a).

Finally, although there remains much to be learned about subject specific history teaching methods, a good deal is known. The last 20 years of research have established that teachers tend the curricular-instructional gate based on "practical" criteria, but why do they embrace some methods as "practical" and reject others? Why is the best of what is known only erratically employed in most practice? In sum, perhaps the main challenge confronting policy-makers, researchers, and teacher educators is to find more effective means of disseminating and institutionalizing what we already know.

REFERENCES

Amherst Project (c. 1961–1972). *Records*. New York: Special Collections, Milbank Memorial Library, Teachers College, Columbia University.

Banks, J. A. (1970). *Teaching the black experience: Methods and materials*. Belmont, CA: Fearon.

Barton, K. C. (1996). Narrative simplifications in elementary students' historical thinking. In: J. Brophy (Ed.), *Advances in Research in Teaching. Vol. 6: Teaching and Learning History* (pp. 51–83). Greenwich, CT: JAI.

Bennett, P. (1967). *What happened on Lexington Green?* (Amherst Project). New York: Special Collections, Milbank Memorial Library, Teachers College, Columbia University.

Beyer, B. K. (1971). *Inquiry in the social studies classroom: A strategy for teaching*. Columbus, OH: Merrill.

Black, M. H., Christenson, B. M., Robinson, R. J., & Whitehouse, L. H. (1962). Critical thinking and problem solving. In: J. U. Michaelis (Ed.), *Social Studies in Elementary Schools* (pp. 150–175). Washington, D.C.: National Council for the Social Studies.

Brinton, C. (1965). *Anatomy of revolution* (rev. ed.). New York: Vintage.

Brophy, J. (1990). Teaching social studies for understanding and higher-order applications. *Elementary School Journal*, *90*, 351–417.

Brophy, J., & Alleman, J. (2000). Primary grade students' knowledge and thinking about Native American and pioneer homes. *Theory and Research in Social Education*, *28*, 96–120.

Brophy, J., Alleman, J., & O'Mahony, C. (2000). Elementary school social studies: Yesterday, today, and tomorrow. In: T. L. Good (Ed.), *American Education: Yesterday, Today, and Tomorrow* (pp. 256–312). Chicago: University of Chicago Press.

Brown, R. H., & Traverso, E. (n.d.). *A Guide to the Amherst approach to inquiry learning*. New York: Special Collections, Milbank Memorial Library, Teachers College, Columbia University.

Bruner, J. S. (1960). *The process of education.* Cambridge, MA: Harvard University Press.

Buckingham, B. R. (c. 1935). *Rugg course in the classroom: The junior-high-school program.* Chicago: Ginn.

Carr, E. H. (1961). *What is history?* New York: Vintage.

Cronbach, L. J. (1966). The logic of experiments on discovery. In: L. S. Shulman & E. R. Keislar (Eds), *Learning by Discovery* (pp. 76–92). Chicago: Rand McNally.

Davis, B. H. (1971). Conceptual teaching in social studies. In: B. K. Beyer & A. N. Penna (Eds), *Concepts in the Social Studies* (pp. 45–48). Washington, D.C.: National Council for the Social Studies.

Dewey, J. (1916). *Democracy and education.* New York: Macmillan.

Dewey, J. (1938). *Experience and education.* New York: Macmillan.

Dewey, J. (1971). *How we think.* Chicago: Gateway.

Eisner, E. W. (1979). *The educational imagination.* New York: Macmillan.

Eisner, E. W. (1985). *The art of educational evaluation.* Philadelphia: Falmer.

Eisner, E. W. (1986). A secretary in the classroom. *Teaching and Teacher Education, 2,* 325–328.

Fenton, E. (1967). *The new social studies.* New York: Holt, Rinehart & Winston.

Fulbrook, M. (1992). *A concise history of Germany.* (rev. ed.). Cambridge, England: Cambridge University Press.

Glade, M. E., & Giese, J. R. (1989). *Immigration: Pluralism and national identity.* Boulder, CO: Social Science Education Consortium.

Goodlad, J. I. (1984). *A place called school.* New York: McGraw-Hill.

Hanna, P. R., & Lee, J. R. (1962). Generalizations from the social sciences. In: J. U. Michaelis (Ed.), *Social Studies in Elementary Schools* (pp. 62–89). Washington, D.C.: National Council for the Social Studies.

Hertzberg, H. W. (1971). *Historical parallels for the sixties and seventies: Primary sources and core curriculum revisited.* Boulder, CO: Social Science Education Consortium.

Hertzberg, H. W. (1988). Are method and content enemies? In: B. R. Gifford (Ed.), *History in the Schools* (pp. 13–40). New York: Macmillan.

Holt, T. (1990). *Thinking historically: Narrative, imagination, and understanding.* New York: College Entrance Examination Board.

Illinois Curriculum Program (1971). Students develop the concept of "imperialism." In: B. K. Beyer & A. N. Penna (Eds), *Concepts in the Social Studies* (pp. 67–75). Washington, D.C.: National Council for the Social Studies.

Jackson, P. W. (1968). *Life in classrooms.* New York: Holt, Rinehart & Winston.

Johnson, H. (1940). *Teaching of history* (rev. ed.). New York: Macmillan.

Kleg, A. A., Jr. (1991). Games and simulations in social studies education. In: J. P. Shaver (Ed.), *Handbook of research on social studies teaching and learning* (pp. 523–529). New York: Macmillan.

Kline, W. A. (1974). *The "Amherst Project": A case study of a federally sponsored curriculum development project.* Unpublished doctoral dissertation, Stanford University, Stanford, CA.

Levstik, L. S. (1990). From the outside in: American children's literature from 1920–1940. *Theory and Research in Social Education, 18,* 327–343.

Levstik, L. S. (1993). Building a sense of history in a first-grade classroom. In: J. Brophy (Ed.), *Advances in research on teaching. Vol. 4: Case studies of teaching and learning in social studies* (pp. 1–31). Greenwich, CT: JAI.

Levstik, L. S. (1996). NCSS and the teaching of history. In: O. L. Davis, Jr. (Ed.), *NCSS in retrospect* (pp. 21–34). Washington, D.C.: National Council for the Social Studies.

Levstik, L. S., & Barton, K. C. (2000, April). *Committing acts of history: Mediated action, humanistic education, and participatory democracy.* Paper presented at the annual meeting of the American Educational Research Association, New Orleans.
Martorella, P. (1985). *Elementary social studies: Developing reflective, competent, and concerned citizens*. Boston: Little, Brown.
Noddings, N. (1983). Why is Piaget so hard to apply in the classroom? *Journal of Curriculum Theorizing*, *5* (2), 84–103.
Noddings, N. (1998). *Teaching for continuous learning*. Paper presented at the annual meeting of the American Educational Research Association, San Diego.
Noddings, N. (1999). Caring and competence. In: G. A. Griffin (Ed.), *The education of teachers* (pp. 205–220). Chicago: University of Chicago Press.
Noddings, N. (in press). The care tradition: Past, present, future. *Theory into Practice*.
Noddings, N., & Shore, P. J. (1984). *Awakening the inner eye: Intuition in education*. New York: Teachers College Press.
Neustadt, R. E., & May, E. R. (1986). *Thinking in time: The uses of history for decision-makers*. New York: Free Press.
Oliver, D. W., & Shaver, J. P. (1966). *Teaching public issues in the high school*. Boston: Houghton Mifflin.
Parsons, T. W. & Shaftel, F. R. (1967). Thinking and inquiry: Some critical issues-instruction in the elementary grades. In: J. Fair & F. R. Shaftel (Eds), *Effective thinking in the social studies* (pp. 123–166). Washington, D.C.: National Council for the Social Studies.
Riebsame, W. E. (2000). The topography of geography: Some trends in human geographic thinking. In: S. W. Bednarz & R. S. Bednarz (Eds), *Social science on the frontier: New horizons in history and geography* (pp. 17–35). Boulder, CO: Social Science Education Consortium.
Rugg, H. O. (1939). Curriculum design in the social sciences: What I believe. In: J. A. Michener (Ed.), *The future of the social studies: Proposals for an experimental social-studies curriculum* (pp. 140–158). Cambridge, MA: National Council for the Social Studies.
Rugg, H. O. (1996). Reconstructing the curriculum: An open letter to Professor Henry Johnson commenting on committee procedure as illustrated by the Report of the Joint Committee on History and Education for Citizenship. In: W. C. Parker (Ed.), *Educating the democratic mind* (pp. 45–60). Albany, NY: State University of New York Press.
Salmon, L. M. (1902). *Some principles in the teaching of history.* Chicago: University of Chicago Press.
Schools Council (1976). *What is history? 4. Problems of evidence: The case of Richard III and the missing princes*. Edinburgh: Holmes McDougall.
Schools Council. (1976). *What is history? Teacher's guide*. Edinburgh: Holmes McDougall.
Seixas, P. (1997). Mapping the terrain of historical significance. *Social Education*, *61*, 22–27.
Shemilt, D. (1980). *History 13–16 evaluation study.* Edinburgh: Holmes McDougall.
Sosniak, L. A. (1997). The taxonomy, curriculum, and their relations. In: D. J. Flinders & S. J. Thornton (Eds), *The curriculum studies reader* (pp. 76–91). New York: Routledge.
Sosniak, L. A. (1999). Professional and subject matter knowledge for teacher education. In: G. A. Griffin (Ed.), *The education of teachers* (pp. 185–204). Chicago: University of Chicago Press.
Stodolsky, S. S. (1988). *The subject matters: Classroom activity in math and social studies*. Chicago: University of Chicago Press.
Taba, H. (1962). *Curriculum development: Theory and practice*. New York: Harcourt, Brace & World.

Tanner, L. N. (1997). *Dewey's laboratory school: Lessons for today.* New York: Teachers College Press.

Thornton, S. J. (1988). Curriculum consonance in United States history classrooms. *Journal of Curriculum and Supervision, 3*, 308–320.

Thornton, S. J. (1991). Teacher as curricular-instructional gatekeeper in social studies. In: J. P. Shaver (Ed.), *Handbook of research on social studies teaching and learning* (pp. 237–248). New York: Macmillan.

Thornton, S. J. (1993). Toward the desirable in social studies teaching. In: J. Brophy (Ed.), *Advances in research on teaching. Vol. 4: Case studies of teaching and learning in social studies* (pp. 157–178). Greenwich, CT: JAI.

Thornton, S. J. (1994). The social studies near century's end: Reconsidering patterns of curriculum and instruction. In: L. Darling-Hammond (Ed.), *Review of research in education, 20* (pp. 223–254). Washington, D.C.: American Educational Research Association.

Thornton, S. J. (1997). Matters of method. *Theory and Research in Social Education, 25*, 216-219.

Thornton, S. J. (in press, a). Educating the educators: Rethinking subject matter and methods. *Theory into Practice.*

Thornton, S. J. (in press, b). Legitimacy in the social studies curriculum. In: L. Corno (Ed.), *Education across a century: The centennial volume.* Chicago: University of Chicago Press.

Totten, S. (2000). Diminishing the complexity and horror of the Holocaust: Using simulations in an attempt to convey historical experiences. *Social Education, 64*, 165–171.

Turner, M. J., & Haley, F. (1975). *Utilization of new social studies curriculum programs.* Boulder, CO: ERIC Clearinghouse for Social Studies/Social Science Education, and Social Science Education Consortium.

VanSledright, B. A. (1997). Can more be less? The depth-breadth dilemma in teaching American history, *Social Education, 61*, 38–41.

Verducci, S. (2000). A conceptual history of empathy and a question it raises for moral education. *Educational Theory, 50*, 63–80.

Wineburg, S. S. (1991). Historical problem solving: A study of the cognitive processes used in the evaluation of documentary and pictorial evidence. *Journal of Educational Psychology, 83*, 73–87.

Wineburg, S. S., & Wilson, S. M. (1991). Subject-matter knowledge in the teaching of history. In: J. Brophy (Ed.), *Advances in research on teaching, Vol. 2* (pp. 305–347). Greenwich, CT: JAI.

Zimmerman, J. (1999). Storm over the schoolhouse: Exploring popular influences upon the American curriculum, 1890–1941. *Teachers College Record, 100*, 602–626.

Zumwalt, K. K. (1988). Are we improving or undermining teaching? In: L. N. Tanner (Ed.), *Critical issues in curriculum* (pp. 148–174). Chicago: University of Chicago Press.

TEACHING METHODS IN PHYSICAL GEOGRAPHY: BRIDGING TRADITION AND TECHNOLOGY

Rolland Fraser and Joseph P. Stoltman

INTRODUCTION

Physical geography was an important topic of study in the earliest civilizations, most notably Greece and Egypt. For example, Eratosthenes estimated Earth's circumference using geometric calculations applied to physical geography. The circumference was calculated to be 28,738 miles, compared to the actual distance of 24,902 miles. To make this calculation, he completed a field study in order to observe and obtain measurements of sun angles at Syene and Alexandria in Egypt. He then mapped the absolute and relative locations of the field study sites, and applied the field study observations to a model, or simulation of Earth relative to the Sun. This accomplishment in 247 BC demonstrated the early interest in physical characteristics of Earth, and the importance of field study, mapping, and simulation as methods in geography research and teaching.

The history of geography as a discipline and its transitions as a pedagogical field have set the context for physical geography as it is taught in K-12 education. Some of that history clearly helps place geography within the broader context of the educational curriculum. Other aspects of the discipline reveal that geography's disciplinary structure enables it to bridge the social and physical sciences, and be recognized as both (National Council for the Social Studies, 1994).

Subject-Specific Instructional Methods and Activities, Volume 8, pages 315–345.
2001 by Elsevier Science Ltd.
ISBN: 0-7623-0615-7

In Chapter 1 of this volume, Jere Brophy presents twelve principles of successful pedagogy. The authors, in reviewing those principles relative to the teaching of physical geography, conclude that each of them is an important element for the teaching of physical geography. There is evidence in the research literature on geographical education that validates the principles, such as the research the effects of cooperative learning on students' collective completion of tasks that required map making (Leinhardt, Stainton & Bausmith, 1998). Similarly, other research could be cited that corroborates the principles presented by Brophy from a recent bibliography on research in geographic education (Forsyth, 1995). In the development of this chapter, the authors selected three instructional methods that complement the principles extended by Brophy. They are field study, map based instruction, and simulation/modeling. Each has a long tradition within geographic research and education, and each has a fitting role to play in teaching and learning in the 21st century.

The example of Eratosthenes and physical geography is ancient, but still significant since much of what we do today, from flying in airplanes to growing crops, is based on Earth's dimensions and orientation to the sun. There are other chapters in the development of physical geography as a school subject, and in order to examine them, we must depart ancient Egypt and fast forward to Colonial America. We immediately observe two things. First, we have missed entirely the Age of Exploration during which time much of the world's physical geography was both mapped and recorded in manuscripts and drawings, and which had a major influence on the importance of the discipline. Second, the fast forward enables us to enter and engage a period of great change in North America as formal education was established. Just what was the methodology of teaching in colonial North America? Physical geography had become an important and respected part of the 17th century educational curriculum. Harvard University taught a course in map and globe study (Warntz, 1964). Harvard's course affected the teaching of geography in the elementary and secondary schools of the time with the expectation that students admitted to Harvard would have prior knowledge of physical geography. Those expectations resulted in a number of geography textbooks being published during the 18th and 19th centuries. *Geography Made Easy* was published in 1784 and was the first in a steady production of textbooks by Jedidiah Morse. The Morse geographies were structured on the premise that almost anyone could use the textbook to teach geography, and promoted read and recite teaching methods. The universal ease of textbook usage made geography an important school subject for all students to study and all teachers to teach (Morse, 1814).

Physical geography continued to gain prominence in the curriculum during the 19th century and the first American physical geography textbook, published in 1855 (Rumble, 1946), introduced several important changes to teaching methods. This was largely the result of research on learning by three prominent educators: Johann Pestalozzi, Arnold Guyot and William Morris Davis. Pestalozzi was a self-directed epistemologist and recognized that direct observation and sense perception were essential to meaningful learning by children. Further, he observed that maps and line drawings included in the geography books along with curricular attention to the study of the home region provided students with the opportunity to directly observe physical geography and compare the empirical evidence with information in the textbook. These were important suggestions and resulted in direct observation and the use of globes and maps being recognized as the best ways to teach geography (Rosen, 1957).

Arnold Guyot, a disciple of Karl Ritter the great German geographer, believed that teaching physical geography was teleological, showing that the great divisions of the world were not random or fortuitous occurrences, but the work of the Creator. Guyot was further influenced by Pestalozzi and postulated that direct observation of the wonders of the environment was the best way to teach physical geography. However, while Guyot's textbooks became especially popular and were widely used in secondary schools, the books contained encyclopedic lists of facts that were not validated by student observation and simulating the physical world, but simply memorized by the students. Guyot thus defied his basic belief regarding the importance of observation (Fairbanks, 1927).

Among the three, it was Davis who relished teaching physical geography. William Morris Davis was educated as a geographer and believed that geography should seek explanations and predict effects, consequences, and conditions of Earth's physical characteristics. Davis became one of the most influential geographers and educators in the history of American geography. He wrote textbooks and teacher's guides on physical geography, including geographic interpretation of the relationship between people and the natural environment (James & Martin, 1979). Due largely to the research by Davis in physical geography and the teaching of the subject, the discipline matured significantly in content, scope, methods, and philosophy. The vague descriptions that were typical of physical geography in the 1700s became intricate descriptions and explanations of the physical environment by 1892 (Stowers, 1962). The scholarly guidance by Davis moved physical geography from the recitation of facts during the Guyot period to a systematic plan of study as an earth science. The methodology promoted by Davis engaged students in discussing the consequences of physical geography for people and the more general environment.

The result was that two classroom methodologies clearly set the course for physical geography during the early years of education in the United States. First, as the prominence of the subject increased, so did the attention to direct observations by students, and the important role of field study. The observations of the geographic attributes within the locale or region became common. Second, the use of diagrams and drawings within physical geography materials became more common (Hill & LaPrairie, 1989). The visual power of geography was recognized both through field study and through graphics. The use of models, diagrams, maps, and drawings and later photographs becomes a strong orientation for teaching methods in physical geography. The catechism form of student recitation changed to the discussion methods for teaching geography by the 1850s and the discussion method was well established by the 1870s as a preferred method suggested in textbooks (Arnold, 1991).

Two significant national reports further elevated the dominance of physical geography at the end of the 19th century. The first was the Committee of Ten Report (National Education Association, 1894), which was the first major reform movement in American education. The physical geography and geomorphology as proposed by William Morris Davis were given sweeping acceptance by the report. The second report was the 1897 Science Section evaluation of physical geography (National Education Association, 1898) which lauded the importance of physical geography and set the stage for its dominate role within geography education. However, at this same time the human and regional elements of geography began to take on greater significance as social and cultural changes swept the United States. Physical geography was relegated to a minor role in geographic education as the discipline changed and as the social studies movement of the 20th century emerged (United States Bureau of Education, 1916).

While the physical geography of the William Morris Davis tradition had been a holistic application of physical principles to explain and predict Earth's processes and the resulting surface features, the larger discipline of geography was undergoing a transformation. The discipline was dividing into many specializations, such as urban, social, political, rural, settlement, and medical geography (James & Jones, 1954). These changes were also reflected in the 1959 Yearbook of the National Council for the Social Studies on geography, which devoted a chapter to physical geography (Kennamer, 1959). While a residual interest was evident, there was no major thrust in favor of physical geography within the social studies. With its disappearance, the teaching methodologies of field study, mapping, and simulation/model building that were particularly well suited to physical geography also declined as common instructional practices.

Such an ebb and flow of an academic discipline relative to social and educational needs is not uncommon. There is a well documented relationship between the academic and research paradigms of physical geography and the teaching of physical geography in K-12 education. As physical geography declined as a disciplinary and research focus, there was a subsequent decline in its inclusion within the curriculum. There was little physical geography included within the social studies curriculum by the late 1970s (Hill & LaPrairie, 1989). Much of the content of physical geography has been subsumed within the earth science curriculum during 1965–1975 (Earth Science Curriculum Project, 1965). Topics that had once been prominent in physical geography, such as Earth materials, Earth changes, climate, and Earth cycles, were incorporated into the newly emerged earth science curriculum. The tradition of geography as an earth science (Pattison, 1964), had been largely neglected by geographers.

The High School Geography Project (HSGP) which was comparable to the Earth Science Curriculum Project, did include physical geography (Pratt, 1970) integrated into a unit entitled Environment and Resources (High School Geography Project, 1979). Despite the inclusion of physical geography within HSGP, the discipline had a greatly reduced profile as either a physical or social science during the period of the 1970s (Libbee & Stoltman, 1988; Vuicich & Stoltman, 1974). The publication of the *Guidelines for Geographic Education* (Joint Committee on Geographic Education, 1984) revitalized the importance of "physical characteristics of place" and once again institutionalized physical geography within the discipline as well as within K-12 teaching. However, a new problem compounded the reintroduction of physical geography. Most social studies teachers who were going to teach physical geography as part of the social studies curriculum had little or no academic preparation with the content of the discipline (Spetz, 1988; Winston, 1984). The strongest case for including substantive physical geography recommended by the *Guidelines for Geographic Education* was within the study of human-environment relationships.

The late 20th century witnessed a major rekindling of interest in the study of human-environment interactions in geography, and subsequently resulted in greater attention to physical geography to the K-12 curriculum (Gaile & Willmott, 1989). The nature-science and nature-humanities issues emerged as scholarly discussions that would have a positive impact on physical geography. For example, *The Earth As Transformed by Human Action* (Turner et al., 1990) examined the significant changes in the geoshpere and biosphere and their effects on people. *Rediscovering Geography* (National Research Council, 1997) further supported physical geography's significant role in studying how people affect and are affected by their physical surroundings.

An unexpected effect of the renewed interest in physical geography with its field study, mapping, and simulation/modeling research methodologies was the application to grades K-12 teaching. Methodologies used by research geographers to collect, process and analyze field study data could also be used by students. The physical processes and human-environment interactions were recognized as potentially meaningful hands-on learning for students. Mapping could be used to study both local and global issues, bringing the power of spatial analysis and informed discussions to the classroom. Students could analyze and simulate/model actual environmental conditions and processes to study their impact on Earth systems just as geographers use simulations and models to investigate interesting environmental changes, problems and issues. Physical geography was demonstrating both a strong intellectual as well as a practical role in learning for students in K-12 education.

PRINCIPLES OF PHYSICAL GEOGRAPHY FOR THE K-12 CURRICULUM

The importance of physical geography for the K-12 curriculum was reaffirmed in the national content standards in geography (Geography Education Standards Project, 1994). The content standards represent the big ideas, or principles of physical geography and physical processes as well as the relationships among and between nature and society. They are key ideas for explaining the past and present issues facing the environment and people, including such issues as natural disasters, resource depletion, and conflicts about land and water, to name just a few (Riebsame, 2000).

The National Geography Content Standards: The national geography content standards have identified the content and principles of physical geography that are important for K-12 education (Geography Education Standards Project, 1994). There are two benefits to the inclusion of physical geography within the national standards. First, the content focus for K-12 education from among a diverse arena of physical geography content is clearly established. This is beneficial since it enables teachers and curriculum designers to more directly address that which is deemed important by the discipline. Second, the content standards clearly develop the linkages between physical geography and society, a missing component in earlier treatments of the subject within the curriculum. Six statements of content and/or principle are identified for inclusion within the curriculum.

(1) Physical characteristics of places: These include climate, landforms, soils, hydrology, animals, and vegetation.
(2) The physical processes that shape the patterns on Earth's surface: There are four physical processes that shape patterns on Earth. They are: atmosphere, lithosphere, hydrosphere, and biosphere. The interactions between those four processes are instrumental in understanding why Earth or some part of it is like it is.
(3) The characteristics and spatial distribution of ecosystems on Earth's surface: Ecosystems form distinct regions based upon climate, vegetation, latitude, wind patterns, ocean currents, and terrain. Ecosystems present both opportunities and limitations for people.
(4) How human actions modify the physical environment: Environmental modifications are made for social, economic, and/or political implications. The geographic context and people's understanding of that context, technical abilities, and the inclination to change the environment are important aspects of environmental modification.
(5) How physical systems affect human systems: The natural environment presents people with many choices depending on their cultural and technological resources. To live in a given environment, people must develop a means to offset natural conditions. When natural conditions prevail, then people must adjust their way of living, or relocate.
(6) Change which occurs in the meaning, use, distribution, and importance of resources. A resource is a part of the natural environment that is given value by the cultural preferences of people. Cultural preferences set the value for a resource, and the natural environment makes its available at a particular location.

These principles of physical geography are interwoven to a large degree within environment and society. Both environment and society have many elements that are observational, the elements may be mapped, and many of the elements may be simulated or modeled. Those teaching methods incorporate classroom practices that enable teaching of the principles of physical geography as well as incorporating sound practices in teaching and in methods used by professional geographers.

THE TEACHING OF PHYSICAL GEOGRAPHY

The teaching of physical geography and the content that is recommended by the national content standards align closely with the principles of pedagogy presented by Brophy in Chapter 1. Geography's traditions as an empirical science are highlighted when students are engaged in field study and observing information in the field and on maps and diagrams. Methodologies that incorporate those techniques and skills demonstrate a firm foundation with the content and theories of the discipline. Syntax geography, or doing and studying geography much as a professional geographer does, links the structure of the discipline closely to instructional strategies.

There are three instructional methods that the authors judge to be important in the teaching of principles, concepts and methods from the field of physical geography. They are: Field Studies; Map Based Instruction; and Simulations/ Models. The selection of those three methods is not intended to convey the notion that other methods are not useful or important, and one could devote discussion space to a range of methods, such as exposition, concept teaching, inquiry, etc. Those methods are applicable across a wide range of school subjects, including geography. The authors believe that the three methods selected and discussed in tis chapter are unique to physical geography. They have been practiced in teaching physical geography since it earliest presence in the curriculum. These methods are similar to the research methods that are used by physical geographers as they investigate the physical geography of Earth. The three methods of teaching discussed include both real life as well as vicarious experiences for the student interacting with the content.

FIELD STUDY AS A TEACHING METHOD

Field study involves students observing, collecting, processing, and analyzing information gathered in the field (outside of school) and using it to explore and possibly answer a persistent problem. Learning does not cease with the analysis of facts in field study, but reexamining the data and proposing reasons for observed relationships are then followed-up in the classroom. The use of field study as an effective method to teach physical geography has been cited in bibliographies (Ball, 1969; Lukehurst & Graves, 1972), local community study guides (Arnold & Foskett, 1979; Committee on Local Geography of the High School Geography Project, 1971), and over a long period in teaching methods publications for teachers at various levels of education (Bland, Chambers, Donert & Thomas, 1996; Everson, 1973; Fairgrieve, 1926; Glynn, 1989; Gold et al., 1991; Gopsill, 1966; Graves, 1965). Most recently, the National Council for Geographic Education has addressed the importance of using field study in the teaching of geography and is encouraging teachers to implement field study as a teaching methodology (Rice & Bulman, 2000). Field study complements the principles of teaching cited by Brophy in Chapter 1. The principles of teaching provide important suggestions that may be used to enhance field study as a teaching methodology.

There are three main ways that field study is incorporated in teaching physical geography. The first is fieldwork in which students actually enter the field and collect information or data directly through observations and recordings. Fieldwork was once common to geography education in the United States, and is still a common element in teaching geography in European, Australian, African, Asian, and South American schools, especially from middle school through high school. In the United States, fieldwork in both the local area as well more traditional residential field camps is being encouraged. A basic premise of fieldwork is that it may be carried out in urban, suburban, and rural schools equally well. An example is the Schoolyard Habitat Project sponsored by the National Wildlife Foundation in which areas adjacent to the urban schools are designed as fieldwork sites (National Wildlife Federation, 2000). The project demonstrates that fieldwork in several subjects, including physical geography and the environment can be completed without large tracts of open land or long travels to distant field study sites.

The second, and much more prominently observed type of field study in the United States is the field trip. The field trip usually takes students to a site away from the school where they may observe, participate, and be informed by experts regarding topics in physical geography. A museum, farm, nature center, wildlife refuge, urban park, school outdoor education center, for example, may be the focus of a field trip. Field trips are highly popular among students and can provide important information not readily available in other ways (Nickelsburg, 1966). On the other hand, from a teaching methodology point of view, it is easy for a field trip to take on what geographers call the "talking head," in which an expert presents long narratives of information, but the students are only marginally engaged either as hands-on or minds-on with the content of the trip.

The third kind of field study is the virtual field trip in which students never leave the school or the computer terminal. Electronic programming and internet connections take students on life-like visits to practically any location on the earth for observing the physical geography of places and the relationships between people and the environment. This vicarious fieldtrip is a modern version of the travel film, photographs, and sound filmstrip combined. While those earlier media have been largely replaced by video tape, CD, and DVD, on line and real time observations have become accessible through the world wide web (www) and the internet. The methodology for the virtual field trip may provide opportunities for students to plan their own virtual excursions. The traditional practice of standing in front of an interesting object in the out-of-doors for an extended explanation and discussion may be replaced by an on screen view of a physical feature, for example. To extend the information beyond virtual observation, the student may link to another www site that has an extended explanation of the physical feature in animated format. Thus, early elementary students studying physical systems may observe a volcano. In the middle grades, the world distribution of volcanoes may be mapped and the spatial relationship with the edges of continental plates may be examined. Models may be called on to demonstrate how the edges of tectonic plates subduct and volcanoes are one result. Volcanoes are still the focus, but the complexity of the explanation from the virtual field trip has been enhanced. At the high school level, the effects of volcanoes as natural events, their role in natural disasters, and the ways that people are affected may be examined using virtual field trips. A well planned virtual field trip provides many of the opportunities to observe conditions in the world away from the school, but it must be carefully planned just as a field trip by bus to a local site of geographic importance must be planned.

Urban schools can often complete walking field trips and engage in more extended field work in the vicinity of the school by using the recreation field or small opens spaces where vegetation, animal life, and weather patterns may be observed. For example, the daily duration of direct sunlight differs between the different sides of a building, and makes a splendid laboratory for taking measurements of length of time for direct sunlight, the types of plants that grow in the micro-ecosystem, and how quickly they grow. Temperature and humidity recordings may be made as well and correlated with the length of direct sunlight and shade over an a period of several months. The resulting micro-climate data may be analyzed to reveal the effects of human activities on biogeography and physical processes. Similarly, students in urban areas may have access to museums, research laboratories, water treatment facilities, and landscape supply firms that will provide opportunities for field trips where they observe or participate in data collection, analysis and reporting.

The principal reason for engaging students in field study in physical geography is the direct involvement in learning about the environment. The students have the opportunity to observe and record their observations as they occur in the world beyond school. Field study is an organized learning activity that engages students in recording information as they stop and make observations at various stations during the field study. The field study should entail hypothesis building and testing, or the testing of "best hunches" that is closer to brainstorming by the students in search of explanation. Whatever the procedure used to raise significant questions from their observations, the students must process the information and construct their own meaning from the observations. The information observed and recorded during a field study needs to be processed in detail and discussed by the students following the field study. In field study, the environment becomes the source for information rather than the textbook or teacher, and time must be spent following the observations to put the information together into meaningful contexts. Textbooks, teacher exposition, and student discussions serve as materials and important resources in good field study practices (Anderzohn & Newhouse, 1959).

BENEFITS OF ENGAGING STUDENTS IN FIELD STUDY

What are the educational benefits of doing fieldwork outside in the environment in a traditional sense? Most traditional fieldwork experiences include the use of instrumentation, or the tools of field research necessary to scientific inquiry. Students gain experience using the instruments for field study, such as measuring tapes, surveying strings, a hand level, topographic maps, altimeter, thermometer, and data sheets for recording observations. Fieldwork may also employ geographic positioning systems (GPS) and laptop computers with geographical information systems (GIS) software. Students engaged in fieldwork get experience using those instruments in authentic situations as they interact with physical the physical environment (Stoltman & Fraser, 2000).

It is generally agreed upon that students engaged in out-of-doors fieldwork receive the following benefits to their general learning (Gold et al., 1991).

(1) Developing observation skills: All students have some skill of observation using their five senses when they arrive in geography class. Fieldwork takes them back into the same or a different environment and demonstrates the importance of those senses. More importantly, fieldwork provides the student with experience in developing a geographic perspective in viewing the landscape, both natural and human.
(2) Experiential learning: Fieldwork provides opportunities to learn through direct, concrete experiences. The experiences one gains in fieldwork are closely related to the experiences one will have in a job or life's general experiences.
(3) Direct involvement of students in responsibility for learning: Fieldwork requires that students plan and carry out learning in an independent manner. While fieldwork may occur in groups, each person has responsibility for a particular component, thus being responsible for the collection, verification, analysis, and reporting of information.
(4) Developing and applying analytical skills: Fieldwork relies upon a range of skills, many of which are not used in the classroom. Observations of patterns in nature, or in the built environment, developing plans to obtain samples that are not readily accessible, taking measurements of dynamic

processes, and using classification, mathematical calculations, and theoretical models to organize and analyze the observed data are essential in fieldwork.

(5) Experiencing real-life research: Fieldwork exposes students to the problems and issues that are faced by people on a regular basis. Not all those problems or issues are negative and fieldwork can focus on processes working successfully in an environment. The important thing is that the fieldwork engages students in the use of messy information that they then have to sift through, organize, and use in testing their hypotheses or comparing their research to models of explanation.

(6) Developing environmental ethics: Working in the environment and analyzing the dynamics of the environment sends a powerful message to students about environmental ethics. The ethical questions underlying the ultimate responsibility for the environment, its quality, and its uses can be a lasting effect of learning through fieldwork.

(7) Teamwork: Fieldwork most often relies upon being a member of a team destined to learn more about a problem or issue, and perhaps to develop a plan to resolve it. Community action groups, public policy groups, and neighborhood watchdog groups all rely on teamwork. Important experiences are carried from fieldwork to community activism on a range of topics in physical geography ranging from environmental aesthetics to sustaining water resources. Fieldwork experiences provide an important teamwork element.

(8) Skills: A powerful set of skills and their application are developed and honed during fieldwork. Beyond the skill of observation are those of synthesizing information, evaluating information, and making reasoned decisions about the efficacy of information. There are also the instrumentation skills that once learned, become life long skills. These include using a compass and topographic map, field sketching and field mapping, using aerial photographs and hand held cameras to obtain a photographic record, and the skills of interviewing people. These are skills developed largely through interaction in the world outside the classroom. Practical problem solving, adaptability to new demands that call upon creative solutions while doing fieldwork, and thinking on the move while making observations and collecting data are important skills that are enhanced through fieldwork.

(9) Uses of Technology: Fieldwork provides the opportunity to apply technology to investigating problems and issues. Laptop computers with spreadsheet software are easily transported to the field. Mapping software permits recording and mapping observations directly while in the field.

> Different types of probes for measuring temperature of water and air, chemical analyses of water, and soil moisture content may be connected directly to a computer and recorded using software. Global positioning systems (GPS) permit recording relatively precise locational information for observation or instrument stations. GPS information may be coordinated with topographic, soil survey, and ecology maps to study relationships within a fieldwork area. Information may be collected photographically using digital cameras and recorded in numeric and textual form using palm pilots and downloaded to a laptop for processing and analysis. Remotely sensed images and aerial photographs may be used to develop base maps of the fieldwork area and to provide information about the character of the landforms, vegetation, drainage pattern, and the land use.

Local fieldwork does not have to end locally for students. The observations made within the local community are the small pieces of a the much larger Earth system. Participation in local fieldwork that complements the information collected and analyzed elsewhere about the larger global environment is within the reach of most schools. In this way, technology and fieldwork make a powerful combination as demonstrated by the Globe Program (GLOBE, 2000). The Globe Program has the participation of 8,500 school is different regions of the world. Students make fieldwork observations in the vicinity of their schools and homes and enter the data from the observations into a computer database. Students from across the earth all complete similar observations. The database is then available to scientists, students, and others to access and analyze the information to search for global patterns and trends that emerge from the 8,500 locations where data are collected by students.

The effectiveness of fieldwork as a teaching method in which students learn equally well or become more proficient than when using a textbook or classroom experience is not well documented in the research. One research report compared the benefits of active, passive, and no fieldwork for groups of 8th and 9th grade students. It was observed that the active fieldwork involvement was superior for the retention of content and concepts (Mackenzie & White, 1982). The effects of experiences in the actual environment on geography learning through field study and excursions are often cited as important research questions in geographic education research (Forsyth, 1995). However, there has been little direct research on the topic as it applies to physical geography and human-environment interaction.

Principles of teaching and conventional wisdom do suggest that hands-on, minds-on study outside the normal classroom and in the natural or human landscape do have compelling benefits. Of the benefits that can be readily observed and recorded are those related to inquiry as a process for teaching and learning. In fieldwork, the students begin with the questions, or hypotheses, and then engage in a process of inquiry that leads them closer to answering the question or testing the hypothesis. The application and immersion within the inquiry model of learning may be the most obvious learning benefit from fieldwork. That immersion also brings other questions and problems to the forefront, and is a persuasive context for the synthesis of ideas and information. Fieldwork allows and requires the participants to view the world around them as a system of interwoven parts, like the texture of a fabric with its interlaced strands.

Planning the field study is an opportunity for students to assist with the procurement of supplies, materials, and equipment they will need to make and record their observations. Route planning and a logbook of observation points may also be completed with the assistance of the students. The planning for the content and information also serves as an advance organizer for the activities the students will complete in the field, the data they will collect, and the way they will process and analyze the information.

VIRTUAL FIELD STUDY

While the learning benefits of actually taking students to field study sites both near and distant from the school are offset by the problems encountered in arranging transportation, liability issues, adequate supervision, and adequate field study equipment, many of those practical headaches are avoided with the newly emerging virtual field study. Virtual field studies are relatively new to physical geography, having entered the world of K-12 education since the mid-1990s. The authors believe that it will be some time before student learning from virtual reality fieldwork is documented to show if it effectively replaces the actual experience of being in the field. However, it promises to greatly enhance the access by students to field study experiences that rest somewhere between vicarious and direct in physical engagement.

Virtual field study is of two kinds. First, there is the remote camera that enables a person to view and record observations at some distance from the site. Television stations often use remote cameras to show the weather conditions outside, or to monitor the traffic on a particular roadway. Using this same technique, it is possible to virtually collect data in real time to study physical geographic processes from a distance. For example, students may study the physical processes of soil stabilization and natural re-vegetation on Mount St. Helens remotely. Furthermore, newly developed virtual excursions demonstrate considerable creativity in studying such environmental problems as landfills and the long- term storage of nuclear waste (Kelly, Ott, Tashiro & Semken, 2000). In both of these topics the problems faced in obtaining access to and real time data about landfills and nuclear waste sites are overcome by the virtual field study. Similarly, there are many places where physical access is not possible, but which are inherently interesting to students, such as viewing the interior of a volcano. To make information accessible and add some of the adventure and discovery of field geography, computer based, virtual fieldtrips to have been developed and are being explored by students (Crampton, 1999).

The second type of virtual field study incorporates virtual reality. The main difference between the virtual and virtual reality is that the virtual field trip remotely takes the student to the site, but the virtual reality fieldwork uses programmed information to design a wide range of scenarios that can be altered between the real and the reconfigured. In virtual reality, the essential information is programmed directly into a computer, somewhat like a flight simulator is programmed for pilot training. The images one sees are often based on real data and the computer simulates physical processes, but it is not real time as one would observe if actually present at the site. A virtual reality helicopter flight thorough the Grand Canyon reveals the physical geography, but it is stored in the computer's memory and plays back as a field of vision on the screen showing the canyon walls in nearly 360 degree fields of vision. The interesting thing about virtual reality is that it sometimes enables the student or the programmer to change the variables. A lake and its shoreline represents the interaction of the physical processes of erosion and deposition, as well as plant succession and land use. For example, fertilization of lakeside, residential landscaping is usually detrimental to the water quality in the lake. The physical process of runoff and contamination can be determined by slope, precipitation, soil porosity, and vegetation based on fieldwork. The value that residents place on landscape aesthetics, the tradeoffs they are willing to make for green grass versus lake water with naturally balanced aquatic growth, and the willingness or resistance to changing values are critical to the same problem. The physical processes are well known and may be demonstrated using virtual reality. However, the human-environment equation depends upon the choices of the people to either fertilize their lawns less or have a vegetation-clogged lake. Many of the human-environment interaction variables incorporated within virtual reality field study involve choices and consequences that may be examined several different ways (Educational Web Adventures, 1998).

MAP BASED INSTRUCTIONAL METHODS

The map is often referred to as the geographer's art (Haggett, 1990). Maps are used to investigate most of the problems and issues that geographers are interested in pursuing. Maps also provide the basis for much geographic instruction, which may be described as inquiry using maps. As such, maps meet three important criteria for enhancing teaching and learning. First, maps are graphic representations of information. Their graphic nature allows the map user to view the non-textual information and relate it to Earth's surface. Thus, there is a close correspondence between the map and reality once the scale of the map is considered, but small landmarks may not appear on a small-scale map. Second, the maps used in map based instruction may include those designed and produced by the learners as map makers. The process of making maps places the student in the role of producer or presenter of information or knowledge as a result of making a map. Third, while one map usually shows discrete information, it is the combination of several maps that has the power to truly investigate an issue about the earth. This entails the production, comparing and contrasting of spatial patterns and locational information on maps to identify spatial relationships.

AN APPROACH TO MAP BASED INSTRUCTION

Maps may be used as a teaching methodology for physical geography in several different ways. Maps may be arranged in sets showing the same area, but with each map showing different information, such as climate, vegetation, terrain, etc. The maps are then compared to determine which of the patterns or locations are similar and which are dissimilar. This teaching methodology is called a map autocorrelation technique. The process of using several maps was once dependent upon available printed maps or maps that were produced by the students, but it is now common to use computer maps and Geographic Information Systems to perform the same procedure electronically. The maps become the means to inquire about spatial patterns and their relationships. The most complex reasoning occurs when it is possible to use the spatial patterns on maps to suggest cause-effect relationships in the data that are shown.

Another aspect of map-based teaching methodology is the use of mental maps, or those graphic images of the environment that people carry around in their heads and that can be drawn on a sheet of paper when the person is requested to do so. In this case, the human mind functions as the geographic information system and overlays new information on prior mental maps in the construction of new knowledge. The mental maps are the product of direct experience with the environment at the local scale. Students traveling to and from school develop a mental map of the route in their minds, knowing where to turn off one street to another in order to locate their school or their home. Drivers of cars develop similar mental maps. Other mental maps are the result of using maps as models of larger areas where one has never traveled. For example, students who have not traveled to the Rocky Mountains may often locate the Rocky Mountains quite accurately on their mental map of North America. This is the result of their repeated use of maps that model location relative to the Pacific Coast, Great Plains, and other features that are included on physical maps that they use in class.

Other aspects of map-based instruction focus on their power as graphic organizers for information, as a means to process information so an idea is communicated clearly, and to present information that is not represented as efficiently or clearly in a written or verbal form. The three steps of organizing, processing, and presenting information as maps is a long-standing teaching methodology within physical geography. However, too often the map is used strictly for presenting information, such as the location of states or major cities. While location is important and maps serve a prominent function in that endeavor, maps do have considerably more analytical use than just as objects for coloring or labeling.

An initial emphasis on using maps as an instructional methodology in United State's schools began during the development of the High School Geography Project (High School Geography Project, 1970). Maps were used as the basis for teaching about and analyzing human-environment relationships, and discovering the principles of physical geography that helped explain those relationships. For example, major questions in physical geography were posed for the student, such as "How are different habitats used differently by different cultural groups?" That form of question was quite a contrast to asking, "What is this habitat like?" The first question required the students to consider settlements, road patterns, and other cultural indicators on the map along with physical geography in constructing their response (Gunn, 1972), while the latter question asked for a descriptive statement of environmental conditions.

In the 1990s, the Association of American Geographers developed a curriculum project called ARGUS that was based in large part upon map based instruction (Association of American Geographers, 1997). The materials were designed on the premise that visual information such as maps present enables students to reason from a graphic as well as a textual or verbal basis. The materials development was based largely upon two research studies. The first compared the immediate and delayed recall of information from texts passages that were either enhanced with maps or were devoid of maps, and students more accurately recalled information when maps were included as part of the text (Gillmartin, 1982). The second study compared the design of maps for two areas, one with a rectangular grid and the other with an irregular street pattern. The study reported that students used the information from the rectangular pattern more successfully than from the irregular pattern (Freundschuh, 1992). Physical features on maps tend to be irregular in pattern. Therefore, as a trade-off to adjust for the irregularity of patterns on maps with physical geography information, special maps were designed so that the categories of information were more easily discernible to the map user.

The complexity of physical geographic information was accommodated by designing maps with a limited number of classifications for information and a similarly limited number of gradations of information within any specific classification. For example, the key for ecosystem maps showed large categories, such as deserts, grasslands, forests, mountains, and tundra, as opposed to sub-classifying forests into sub-species or latitudinal ranges for various types of forests which result in a more complex map. Thus, the maps for ARGUS were designed to be less complex, yet to present information necessary so that students could effectively use them in a map based methodology in teaching physical geography. ARGUS also proposed an instructional methodology that used map autocorrelation in determining the relationships between sets of information that were shown by the maps.

Following is an example of a map based instructional methodology from ARGUS. Three sets of physical geography maps of the 48 contiguous states are used in the instructional method (Figs 1a, b, c). (Note that Alaska and Hawaii are not included on the map since the ranges of agricultural conditions are comparatively small.) The students are also provided eight sets of mapped data showing the pattern for agricultural activities (Figs 2A, B, C, D, E, F, G, H). The instructional methodology is map based in order to address the geographic question: How do physical systems affect human systems?

The instructional methodology incorporated within ARGUS engages the students in studying the information and patterns for local relief, growing season, and precipitation and using that information to explain the patterns for the eight important crops shown by the second set of maps. The task for the student is to propose and explain the crop pattern that is most feasible within the environmental conditions shown by the maps (Fig. 3).

Map based instruction engages students in using spatial information, looking for spatial patterns, comparing patterns for spatial relationships, and using one or more patterns to explain the presence of yet another pattern (Cruz & Bermudez, 2000). For example, in the case of corn production, there are places in the United States where the local relief and precipitation are well suited for

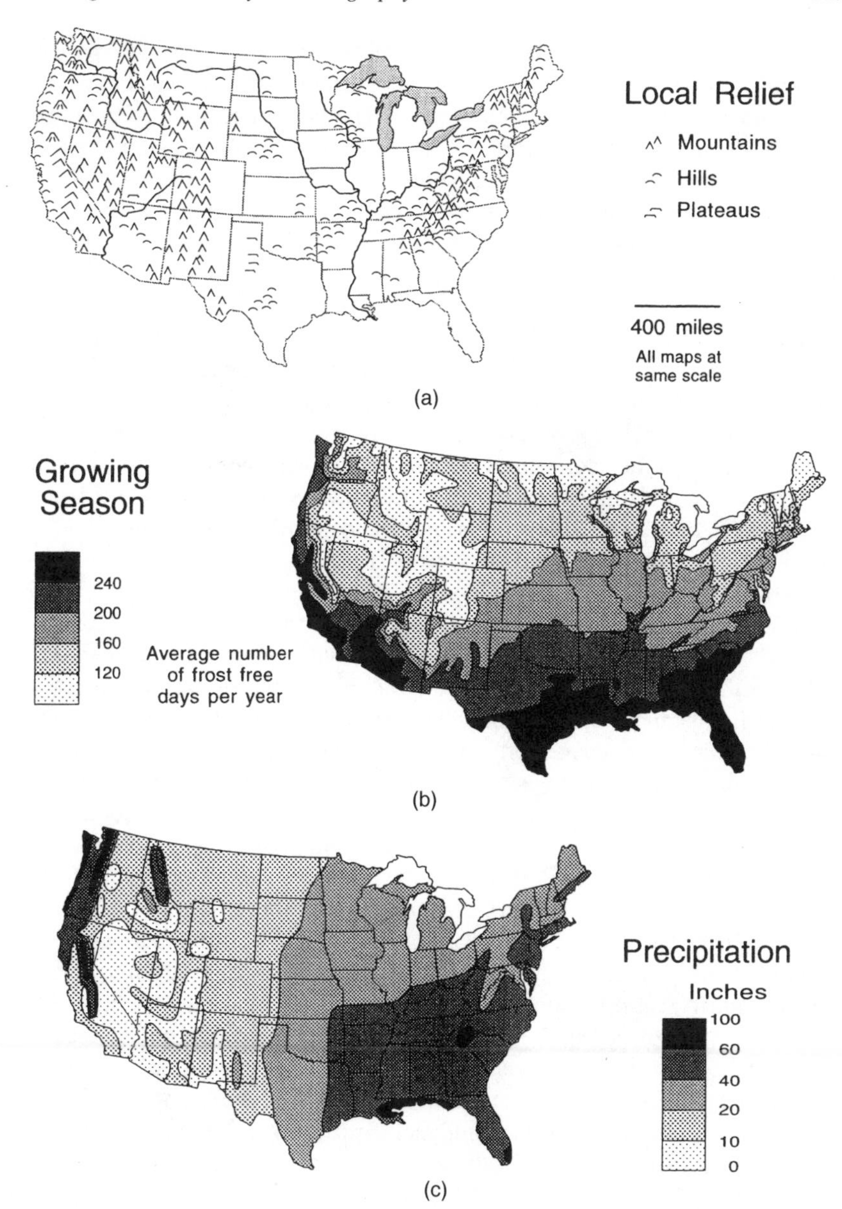

Fig. 1a, b, c. Environmental conditions that affect crop production. The following physical features and environmental conditions affect where particular crops may be grown: (a) Local Relief; (b) Growing Season; (c) Precipitation.

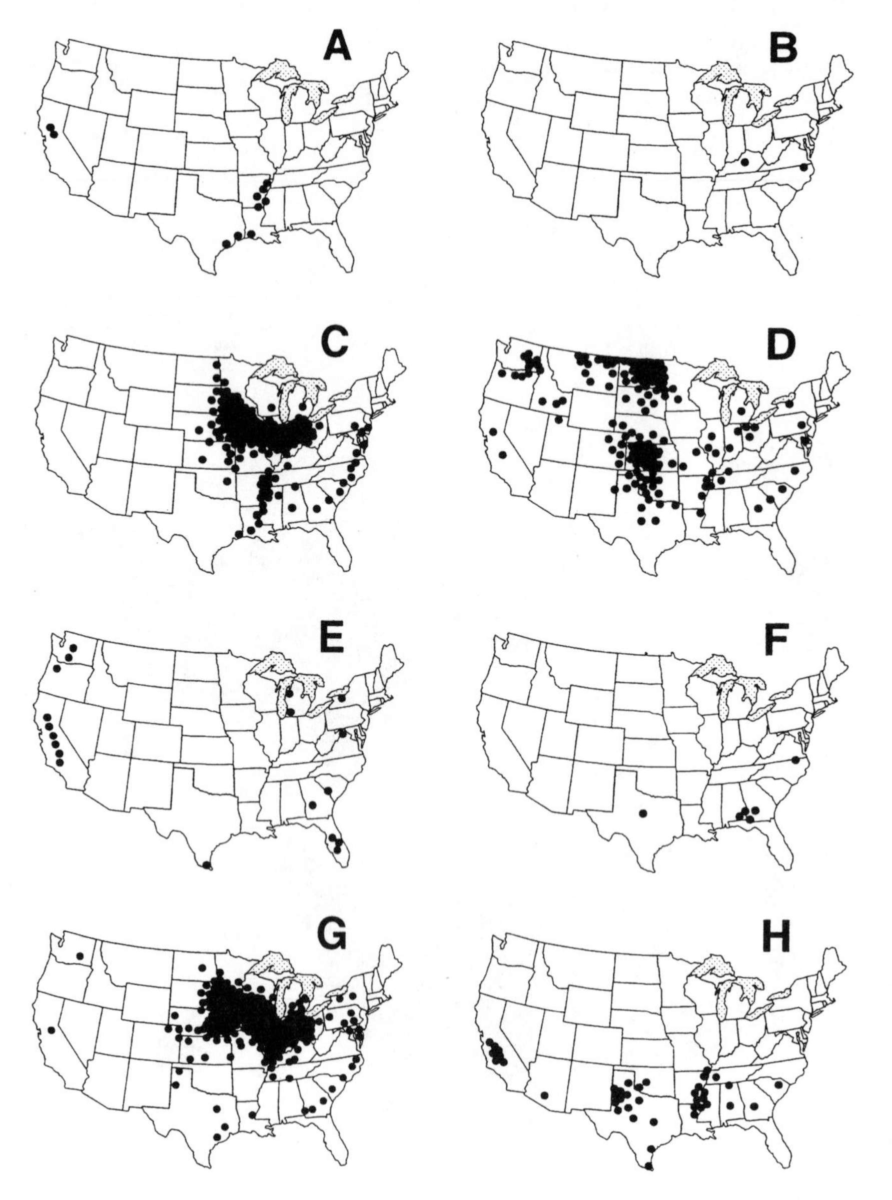

Fig. 2A, B, C, D, E, F, G, H. Eight important crops in the United States. The maps show the distribution of eight important crops.

What is the relationship between the environmental conditions and the pattern for each crop? The number in parentheses (40) tells how many thousands of square miles are used for the crop. You are to match the letter of the crop map with the description that best suits the environment for that crop. Use the environmental conditions map to help you do this. Write the letter of the map before the crop that has a spatial distribution best explained by the brief written description and by the patterns on the Environmental Conditions maps.

_____ Corn (90) is used mainly for hog feed; comparatively small amounts are eaten by people, turned into gasohol, and made into corn syrup for use in candy, soft drinks, and cooking. Corn needs good soil, at least three frost-free months, and about 22 inches of precipitation. Erosion is a serious problem in hilly areas; weeds cause problems for corn in low and wet places.

_____ Cotton (15) cloth has gained in popularity, but the crop is grown mainly for its seed that is used to make cooking oil and is added to some chemicals. Cotton needs seven frost-free months and about 30 inches of precipitation. It may be irrigated if there is not adequate natural precipitation during the growing season. In the early 1900s the boll weevil invaded the Southeastern United States, nearly destroyed the crop, and eventually drove many farmers from the land because their main crop, cotton, was no longer produced.

_____ Orchards (7) include citrus fruits in frost-free places and apples and other tree fruits and nuts. These crops sometimes grow in mountain foothills and near lakes.

_____ Peanuts (2) do best in sandy soil with a long growing season. President Jimmy Carter owned a peanut warehouse in his home state.

_____ Rice (4) needs flat land, plenty of water, and at least 6 frost-free months of growing season. In some locations rice is irrigated and in others it receives enough precipitation.

_____ Soybeans (86) are used for animal feed, cooking oils, and industrial chemicals. Their roots restore nitrogen to the soils. They are a good crop to grow with corn and cotton.

_____ Tobacco (1) growing is limited to a small area with a growing season of 200 days, and 40 to 60 inches of precipitation. The use of tobacco products is known to cause cancer and other health problems.

_____ Wheat (83) is used for cereal, bread, and pasta. It can grow in a drier climate than corn, but has a less valuable to the farmer than corn.

Fig. 3. Task response sheet.

growing corn, but the growing season is too short. The pattern, in this case the absence of corn, is explained by a growing season.

USING THE COMPUTER FOR MAP BASED INSTRUCTION

The computer adds new dimensions to map based teaching methodologies. There are maps of the world available over the www that have underlying electronic grids that enable several different sets of information to be shown on the same map (CIESIN, 2000). For example, worldwide water pollution information can be overlain on the population map to reveal the spatial relationship between population density and safe, clean supplies of water. In a similar way, environmental quality patterns may be compared for two or more regions that are under population density stress, such as Northern India and Central Mexico. In its electronic form, the map based instruction methodology has the potential to bring the world and nearly all the information we have regarding physical geography and human-environment interactions to become useful classroom resources in learning and teaching. While this availability presents an opportunity on the one hand, on the other hand it requires careful planning by the teacher to assure that intended learning outcomes result from using the mapping technology.

Both paper and pencil maps as well as computer map teaching methodologies may be used with students ranging from K-12. Very young students can compare maps of vegetation for a region, their local community for example, with the presence of wild animals, birds, terrain features, and streams and raise questions and seek explanations about the relationship between the physical environment and how people modify it or are influenced by the local environment. Older students should apply the same basic mapping skills to explore similar human-environment issues, but in a more complex way and which build upon prior knowledge. For example, middle school students may investigate the global demands for more crop and pasture lands in relationship to the clearing of tropical forests. At the high school level, the students might use maps to investigate the spatial relationships between precipitation, temperature, soil erosion, and clearing of the tropical rainforest. Scientists have made many of these data from study sites and research stations available over the world-wide-web. The spatial relationships observable on maps are a powerful means to engage in discussions and perhaps take action to influence or change public policy in places where negative human-environment interaction are occurring. Students at the high school level should also consider alternatives to land use practices within the economic and cultural contexts of the populations affected.

USING SIMULATIONS/MODELS IN PHYSICAL GEOGRAPHY

Simulations are models that replicate accurately the interaction between content or topics and processes by which they occur. For example, computers may simulate the development of hurricanes based on data that have been stored in a databank. There is no actual hurricane, but all of the conditions necessary are simulated and the development of the tropical storm can be viewed as a simulation or model on the computer screen. Other simulations directly involve people, such as a simulation game entitled the Oregon Trail (MECC, 1996). In the latter case, students make decisions and interact with the information they have available that recreates conditions in the mid to late 1800s along the trail. The method used in the Oregon Trail incorporates physical geography as a challenge to the migrants traveling westward across North America by wagon train. Oregon Trail, like many simulations, stresses the involvement of students actively rather than passively. This does not imply that it is physical activity, since simulations often focus on active thinking and problem solving as opposed to physical activity. Some simulations may be computer based while others will rely on paper and pencil or game boards. In general, simulations usually engage students in discussion with other students and utilize both traditional as well as electronic educational resources (Educational Web Adventures, 1998).

Geography was an early participant in the development of simulations and models as methods of instruction. For example, the Game of Farming from the High School Geography Project was the simulation of human-environment interaction using actual climatology and agricultural economy data from the periods prior to, during, and following the Dust Bowl (High School Geography Project, 1970). In the Game of Farming, students played the role of farmers and used the data about past patterns of the physical environment (precipitation, temperature) and market conditions for crops (field preparation costs, market prices) to make choices about what crops to plant and how to manage their land. The uncertainty of the conditions that farmers regularly face presented major challenges to each student's survival through years of less than adequate growing conditions, drought and economic depression. The Game of Farming was widely used since it exhibited a high degree of authenticity. It also challenged students in data processing, decision-making, and the evaluation of past choices regarding farming prior to making new choices. Students engaged in discussions of their options, the consequences of each decision, and in making plans for the future.

There are strong arguments for using simulations/models as teaching methodologies.

(1) They are novel approaches to content, concepts, and skills that are unlike most other ways students approach learning.
(2) They require active student participation, using hands-on and minds-on tasks.
(3) They engage students in the application of content (and the principles underpinning the content) that enhance learning.

(4) Simulations often have a game aspect (competition) and thus evoke an emotional response.
(5) Simulations may include practical examples of abstract ideas and concepts and enhance the meaningfulness of those ideas for learners.
(6) They increase interpersonal and group communications, setting the stage for the development of skills in listening, conversing in an informed manner, and civil discussion over issues that emerge during the simulation (Bartlett & Cox, 1982).

Physical geography simulations and models also have elements that make them distinct as instructional methodologies and that provide a strong rationale for incorporating them within classroom instruction. Foremost, a simulation involves authentic leaning in which a real situation is duplicated for its essential features. Those features may be specific to the content of physical geography, or the processes that occur with the physical environment. In the case of human-environment interaction, the simulation generally entails working with environmental and human advantages and constraints in arriving at a solution to a realistic problem or issue. There are usually two or more student players. However, it may be just one student and a computer. Students are normally divided into groups or teams, and have roles within the simulation. Rules are clearly spelled out to inform the students what they are expected to do, or their capacity to take action, in the simulation. Outcomes for the simulation are usually specified, or a system is clearly defined which enables student players or teams to determine their points gained for specific actions or behaviors during the simulation. Lastly, there is a process for disseminating the information gained in the simulation or to inform others of the decisions taken and their intended consequences (Walford, 1973).

Numerous simulations, including those focused on physical geographic processes and human environment interactions have been developed (Muir, 1996). Whereas simulations were initially developed either as components of curriculum projects and linked to specific content or issues, or developed by teachers to align with their instructional programs (Gress & Scott, 1996), a common trend during the 1990s was for simulations to be developed by commercial publishers. The result has been that many simulations are fun to do, keep the students occupied for an extended time, and have elaborate pre and post classroom components. However, commercially developed simulations are not designed to complement the context of the curriculum, nor aligned with the curriculum. Simulations acquired from commercial sources or the www, for example, require a careful evaluation of their benefits to the curriculum and learning of students prior to their use.

The teaching methodology underlying the use of simulations first requires a basic familiarity by the teacher with the content, concepts, and skills developed by the simulation. Once those conditions have been met, then the following criteria for selecting and using simulations should be followed.

(1) The simulation should be authentic for the students? There should be a discussion among the students and teacher about the purposes for the simulation and its objectives.
(2) Simulations that are especially adapted to the class should be used. Complex simulations should occur after the class has gained some familiarity with the operation of simulations. Simulations that go on for several days should be used later in the school year after using several that have taken a shorter time to complete.
(3) A thorough debriefing and review of the simulation and its context within the larger unit of instruction should be completed following the simulation.
(4) Manage the class in an unobtrusive fashion. Simulations require direction and redirection by the teacher just as any lesson requires keeping track of the pace and movement towards the goals (Bartlett & Cox, 1982).

Simulations in physical geography and human-environment interaction have been used as instructional methods and learning materials for a number of years. However, their effects on learning have not been widely researched and reported. The development of simulations in physical geography began in earnest in the 1960s and peaked during the early 1980s. Many of the simulations were developed by classroom teachers and special textbooks and handbooks on designing simulations, models, and games were published that encouraged teachers to design their own simulations (Livingston & Stoll, 1973; Stadsklev, 1974). Pierfe completed a study of fifth grade students using simulated global wind and ocean current patterns on a game board to show the effects those physical processes had on the routes of 15th century ocean explorers (Pierfy, 1972). Students who participated in the simulation retained more information about global wind patterns and ocean currents as well as being better able to demonstrate the effects of those physical processes on exploration than students who followed a more traditional map study approach to the same information. Clegg reviewed simulations developed in social studies and reported that over 800 articles, documents, and books on simulations and games had been published between 1955 and 1989 (Clegg, 1991). The sheer quantity of attention to simulations revealed by Clegg's review suggests both a high level of interest as well as an underlying belief that they were a beneficial methodology for teaching social studies. Clegg's review of simulations was published just as the development of simulations using computers was beginning to impact instruction using simulations (Levin, 1991). The movement of simulations to computers reduced the number of teachers involved in producing simulations and increased their role as consumers of simulations, models, and games that were computer based. This occurred largely because the number of teachers with proficiency in software design and programming was limited, and that was a necessary component for developing simulations. In addition, the commercial value of computer based simulations greatly increased as schools began to acquire computers and computer laboratories, and teachers were in search of opportunities to use the computers. Simulations were a logical way to apply technology to the teaching of physical geography.

The ARGUS Project has adapted a paper and pencil simulation to a computer based simulation. This was accomplished using the data from the map based instruction focusing on agricultural conditions presented earlier. An electronic grid was developed in the shape of the United States, and the local relief, precipitation and frost free maps were added as small drop down menus. The students are given the task to select where in the United States they will choose to plant any of the crops available. Depending on the location on the map and the crop, the simulation assigns values. Attempting to grow orange groves in Minnesota is awarded with a negative value, whereas Southern California and Florida receive maximum points. The simulation operates from a compact disk and uses actual climatic and terrain information in a generalized form (Association of American Geographers, 1997).

The simulation is aligned with the national content standards in geography. It addresses the physical characteristics of place and demonstrates how physical systems affect human systems. It complements the study of United State history and geography since it provides physical geographic characteristics and practical applications of what those characteristics of climate and landforms mean to the production of agricultural products that students encounter in their daily lives. The simulation provokes the students to ask the questions, "If I do this on the basis of what I know, then what will be the consequences?" Perhaps the most challenging aspects of simulation use in physical geography is to enable students to take informed risks and make informed decisions relative to the physical environment. While orange growing in Florida is not without risks, orange growing in Minnesota is a wild guess. Environmental information enables students to verify the pros and cons and reduce the risks as much as possible. It is that reasoned decision making process that is a major strength of simulations.

CONCLUSIONS

The content oriented methods of teaching physical geography and human-environmental interaction have both traditional and technological facets. There are strong traditions in the discipline for field study, map based instruction, and simulation. Geography as a discipline has been greatly impacted by modern technology. The traditions of field study, map-based instruction, and simulation are not exempt from those changes. The strengths of physical geography and human-environment interaction topics, issues, and methodologies have been in the authenticity of the content and its application. Initially, field study meant going into the out-of-doors and observing, recording, processing, analyzing, and presenting information that was observed. Today, those same steps are applied to methodology in teaching physical geography, but the students may have a virtual field study in which they observe the elements of physical geography from a distance, have the capabilities to ask similar types of questions, and obtain similar types information as if they were on site in the field. The methodology of field study remains similar, but the technology used to carry it out has changed dramatically. This anticipates that changes in future years will make it feasible for students to participate in field study components of fieldwork, field trips, and virtual reality field study on a regular basis. Similarly, computer mapping programs and databases that are very easy to use are permitting teachers to select map-based teaching methods to investigate how people affect the physical environment and how the environment limits the activities of people. The importance of environmental information in making decisions and addressing problems is a powerful element in the use of maps to examine spatial relationships between the various patterns that show information about Earth's surface. Simulations enable students to address one of humankinds most frequently asked questions, "What if?" Both paper and pencil and electronic simulations for studying physical geography and human-environment interaction permit not only the question to be asked, but also experimentation by changing the conditions invoked by the question. Simulating and modeling have been enhanced greatly by modern technology in physical geography and human-environment interaction, ranging from natural disasters and their effects to the long term effects of global warming on vegetation patterns and on the health of the human population. Field study, map-based instruction, and simulation/modeling each bring traditions of successful teaching methodologies in physical geography as well as promises for advancing technological based instructional methods in the future.

REFERENCES

Anderzohn, M. L., & Newhouse, H. R. (1959). Teaching Geography Out-of-Doors. In: P. E. James (Ed.), *New Viewpoints in Geography* (pp. 177–199). Washington, D.C.: National Council for the Social Studies.

Arnold, P. S. (1991). *A Description of a Content Analysis of Elementary Geography Textbooks From 1789 to 1897*. Unpublished Ph.D., University of Akron, Akron.

Arnold, R., & Foskett, N. (1979). Physical Geography in an Urban Environment: Two Fieldwork Examples. *Teaching Geography, 5*(2), 60–62.

Association of American Geographers (1997). *Activities and Readings in the Geography of the United States*. Washington, D.C.: Association of American Geographers.

Ball, J. M. (1969). *A Bibliography for Geographic Education*. Athens, GA: Geography Curriculum Project, University of Georgia.

Bartlett, L., & Cox, B. (1982). *Learning to Teach Geography*. Milton, Queensland: John Wiley.

Bland, K., Chambers, B., Donert, K., & Thomas, T. (1996). Fieldwork. In: P. Bailey & P. Fox (Eds), *Geography Teachers' Handbook*. Sheffield: The Geogaphical Association.

CIESIN (2000). *Socioeconomic Data and Applications Center*. Columbia University. Available: http://sedac.ciesin.columbia.edu [July 19, 2000].

Clegg, A. A. (1991). Games and Simulations in Social Studies Education. In: J. P. Shaver (Ed.), *Handbook of Research on Social Studies Teaching and Learning* (pp. 523–529). New York: Macmillan.

Committee on Local Geography of the High School Geography Project (1971). *The Local Community: A Handbook for Teachers*. New York: Macmillan.

Crampton, J. (1999). Integrating the Web and the Geography Curriculum: The Bosnian Virtual Fieldtrip. *Journal of Geography, 98*(4), 155–168.

Cruz, B. C., & Bermudez, P. R. (2000). Mapping the Caribbean Region. *Social Education, 64*(2), 83–87.

Earth Science Curriculum Project (1965). *Investigating the Earth*. Denver, CO: Smith-Brooks Printing Company.

Educational Web Adventures (1998; June 1, 2000). *Amazon Interactive (1st), [www]*. Educational Web Adventures. Available: http://www.eduweb.com/amazon.html [July 24, 2000].

Everson, J. (1973). Fieldwork in School Geography. In: R. Walford (Ed.), *New Directions in Geography* (pp. 107–114). London: Longmans.

Fairbanks, H. W. (1927). *Real Geography and Its Place in the Schools*. San Francisco: Harr Wagner Publishing.

Fairgrieve, J. (1926). *Geography in School*. London: University of London Press.

Forsyth, A. S., Jr. (1995). *Learning Geography: An Annotated Bibliography of Research Paths*. Indiana, PA: National Council for Geographic Education.

Freundschuh, S. M. (1992). *Spatial Knowledge Acquisition of Urban Environments from Maps and Navigation Experience*. Unpublished Doctoral Dissertation, State University of New York, Buffalo.

Gaile, G. L., & Willmott, C. J. (Eds) (1989). *Geography in America*. Columbus, Ohio: Merrill Publishing Company.

Geography Education Standards Project (1994). *Geography for Life: National Geography Content Standards 1994*. Washington, D.C.: National Geographic Society.

Gillmartin, P. P. (1982). The Instructional Efficacy of Maps in Geographic Texts. *Journal of Geography, 81*(4), 145–150.

GLOBE (2000). *Global Learning and Observation to Benefit the Environment*. U.S. Government. Available: http://www.globe.gov/ [July 12, 2000].

Glynn, P. (1989). *Fieldwork – Firsthand: A Close Look at Geography Fieldwork*. London: Crakehill Press.

Gold, J. R., Jenkins, A., Lee, R., Monk, J., Riley, J., Shepherd, I., & Unwin, D. (1991). *Teaching Geography in Higher Education*. Oxford: Basil Blackwell.

Gopsill, G. H. (1966). *The Teaching of Geography*. London: Macmillan.

Graves, N. (1965). Teaching Techniques: Direct Observation. In: UNESCO (Ed.), *Sourcebook for Geography Teaching*. London: Longmans/UNESCO.

Gress, G., & Scott, R. W. (1996). Field Trip Simulation: Developing Field Skills in a Junior High Classroom. *Journal of Geography, 95*(4), 154–157.

Gunn, A. M. (Ed.) (1972). *High School Geography Project: Legacy for the Seventies*. Montreal: Centre Educatif et Culturel Inc.

Haggett, P. (1990). *The Geographer's Art*. Cambridge: Basil Blackwell.

High School Geography Project (1970). *Geography in An Urban Age*. New York: Macmillan.

High School Geography Project (1979). *Environment and Resources*. New York: Macmillan.

Hill, A. D., & LaPrairie, L. A. (1989). Geography in American Education. In: G. L. Gaile & C. J. Willmott (Eds), *Geography in America*. Columbus, OH: Merrill Publising Company.

James, P. E., & Jones, C. F. (Eds) (1954). *American Geography: Inventory and Prospect*. Syracuse: Syracuse University Press.

James, P. E., & Martin, G. J. (1979). *The Association of American Geographers: The First Seventy-Five Years, 1904–1979*. Washington, D.C.: Association of American Geographers.

Joint Committee on Geographic Education (1984). *Guidelines for Geographic Education: Elementary and Secondary Schools*. Washington, D.C.: Association of American Geographers.

Kelly, M., Ott, M., Tashiro, J. S., & Semken, S. (2000). *VR Excursions: Exploring Earth's Environment*. Upper Saddle River, NJ: Prentice-Hall.

Kennamer, L. (1959). The Place of Physical Geography in the Curriculum. In: P. E. James (Ed.), *New Viewpoints in Geography* (pp. 211–228). Washington, D.C.: National Council for the Social Studies.

Leinhardt, G., Stainton, C., & Bausmith, J. M. (1998). Constructing Maps Collectively. *Journal of Geography, 97*(1), 19–39.

Levin, S. R. (1991). The Effects of Interactive Video Enhanced Earthquake Lessons on Achievement of Seventh and Eighth Grade Earth Science Students. *Journal of Computer-Based Instruction, 18*(4), 125–129.

Libbee, M., & Stoltman, J. P. (1988). Geography within the social studies curriculum. In: S. J. Natoli (Ed.), *Strengthening Geography in the Social Studies* (pp. 22–41). Washington, D.C.: National Council for the Social Studies.

Livingston, S. A., & Stoll, C. S. (1973). *Simulation Games: An Introduction for the Social Studies Teacher*. New York: Free Press.

Lukehurst, C., & Graves, N. J. (1972). *Geography in Education: A Bibliography of British Sources 1870–1970*. Sheffield: The Geographical Association.

Mackenzie, N. C., & White, R. T. (1982). Fieldwork in Geography and Long-term Memory Structures. *American Educational Research Journal, 19*(4), 623–632.

MECC (1996). *Oregon Trail [CD]*. Minneapolis, MN: MECC.

Morse, J. (1814). *American Universal Geography* (17th ed.). Boston: Buckingham.

Muir, S. P. (1996). Simulation games for elementary and primary social studies: An annoted bibliography. *Simulation and Gaming: An International Journal of Theory, Practice, and Research*, *7*, 41–73.

National Council for the Social Studies (1994). *Expectations of Excellence: Curriculum Standards for Social Studies*. Washington D.C.: National Council for the Social Studies.

National Education Association (1894). *Report of the Committee of Ten on Secondary School Studies*. New York: American Books Company.

National Education Association (1898). *Preliminary Report of the Subcommittee on Physical Geography*. Chicago: University of Chicago Press.

National Research Council (1997). *Rediscovering Geography*. Washington, D.C.: National Academy Press.

National Wildlife Federation (2000). *Resource News*. Washington, D.C.: National Wildlife Federation.

Nickelsburg, J. (1966). *Field Trips*. Minneapolis, MN: Burgess Publishing.

Pattison, W. D. (1964). The four traditions of geography. *Journal of Geography*, *63*.

Pierfy, D. A. (1972). *The Effects of a Simulation Game on the Learning of Geographic Information at the Fifth-grade Level*. Unpublished Doctoral Dissertation, University of Georgia, Athens, GA.

Pratt, R. B. (1970). *A historical analysis of the High School Geography Project as a study in curriculum development*. Unpublished Doctoral Disertation, University of Colorado, Boulder.

Rice, G., & Bulman, T. (2000). *Fieldwork in the Geography Curriculum: Filling the Rhetoric Reality Gap*. Indiana, PA: National Council for Geographic Education.

Riebsame, W. E. (2000). The Topography of Geography: Some Trends in Human Geographic Thinking. In: S. W. Bednarz & R. S. Bednarz (Eds), *Social Science on the Frontier: New Horizons in History and Geography* (Vol. 1, pp. 17–35). Boulder, CO: Social Science Education Consortium.

Rosen, S. (1957). A Short History of High School Geography. *Journal of Geography*, (December), 405–412.

Rumble, H. E. (1946). Early Geography Instruction in America. *The Social Studies*, (October), 266–268.

Spetz, D. (1988). The Preparation of Geography Teachers. In: S. Natoli (Ed.), *Strengthening Geography in the Social Studies* (pp. 51–58). Washington, D.C.: National Council for the Social Studies.

Stadsklev, R. (1974). *Handbook of Simulation Gaming in Social Education*. Birmingham, AL: University of Alabama.

Stoltman, J. P., & Fraser, R. (2000). Geography Fieldwork: Tradition and Technology Meet. In: R. Gerber & G. K. Chuan (Eds), *Fieldwork in Geography: Reflections, Perspectives, and Actions*. Dordrecht: Kluwer Academic Publishers.

Stowers, D. M. (1962). *Geography in American Schools, 1892–1935: Textbooks and Reports of National Committees*. Unpublished Doctoral Dissertation, Duke University, Durham.

Turner, B. L., Clark, W. C., Kates, R. W., Richards, J. F., Mathews, J. T., & Meyer, W. B. (Eds) (1990). *The Earth As Transformed by Human Action*. Cambridge: Cambridge University Press.

United States Bureau of Education (1916). *The Social Studies in Secondary Education: Report of the Committee on the Social Studies (28)*. Washington, D.C.: U.S. Bureau of Education.

Vuicich, G., & Stoltman, J. P. (1974). *Geography in Elementary and Secondary Education*. Boulder, CO: Social Science Education Consortium.

Walford, R. (1973). Models, Simulations, and Games. In: R. Walford (Ed.), *New Directions in Teaching Geography* (pp. 95–106). London: Longmans.

Warntz, W. (1964). *Geography Now and Then*. New York: American Geogaphical Society.

Winston, B. (1984). Teacher Education in Geography in the United States. In: W. Marsden (Ed.), *Teacher Education Models in Geography: An International Comparison* (pp. 133–149). Kalamazoo: Western Michgian University.

TEACHING METHODS IN CULTURAL GEOGRAPHY: MAKING A WORLD OF DIFFERENCE

Stephen F. Cunha

INTRODUCTION

This chapter investigates good practices for teaching about culture at the K-12 level. The focus is on instructional methods and learning activities integral to teaching about culture from a geographic perspective. That said, these techniques are as cross-disciplinary as people are diverse, and educators of all stripes can easily adapt them to their own disciplines. A few practices will be unique to geography in that they reflect our emphasis on spatial inquiry and a penchant for a "local and global" perspective. One important objective of this volume is to select examples that entice students to think about *their* world. In geography, we want them to also ponder *our shared* world with equal enthusiasm.

There are three sections to this chapter. I initially define and discuss the nature of cultural geography over the last half century, and where contemporary culture studies are headed. The second section provides best practices for teaching about culture, emphasizing the National Geography Standards and current themes in pedagogy. The conclusion tackles the issue of teaching about cultures in the Digital Age, and presents two traditional units that can be effective *sans* technology.

Subject-Specific Instructional Methods and Activities, Volume 8, pages 347–372.

ISBN: 0-7623-0615-7

I. CULTURE STUDIES IN A CELL PHONE WORLD

Although we live at the dawn of the "Digital Age," the Jeffersonian ideal regarding school curriculum still stands. That is to say, a balanced agenda incorporating science, language, mathematics, arts and humanities still produces a more capable and enlightened citizenry than a focus on just one or two subjects. Although Jefferson never booted up, he correctly forecast the value of exercising all parts of the human brain.

The frenetic pace of globalization – the emergence of NAFTA, GATT, and WTO into household lexicon – arguably makes culture studies more important today than ever, especially when we include the spatial perspective. The emerging World Wide Web provides unprecedented global information at our fingertips. Within this milieu, culture studies are valuable not just because they inquire and teach about others, but because they also counteract the drift to spend more time on-line and in-cubicle, using our technology to increase productivity and solve problems. However, as corporate leaders the world over are discovering, there is more to a global economy than simply swelling production with cheaper labor and faster shipping. In fact, building a better mousetrap over *there* to then market *here* requires a human face (minimum wages, health care, pollution standards, education, etc.) behind every label. The protests that shadow World Trade Organization, World Bank, and European Union meetings are illustrative. Thus culture studies oblige students and teachers (and we hope their parents) to investigate core-culture concepts such as: (1) diversity: how different groups produce different ideas and perspectives; (2) tolerance: how prejudice, discrimination, and acceptance vary among societies, and how these values evolve over time, and (3) sense of place, territoriality, migration patterns, resource use, etc.

The process of teaching and learning about culture is inherently interdisciplinary. For example, probing the cataclysmic change in Amazonian cultures demands numbers (demographics), biology (declining biodiversity), and history (conquest and plunder). Arguably, one of the benefits of investigating *them* is that we also learn how our society interacts locally and globally. Culture studies helps people explore the world and their place within it. Whatever rubric or methodology we educators employ to teach about culture, it better prepares our students to live and compete in a world – or marketplace if you prefer – that is growing more diverse at every scale.

Defining Culture

Defining culture is much like handling mercury. [The many iterations share common underpinnings that will serve us well in our teaching.] Here is my definition paraphrased and slightly altered from McDowell (1994, p. 148) and Fellmann et al. (1997):

> Culture is a set of ideas, values, customs, power structures, and beliefs that shape people's actions and their production of material artifacts, including their landscape and the built environment. Culture is not inborn or instinctive, but rather, is socially defined and determined, and subject to change over time. The pervasive influence of culture shapes people into a common mold.

The definition is difficult to grasp for students at any grade level because it represents a concept rather than a place or thing. Individual bias further compounds the issue as nearly everyone has some preconceived notion of his or her culture. Furthermore, the emphasis on multiculturalism which began in the late 1970s sparked many popular interpretations or culture. Some of these proponents write, appear on television, market products, and embrace multiculturalism and diversity for political gain. This is all the more reason why those teaching culture studies should intellectually ground their definitions within a discipline that offers an established curricular focus across all grade levels. Armed with both this definition and the caveat, I present one such option below.

Geography and Geographers

Geography is a spatial science that explains the "why of where." Its subject matter is the physical and human phenomena that make up the world's environments and places. Geographers analyze changes in the Earth over space and time drawing on the central themes presented in Table 1. Each year our Earth grows more crowded with people (Baby 6 billion joined us in 1999), the physical environment is further degraded, and raw materials are further

Table 1. Five Core Geographic Themes.

1. **Location**: The spatial relationships between people, places, and environments; expressed by means of a grid (absolute) or in relation (relative) to the position of other places.
2. **Cultural Ecology**: The interaction between humans, acting through their culture, and the Earth.
3. **Cultural Landscape**: The human imprint on the physical environment as created and modified by people.
4. **Movement**: The motion of people, goods, ideas, etc., from a point of origin (cultural hearth) across the Earth.
5. **Regions**: An area with similar cultural and physical traits that make it distinctive from other areas.

depleted. At the same time, an increasingly competitive global economy makes societies, states, and cultures more interconnected.

While the public too often perceives of geography as simply the study of place names or "dead people" (meaning ancient ones), the discipline is overwhelmingly contemporary and alive with everyday insights. To understand where cultural geography is headed and the relevance to the exercises that follow in the next section, I provide a brief summary of how it has evolved over the past 75 years.

The UC Berkeley Geographer, Carl Sauer (1925), put cultural geography on the map. From 1930 to 1970, he trained geographers in the "Morphology of Landscapes" and cultural ecology. He probed how people left their imprint on the Earth through their productive activities and settlements. Saur's students reconstructed the environmental and human forces that shaped the landscape. They identified cultural regions by material artifacts and other cultural attributes such as language or diet.

The 1960s "Quantitative Revolution" challenged Sauer's "Berkeley School" (Burton, 1963). This new approach demanded a strident theoretical base and used more empirical evidence to support their claims. They applied statistical and mathematical methodology to geographical systems. They also engaged in communal construction of formal theories of spatial organization.

In the 1990s, a "new geography" emerged and today cultural studies are one of the most exhilarating areas of the discipline. This most recent emphasis analyses everyday objects, explores different views of nature in art or film, and studies the meaning of landscapes and the social construction of place-based identities" (McDowell, 1994, p. 148). We find cultural geographers investigating how minorities are portrayed in films, labor use in Tijuana *maquiladores*, refugee assimilation in Turin, and the impact of forest clearing in Borneo. Today's geograhers are armed with Geographic Information Systems (GIS), a data-rich series of digital map overlays that becomes a predictive tool used to more closely scrutinize everything from rainforest clearing (with satellite imagery) to neighborhood migration (using detailed census data). Politicians and corporate leaders now realize that interpreting cultural patterns over time and space is paramount to comprehending today's world. Academics too are finding that the movement across time and space of humans, commerce, ideas, and natural resources provides insights that complement other disciplines.

Recent K-12 Developments

Realizing the centrality of culture studies, the 1989 *Goals 2000: The Educate America Act* – passed by Congress in 1994 and signed by then-President Bush, included geography as one of five core subjects (along with reading, writing, mathematics, and social science). That same year Congress authorized the National Assessment of Educational Progress (NAEP) to conduct an assessment of geography in the schools. "In framing the assessment, NAEP's Governing Board stated categorically that it was guided by the conviction that a broad knowledge of geography is an essential part of a full education" (Murphy, 1998, p. A64). During the previous 50 years geography had slipped off most K-12 radar screens as shortened school days, declining funding, and competition from more aggressive subjects elbowed into the curriculum (a situation unique only to the United States). A 1988 Gallup poll commissioned by the National Geographic Society revealed that U.S. students were seventh out of eight industrial countries in their knowledge of geography. Most schoolkids could not identify the world's major culture regions or trace migration routes, or even locate the continents (Gallup, 1988).

A flurry of initiatives returned geography to America's classrooms, albeit with uneven acceptance. Among these, the adoption of the National Geography Standards and the emergence of a Geographic Alliance Network to support them are the most significant. One continuing byproduct is an avalanche of new classroom activities, pedagogy, and assessment tools to promote geography education. Most support teaching about culture, the best of which we investigate below.

II. BEST PRACTICES FOR TEACHING ABOUT CULTURE

This section explores best practices for teaching about culture. Brophy's Twelve Practices presented in Chapter One are integrated throughout, although in a different order to create a more culture-specific discussion. I agree with Sosniak's suggestion that, "what is needed to help teachers is a small number of big subject-specific ideas and concepts, well articulated that have broad consequences for teaching and learning (Sosniak, 1999, p. 198). Thus I present a few relevant and contemporary classroom exercises to lead faculty and credential students into the wealth of materials being created and utilized to teach students about culture, primarily from the geographic perspective discussed above.

Curricular Alignment: Teaching to a Higher Standard

The National Geography Standards (GESP, 1994) define an ascending curricular focus for K-12 students. They are an effective roadmap for teaching about culture, and for guiding interdisciplinary efforts. School subject standards received much publicity from President Clinton, and are frequently hailed as a reliable mechanism to distance schools from mediocrity (Tucker & Codding, 1998).

The Geography Standards emerged from a cooperative effort of the National Geographic Society, higher education, K-12 schools, and private entities. All three national academic organizations in geography – the Association of American Geographers (AAG), the American Geographical Society (AGS), and the National Council for Geographic Education (NCGE) – were vital players. The fully coordinated scheme for teaching and learning geography is designed to:

(1) Build an understanding of our spatially interconnected and multicultural world;
(2) Identify the most important and enduring ideas in geography; and
(3) Develop a framework that will allow students to learn about geography in a logical, coherent, and accessible progression starting in kindergarten and ending in the twelfth grade (GESP, 1994).

The eighteen National Geography Standards are grouped under the six essential elements presented in Table 2. They target three grade levels: K-4, 5–8, and 9–12. In addition, each standard:

(1) Is identified by a number and applicable grade level (e.g. Geography Standard 2 – Grades K-4).
(2) Is grouped within an essential element (e.g. Places and Regions).
(3) Has a title summary that clearly identifies what the student needs to know and understand about **a** specific set of ideas and approaches (e.g. By the end of the eighth grade the student knows and understands the demographic structure of a population).
(4) States exactly what the student should know and understand after completing a particular grade level (e.g. Therefore, the student is able to explain changes that occur in the structure – age and gender – of a population as it moves through the different stages of the demographic transition).
(5) Spells out what the student should be able to do on the basis of this knowledge (e.g. *Create population pyramids for different countries and organize them into groups based on similarities of age characteristics*).

Each standard offers questions and activities structured around central ideas that are connected to clearly defined goals and intended instructional outcomes. Each grade level skill is fully integrated into the knowledge content. They

Table 2. The Eighteen National Geography Standards.

The *Geographically Informed Person* knows and understands . . .

THE WORLD IN SPATIAL TERMS	
STANDARD 1:	How to use maps and other geographic representations, tools, and technologies to acquire, process, and report information.
STANDARD 2:	How to use mental maps to organize information about people, places, and environments.
STANDARD 3:	How to analyze the spatial organization of people, places, and environments on Earth's surface.
PLACES AND REGIONS	
STANDARD 4:	The physical and human characteristics of places.
STANDARD 5:	That people create regions to interpret Earth's complexity.
STANDARD 6:	How cultures and experience influence people's perception of places and regions.
PHYSICAL SYSTEMS	
STANDARD 7:	The physical processes that shape the patterns of Earth's surface.
STANDARD 8:	The characteristics and spatial distribution of ecosystems on Earth.
HUMAN SYSTEMS	
STANDARD 9:	The characteristics, distribution, and migration of human populations on Earth's surface.
STANDARD 10:	The characteristics, distributions, and complexity of Earth's cultural mosaics.
STANDARD 11:	The patterns and networks of economic interdependence on Earth.
STANDARD 12:	The process, patterns, and functions of human settlement.
STANDARD 13:	How forces of cooperation and conflict among people influence the division and control of Earth's surface.
ENVIRONMENT AND SOCIETY	
STANDARD 14:	How human actions modify the physical environment.
STANDARD 15:	How physical systems affect human systems.
STANDARD 16:	Changes in the meaning, use, distribution, and importance of resources.
THE USES OF GEOGRAPHY	
STANDARD 17:	How to apply geography to interpret the past.
STANDARD 18:	To apply geography to interpret the present and plan for the future.

Source: GESP (1994, pp. 34–35).

guide curriculum to promote balanced depth and coherent content. Teachers can be confident their subject matter will be relevant for both the next grade level and standardized tests. Textbook authors and publishers benefit because standards-based content presents a clear grasp of topics, depth, and linkages to the next level. The standards also focus the ancillary learning materials (e.g. workbooks, study guides, teacher's aids, and overheads).

At first glance, it *may* appear that some standards have little to do with teaching about culture. However, teaching about culture from a geographic perspective involves spatial tools such as maps (Standards 1, 2 and 3), as well as comprehending the physical environment (Standards 7 and 8). For example, understanding African diaspora during the 17th and 18th centuries requires a spatial perspective to fully grasp the distance between the old and new worlds (which thwarted return), and for understanding the geographic distribution of Africans in the new world (which accounts for contemporary North and South American diversity). Geography also explains the environmental constraints of crop labor demands (hence the lack of plantations in the U.S. North), and wind belts (which made the entire slave trade operation so favorable in the first place). From this geographic perspective, teaching about culture involves multiple standards in each unit, just as understanding slavery involves meteorology, spatial representation, climate, human migration, and literature.

Standards-based instruction emphasizes coherent networks of connected content structured around powerful ideas. To varying degrees the standards are incorporated into numerous state curricular frameworks (Bednarz 1997). The principle benefit to students, particularly as they progress through the curriculum, is that at each stage they should be able to integrate acquired information and skills with other academic disciplines. Even a second grader who learns to apply new cognate spatial skills with human and environmental attributes, will derive a deeper understanding of the Hundred Acre Wood so familiar to Pooh and Piglet. Similarly, a twelfth grader will benefit from the ability to examine "the sequent occupancy of a specific habitat" (e.g. the impact on the Sierra Nevada foothills by indigenous peoples, European explorers, gold miners, loggers and dam builders, and hikers and suburbanites). In this case, integrating literature from high school reading lists (*Ishi in Two Worlds, Storm, California History*, etc.), and spatial skills (interpreting on-line satellite imagery), and mathematics (census data) results in a more sophisticated comprehension of geography, literature, history, and mathematics.

While the Standards establish a framework to teach a geographic perspective on culture, the Geographic Alliance Network prepares and supports teachers in this task. During the 1980s the National Geographic Society promoted "Alliances" in all 50 states; these are teacher professional organizations that:

(1) Sponsor professional growth workshops and institutes;
(2) Develop curricular materials to support geography teaching;
(3) Initiate school-university partnerships; and
(4) Integrate the National Standards within state curricular frameworks.

Each Alliance provides lesson plans and activities with advance organizers or previews to accompany the Standards (e.g. TAGE, 2000). More importantly, they help teachers prepare and tailor their own materials through interdisciplinary pedagogy. The Alliances have a multiplier effect, in that institute graduates then train colleagues in their home districts, and educational materials (lesson plans, literature guides, audio and visual produces) developed both during and after Alliance activities are published (in hard copy and online) for nationwide access (NGS, 2000).

Make the Census Count: The Standards in Action

Using readily available current and past census data to teach about culture is a new avenue for most teachers. Yet this data – indeed all demographic information – reveals changing patterns of immigration, population age structure, economic trends, and even "Great Events" such as epidemics, warfare, political upheavals, and environmental disasters. Within the United States, census data is a window into a society that grows more diverse, mobile, and globally interconnected each year. With massive immigration the current reality, teachers will play a vital role in shaping the new America (Rong 1998). The journal *Social Education* (1998) devoted an entire issue to teaching about immigration. Moreover, comparing former data with emerging trends enlivens culture studies and illustrates the key idea that societies are dynamic entities. Finally, students at all level enjoy learning about population trends. Employing actual population data at any grade level is relatively simple. According to Boone, 2000, p. 77):

> Second grade students can understand relatively sophisticated population concepts. I have always believed that 'little people can learn big kid stuff' if the material is presented appropriately. This proved to be true with census data. My population instruction is based upon two assumptions basic to elementary education. First, material must be understandable; that is, related to students' prior knowledge and experience. Second, material must be interdisciplinary, that is, clearly connected to students' current and varied interests.

Boone then provides the Curriculum Connections (Geography and Math), the Applicable Geography Standards (Grades K-4 Standards 3, 5, and 9), and the Skills and Objectives: (a) Develop an understanding of large numbers, (b) use manipulatives and maps to represent large numbers, (c) interpret data collected by the U.S. census, and (d) explore the geographic distribution of population.

The author clarifies intended outcomes and cues desired learning strategies through pre-activities such as reading *Stories About a Million* (Schwartz, 1985), paper cutouts, estimation exercises, and kinetic activities. He also clearly prepares students for a discussion about the U.S. population by having them represent each state's population with stacked blocks. The resulting towers become the data for colored choropleth maps that are used to distinguish various population regions. These pre-activities help students to focus on the main ideas and to order their new thoughts about what all those census numbers really mean.

The students then use raw census data – individually or as groups – to represent the population of each state with blocks. Each block represents one million people and the resulting towers are placed on a large floor maps. The U.S. becomes a giant three-dimensional cartogram used to answer questions that illustrate various demographic and physical topics. Ely (2000) and Sharma (2000) also employ census data to construct interdisciplinary and Standards-based lesson plans for grades 4 through 8 and Grades 9 through 12 respectively.

Teachers can use these lesson plans with confidence that they fit into the National Standards. Both the ideas and supporting materials supplied by the Alliance prepare and channel students into thinking about culture in a systematic way (in this case through population). Other examples of teaching with standards include Marra's (1996) integration of geography and literature using children's picture books; Hurt's (1997) Main Street foray; and Merrett's (1997) unit on political geography.

Integrating Technology and Cooperative Learning

Teachers are most effective when they create a climate for molding students into a cohesive and supportive learning community. This is especially true when instructing about culture, where the sensitive subject matter can impose a dynamic tension on the classroom, even in the elementary grades. Teachers of course set the tone, and the importance of good and consistent modeling cannot be overstated. Collaborative group efforts are an excellent strategy to develop both class camaraderie and affective and social benefits such as increased student interest and output. Rapidly evolving digital technologies can foster cooperation and build an effective learning community. The expense and sophistication of most computer technology, which doubles up students at every terminal, virtually dictates a cooperative approach anyway. The following exercise integrates cooperative learning and technology.

The Community Atlas Project

This interdisciplinary project developed by the Environmental Systems Research Institute (ESRI) offers numerous benefits for community learners. It involves collaborative effort at several scales and invites exploration of cultural diversity within the student's local community. Working on a community atlas builds spatial relationships and teaches – in an active learning mode – the relationship between students and their natural environment. The project helps students understand how communities change over time through external (regional development) and internal (human migration) factors. Teachers can maximize their predefined and expected outcomes with a pre-project agenda. Once completed, it can be extended in comparative analysis to other communities, reflecting the geographer's penchant for local and global analysis. Moreover, it can be tied easily to the National Geography Standards and tailored to almost every grade level. Working with the various on-line and mapping parameters also builds technical competence that can be applied to other disciplines.

They begin by asking teachers and students:

> What do small farming communities in New Jersey, Texas, California, and Wisconsin have in common? Is "the community" for high schoolers in Juneau more like "the community" for those in Montpelier or in Honolulu? Do junior high kids in Chicago see things more like kids in Decatur or in Dallas (ESRI, 2000)?

The goal of the Atlas project is to illustrate and define – through student eyes – the nature of their communities through maps and text. Classes create a presentation about their community. Since ESRI provides the necessary software, only a recent web browser and internet connection are requirements to participate. All U.S. schools are eligible and submissions should be from entire classroom groups, not individual students. The results are variable because communities are as dissimilar as the traits various students and teachers select to represent them. Popular defining characteristics include:

(1) Boundaries: size and defining traits for each boundary;
(2) Natural landscape: landforms, climate, water resources, vegetation;
(3) Population: size and density, diversity, trajectory of change;
(4) Historical patterns of settlement and sequent occupancy;
(5) Land use, economic patterns, transportation corridors and networks;
(6) Significant current social, political, and environmental issues.

Students post descriptions and maps on an internet site maintained by ESRI (communityatlas@esri.com). The site is searchable by selected characteristics that invite nationwide comparative analysis for both similarities and differences. The vast array of searchable topics allows participants to tailor for desired grade level, Geographic Standards, curriculum needs, and region.

The project also requires multiple cooperative learning strategies: students help one another accomplish individual learning goals (finding data, checking work, tutoring), group goals (selecting key traits for cartographic representation), and a division of labor (to produce the atlas in electronic and hard copy format). The highly collaborative nature of this project also invites parental assistance with data collection and access to community and business groups. This encourages active involvement in their children's learning and sets the stage for mutual achievement (the project is an Open House tour-de-force). The teamwork develops an effective educational community, as working together to produce a community profile requires significant research and exploration that:

> Requires consensus on what are the bounds of 'the community,' what is it like, and what is the best way to represent its nature within a modest space. Getting an entire class to reach consensus on a vision of 'the community' or a 'conservation issue' can be a challenge. But sharing expertise, communicating vision, and seeking agreement are essential skills in the world at large (ESRI, 2000).

For additional ideas on teaching culture with new technologies, refer to Audet and Paris (1997), who assess the emerging GIS needs of schools; Kirman's (1997) introduction to remote sensing; Keiper's (1999) GIS case study for elementary students; and Risinger's (1998) review of global education and the World Wide Web.

There are many less involved variations of this project that do not require software, but still explore culture from an interdisciplinary perspective (Benson, 2000; Chiodo, 1993; Gould & White, 1974; Hays, 1993; Saarinen & McCabe, 1995). Collaborative student groups of all ages can devise mental maps about their community on blank paper, and then form an overall profile based on pre-set criteria. Conducting a community survey with simple data on ethnicity, family size, cultural origin, etc., provides excellent data for maps and further discussion and analysis. Whatever the task, students almost always benefit from working in pairs or small groups to construct understandings or help one another master skills, provided the teacher monitors their strategies and their effort.

Exploring Cultural Diversity while Building Academic Skills

Students need to practice and apply skills as they learn, and receive constructive feedback. Appropriate task scaffolding perpetuates a challenging learning environment. The teacher must concurrently provide whatever help the students need to keep them productive and confident learners. The necessity of skill building is well documented in education research, but to remain effective, the skills must be practiced and followed up in a variety of contexts. This is as true within a grade level as it is for the entire grade progression (Brophy & Alleman, 1991; Knapp, 1995).

As discussed earlier, the protean nature of culture complicates both teaching and learning about it. During a single student's K-12 progression, both local and global cultures will undergo radical change. Teachers can deepen understanding global cultures by applying three practices at all grade levels:

(1) Maintain – in presentations and assignments – an interdisciplinary pedagogy;
(2) Emphasize skill-building (writing, speaking, analysis, using technology), but only after a base competency has been established; and
(3) Select activities that show culture as an interpretive and malleable concept.

There are many excellent workbooks that build skills in the social sciences. During the last decade a sea change in format has produced assignments that require less "fill in the blank" (after the teacher's presentation) and more research and analysis. These begin with simple writing and public television analysis and quickly escalate to sophisticated internet, CD-ROM, and GIS applications. However, the non-workbook exercise presented here shares these characteristics, offers flexible interdisciplinary use, and can be augmented each year to keep pace with student development. It represents an emerging genre of "real person interactivity" that utilizes new technology while incorporating the three principles stated above.

Exploring Cultural Universals

Payne and Gay (1997, p. 220) admit that "illustrating the complexity and impact culture has on groups of people is a challenging task for classroom teachers." They employ "Cultural Universals" described by Cleaveland, et al. (1979) to help students connect, literally and figuratively, with their contemporaries elsewhere in the world (Table 3). This exercise and others like

Table 3. Examples of Cultural Universals.

Material Culture	Food, clothing, tools, weapons, shelter, books
Arts, Play, Recreation	Standards of beauty and taste, art and recreation forms
Language/Nonverbal	Language and nonverbal communication
Education	Informal education and formal education
World View	Belief systems and religion
Social Organization	Societies, families, kinship systems

Source: Payne and Gay (1997, p. 221).

it are valuable because they require students to think critically about their place both at home and in the larger world. Moreover, they interact with peers abroad, while developing writing, research, group, and analytical skills. Social Science teachers also favor the opportunity for interdisciplinary cooperation and the fact that multiple "correct" answers are possible.

The first objective of this exercise is to orient learning by debunking popular cultural myths (GeoSphere Project, 1994; Kunstler, 1993; Relph, 1976) and modifying the common student perception that "we have a McDonalds and so do they; therefore their culture is just like ours" (Payne & Gay, 1997, p. 220). Working in large groups, students generate their definition of a cultural universal. They then discuss and analyze what constitutes cultural commonalties, before selecting representative universals they wish to explore. A labeled and annotated database illustrating their universals (from hard and electronic sources) completes the first phase.

After this exploratory work, the students charge onto the internet in search of contemporaries from Addis Adaba to Zurich (type internet search <K-12keypals> for international classroom exchange sites). After making contact and wrestling with different time zones and diverse school/holiday schedules, they are ready to collaborate on an atlas of "Cultural Universals" that includes stories and news items.

Cushner's (1992) *Creating Cross-Cultural Understanding Through Internationally Cooperative Story Writing*, and Payne and Gay's (1977) exercise develops four primary skills:

(1) Written: transactional and expressive;
(2) Research: for facts, definitions, cultural differences, images;
(3) Technological: email, internet, electronic searches; HTML; and
(4) Interpersonal: teamwork, patience, work ethic, and independence.

This content-laden endeavor explores culture from multiple sides, and puts students in charge of their lesson plans and progress. Some ongoing exchanges are now eight years old and have expanded to include joint science projects, political action, and actual visits. Payne and Gay (1997, p. 223) conclude:

> Students can benefit tremendously from analyzing and speculating about why other cultures evolved with different characteristics and traits. Exchanging stories and examining how students in other cultures respond to a common cultural event will help them toward an understanding of the complex human systems of which they are part. Researching for clues from images, music, language, dress, customs, and beliefs provides students with examples of how interconnected cultural systems are with world events.

Utilizing "primary materials" is another effective way to build research, writing, presentation, and technology skills that are cross-disciplinary and applicable through all grade levels. Peroco's (1998) students utilize "original" historical documents (those free of interpretation) located on-line, in libraries, and in specialized books, to teach "themselves" about history, culture, and geography. Field trips to historic sites and libraries facilitate collection or viewing of diaries, images (paintings, photographs, drawings, etc.), old newspapers, maps, and other archival materials. Students use this raw material to re-construct the California Gold Rush, American Revolution, Native American settlement, and other events. Hartman and Vogeler (1993) developed another angle following real world events as indicated by travels of the U.S. Secretary of State. Students use the data and events to weave the "Shape of the World" according to visit types (e.g. peace negotiations, conferences, treaty signings, stopovers with or without diplomatic activity), and number of trips per year. They portray this data through maps and essays. This type of exercise also proves to students that history and geography are "made" each and every day.

The ideas presented above move teaching about culture from lecture-discussion-workbook pedagogy, to a more interactive, research, and analytical project-oriented learning. They incorporate the new digital tools that enhance learning, and the realization that culture is best taught as an interdisciplinary and ongoing part of the curriculum. Students should begin within their zones of proximal development, and through diminishing levels of tutoring, progress towards greater autonomy. Faculty will spend more time coaching, organizing and monitoring progress, while students will put more effort into cooperative problem solving, research, and real-world interaction. This becomes all the more effective when content is organized around our next topic; significant content-based discourse.

Thoughtful Discourse Around Big Ideas

Contemporary teachers may find this the "Age of Distraction". There are the usual suspects – television, movies, music, sports, minimum wage jobs, behavior issues, and a societal malaise about teaching and learning. But the education ones – the frenetic and sometimes divisive concern over test scores, pedagogy, bilingual programs, merit pay, and curricular turf battles also distract. While dodging alligators in the swamp we must bear in mind that for our students, it is a sustained discourse structured around powerful ideas that will carry the day.

The following 'Family Migration' exercise can be used at any grade level, offers wide interdisciplinary application, and applies to many National Geography Standards. It is organized around movement, an important geographic theme – or Big Idea – that is a core cultural attribute. The project design involves parents and family members. Teachers can scaffold to any level of archival and digital research, spatial representation, and presentation skills. The finished products can be displayed around both the school and the community. More importantly, the process requires open-ended questions that call for students to apply, analyze, synthesize, or evaluate what they are learning. Although they will share many similar problems gathering information, the individual nature of the data will produce a wide amplitude of "correct" answers that will invite discussion or debate over both the results and the methods. Perhaps most importantly, studying family migration history links students to their own cultural roots, and demonstrates in a very personal way, how they are part of the dynamic forces that shape culture and place.

Family Migration: Why You Can't Go Home

Rodrigue (1996) summarizes this exercise now used throughout California, a state where over 50% of the population was born elsewhere. The task is to investigate various push and pull factors that trigger migration and result in each student having arrived in their current location (Table 4).

Using family information, students reconstruct the spatial and temporal patterns of their own family's migration to California. The migration routes are plotted on maps, and their accompanying essay accounts for the:

(1) Various push and pull factors that triggered both primary and secondary movements;
(2) Different social/cultural traits, material culture, likes or dislikes, and prejudices they may have acquired or left along the way;
(3) Any real or perceived discrimination either inflicted or received, however subtle; and
(4) Any material culture acquired or left behind.

There are many variations from here. Concerned with the lack of ethnic and socioeconomic diversity at her campus, Rodrique has students follow their own migration with an imaginary class partner. The students use historical atlases,

Table 4. Cross-Disciplinary Push and Pull Factors.

Political	War, discrimination, ethnic cleansing, change-of-power
Environmental	Floods, fires, earthquakes, deforestation, insect infestations
Economic	Globalization, poor economy, resource depletion
Demographic	Population pressure, high dependency ratio, disease

encyclopedias, novels and autobiographies (Bing, 1992; Perez, 1991; Tan, 1989) and other sources to acquaint them with someone "of a completely different ethnic and racial background" (Rodrique, 1996, p. 82). Students must learn enough about their fictitious companion to construct a *plausible* history and geography of the migration. In addition to the literary and archival research, she encourages students to interview people of like ethnicity who may have undergone comparable movement, and/or have experienced analogous push/pull and socio-cultural adjustments. The same strategy can be applied to filling in the blank spots or periods left unaccounted for in their initial (and real) migration exercise.

Many instructors incorporate various presentation skills into this assignment. Depending on the grade level these can range from simple oral reports to elaborate posters augmented with maps and duplicated archival family photos. Some classes share their results in city halls and on community billboards.

This exercise and the potential variations engage every student in cognitive processing and construction of knowledge. Although initially addressed to the entire class, each student must, in a very personal way, consider the big ideas of migration, prejudice, discrimination, assimilation, and what it means to be part of a cultural group – and to leave that society for another. In this regard, their journey through this content is clearly more important than the destination.

Here is a sampling of other innovative plans that focus on big ideas. Smith and Brown (1996) develop a sense of place using five core geographic themes that connect students to their own backyards. Wall's (1994) "Friends in High Places" is a group activity designed to integrate geography with other social sciences disciplines. Trilfonoff (1997) applies Amish quilt production to questions about the role of customs and beliefs in the shaping of material culture (and can be easily adapted to specific local material culture). Finally, Byklum's (1994) "Geography and Music: Making the Connection," uses perception to explore the links between culture, geography, and popular music.

Assessment and Achievement Expectations

Assessing assessment is difficult and has two angles: evaluating student efforts, which is largely the teacher's domain, and that which is used to assess the curricular decisions made by teachers, and to evaluate how well they taught their chosen material. Teachers, school officials, politicians, and parents now debate this topic under the rubric of school performance. Nickell articulates the current status:

> The late 1980s gave us a number of highly vocal advocates for more "authentic" assessment, or for tasks that are a better measure of what students can do with what they learn, rather than simply what they can commit to memory for the short term (Nickell, 1999, p. 353).

Alleman and Brophy (1999) offer guidelines for planning assessments. I present here two creative ideas on evaluating student performance, and the overall assessment scheme developed in the National Geography Standards.

Nickell and Wilson (1999) developed a "Teacher Observation Form" for evaluating student projects such as creative skits to portray the court system, devising public opinion surveys, and oral presentations that depict multiple perspectives. Each of these activities requires a high degree of sustained creativity, cooperative effort, and applied skills that in the real world will not be evaluated by multiple choice exams, but rather how well the graduates perform in a dynamic interpersonal environment. Hence the observation tool that evaluates such categories as task preparation, skill development, content gains, and effort within the group. The traditional grade scale is replaced with checked boxes indicating whether the category was observed, and if so, whether the student was an "Expert," made progress, or "Needs Practice." There is also room for *both* teacher and student comments. The authors suggest that teachers develop assessment to fit their needs, and that with practice and thoroughly explaining the method to parents and students, "observation should be used often and with confidence" (Nickell & Wilson, 1999, p. 352).

Hart (1999) advocates "Portfolio Assessment," where annotated comments on diverse student work replace traditional graded assignments. Students then write a self-evaluative letter before meeting their teacher for a quarterly review. The process of self-evaluation may begin in the primary grades as one way to develop less dependence on the teacher for directions, standards, and judgement (Varus, 1990). The many nuances in evaluative forms share at least two characteristics. First, they offer multiple levels of evaluation (self, small group, or entire class). Second, they ask students to evaluate their effort, content mastery, interpersonal skills, presentation, and achievement of objectives. "The real value of teaching students how to evaluate their own performance should be, ultimately, the whole point of assessment" (Hart, 1999, p. 345).

The leading proponents of curricular standards (Tucker & Codding, 1998), advocate state standards for strong and functional assessment components. The National Geography Standards (GESP, 1994) address this aspect for cultural geography. They offer a scheme to assess student progress at grades four, eight, and twelve. Evaluation narratives provide a rubric to help teachers identify specific degrees of student achievement – whether an individual does not meet, meets, or exceeds the standards. The fourth grade norms regarding human systems are illustrative:

(1) Below Standard: tends to ethnocentricity because of an inability to understand or appreciate other cultures.
(2) At Standard: possesses a global sense and has an awareness of economic and political interdependence.
(3) Beyond Standard: uses themes and ideas from many subject areas.

In addition, the standards help teachers develop accurate evaluative procedures to assess their students' geographic knowledge and skills.

III. TEACHING ABOUT CULTURE IN THE DIGITAL AGE

The emerging tools of the Digital Age open new worlds for teaching and learning about culture. Many believe that the World Wide Web, CD-ROM and other electronic connections are prerequisite to linking multiculturalism across the curriculum. These new tools facilitate active learning, international cooperation, and more comprehensive research, all with ease and expediency not possible a generation ago.

However, new technology poses new risks. While few people, including this author, downplay the benefits of electronic learning, even the best interactive multi-media equipment limits students to visual and auditory sensory perception. Thus as logging on grows more pervasive, we should not abandon the "hands on" and "boots wet" pedagogy that has served teachers so well since antiquity. Educators and schools can maximize both funding and results by complementing the full range of classroom efforts with the field efforts discussed below.

Student Field Trips

Peroco (1998) believes that field trips are the "ultimate learning experience." Several authors (Everson, 1969; Kern & Carpenter, 1986; Zant, 1977) provide models for running off campus trips that range from an hour to week-long sojourns. The benefits of cooperative learning, community building, hands-on experience, and multisensory learning (smells, sight, auditory, tactile, and even taste) are both tangible and lasting. Field trips are most effective when:

(1) Advance preparation cues students in to what they will see and do;
(2) Down time allows individual exploration and comprehension during the trip; jam-packed itineraries reduce learning effectiveness;

(3) Diverse activities mix presentations, data collection, kinetic activities, group action, and down time;
(4) Adequate chaperones ensure safety and facilitate numbers two and three above;
(5) The travel-time is less than 50% of the activity time; and
(6) Field data is integrated into classroom projects.

Field trips are a perfect springboard to integrate culture across the curriculum. While visiting historic parks and sites, students can reenact events in period costume and with appropriate props. Alternatively, student surveys conducted in downtown plazas can gather much cultural and socio-economic data for further classroom discussion and analysis. Participating in town meetings, planning processes, and political debates is yet another way to bring real world meaning to the curriculum while leaving the classroom behind. Teachers are also discovering that field study can be seamlessly integrated with new technology to further enrich post-trip classroom learning. Thomas et al. (1998) found that a website for elementary school field trips blossomed into a full-scale "electronic curriculum" in their large urban school system.

Collecting Material Culture to Impart a Sense of Place

In addition to student field trips, those who teach about culture should make every effort to include travel as part of their overall professional growth strategy. Alternatively, the liberal use of items collected in the field can guide students to terrain where computers never journey. Of the many benefits accrued abroad – insights, anecdotes, slides, rich descriptions – collecting material culture is often the most underutilized. Every fall too many teachers return home with great wall hangings, but little to share with students. Carpenter (1996, p. 73) believes "teacher's travels, and summer travel in particular, can be of great use in the classroom."

Material culture consists of physical and visible objects used by different societies to accomplish everyday tasks. This includes everything from dwellings, trucks and horse carts, to clothing, weaponry, cooking utensils and jewelry. These items comprise the cultural landscape that every society creates. When used in cultural context, they provide insights into how humans use natural resources, their level of technology, social relationships, and whom they worship (Burgmann, 1992; Martin, 1996; Small, 1996). Their regional origins reflect human-environment interaction, the diffusion and modification of ideas, and other attributes relevant to completing everyday tasks. The integration of this "old technology" can enliven classroom presentations and effectively integrate into the curricula such core geographic themes as place, diffusion and human-environment relationships (Cunha, 1998).

For example, drinking containers are universal; we all use them. A plastic water jug in industrialized Warsaw takes the form of a yak bladder in Mongolia, a goat skin in the Indus River Plain, a clay pot in Bangalore, an empty squash gourd in the Amazon, and a discarded Clorox bottle in a Lima shantytown. The yak and goat containers reflect the strong reliance on animal husbandry found in Mongolia and Pakistan. Although herding is also very important in southern India, the clay derives from a soil not found in the two other locations. In contrast, despite abundant clay, isolated Amazonians opt for empty gourds. Finally, the higher wage structure allows many Poles to indulge in the "throw-away" penchant of developed economies. The disposable water bottles, along with other containers (Clorox bottles, jelly jars, milk jugs, etc.) are scavenged by shantytown dwellers the world over.

The endless possibilities depend upon a teacher's chosen curriculum, pedagogy, and imagination. Clothing, kitchen items, and toys are student pleasers that also share key common traits:

(1) They are everyday items with counterparts in America;
(2) Their regional characteristics can be plotted on maps;
(3) They illustrate and enliven discussion about core geographic themes (e.g. cultural ecology, diffusion, regionalism, and landscape);
(4) They can be systematically collected at reasonable expense; and
(5) They are multi-sensory teaching aids that students remember.

Teachers can find inexpensive everyday material culture in ordinary shops and markets. In the countryside, locals are often willing to trade their items for ones brought from home (large spoons, pocketknives, and tee shirts are perennial favorites).

Before traveling identify the important curricular units (geography, history, language arts) and geographic themes (environment, culture, place), then target these for collection. Once home, material culture will invigorate the classroom. Students may play with toys, carry water, dig trenches, smell incense, and sample food from far away lands. They can locate artifacts on a map and reason why they are suitable for that region and not another (e.g., clothing and climate, religious icons and culture diffusion). Linking cognate items across the oceans can tangibly illustrate the role development and technology play in our global economy. A collection of music cassettes, Pepsi cans, and computer diskettes – respectively labeled in Spanish, Chinese, and Arabic script is one example. The most effective objects appeal to multiple senses (texture, sound, odor, and sight). They can also withstand years of hands-on student use and can illustrate multidisciplinary curricular themes.

Material culture in the classroom can bring a flat map to life, especially when deployed by the peripatetic teacher who collects and integrates them across the curriculum. As we move deeper into the Information Age, amidst mountains of new hardware and software should come tangible, odorous, noisy pieces of other cultures – and our own – that help students understand how people elsewhere meet the challenges of everyday living.

Traditional Units: Combining the Old and the New

Despite the emphasis in this chapter on emerging technology, we should not abandon traditional ways of teaching about culture. For years, American educators have crossed the Delaware with Washington, saddled up with Lewis and Clark, and plumbed the history and geography of distant peoples from Amazon to Zanzibar. Rather than disregarding these units, teachers should balance and combine the best of the old with what the present offers. The following two examples are illustrative.

The National Geography Standards require students to master the world in spatial terms, including the relative location and size of the continents. Johnson and Brown (1998) designed "How Big is Africa?" to dispel the abundant and inaccurate stereotypes about the size and variety that exists in Africa. They want students to recognize that "it is hard to generalize about Africa because its diversity is great and each of its 53 countries is unique" (Johnson & Brown, 1998, p. 278). The exercise, designed for upper elementary and middle school students, integrates the Geography Standards, requires spatial analysis, literature evaluation, and interactive participation. Students first listen to a story told in the West African traditional mode by a *groit* (oral historian – a student or teacher in this case) who elicits input from his audience. This particular tale, or group of stories, illustrates the misuse of the term Africa as though it applies to a single country (e.g. The President will soon make trips to China, France, and Africa), or reinforces stereotypes (all of Africa is hot; famine persists everywhere). This process continues for several weeks while students use their eyes and ears to gather news items, references, video clips, and photos that illustrate the size and complexity that is Africa. The next step is constructing a "How Big is Africa?" map by placing country cutouts of identical scale of various political states that fit inside an outline map of Africa. This evolves into a host of spatial activities such as comparative country measurement (math skills), topographic, climate, and terrain analysis (scientific inquiry), and linguistic relationships. The follow-up activities include predicting time travel in various conveyances (auto, train, plane), investigating how early explorers (e.g. Ibn Batuta, Vasco da Gama, Zheng He) influenced long held perceptions about Africa, and accounting for why comparative continental-scale projects such as railroads, nationhood, and canals, still remain uncompleted in Africa. This exercise can be repeated with any world region, including the oceans, without utilizing any new technology. On the other hand, the vast internet/CD-ROM resources can enliven the research and *groit* aspects. Finally, it is possible using computerized map programs to construct and alter the "cutout" images digitally. Geography teachers value this multi-level grade exercise because it integrates the Standards, requires copious family and classroom interaction, involves (but does not require) new learning technologies, and is easily adapted to any region of the world on an individual or comparative basis.

Living History is yet another traditional instructional genre that can be augmented but never fully displaced by digital age technology. Although the re-creation of colonial Williamsburg is probably the most well known location to observe living history, the past also comes to life in numerous other national park and historic sites throughout the United States. Many teachers have students act out history, from native encounters to the present.

Although highly demanding from the teacher's perspective, it is possible to combine living history, field study, and analysis of original sources into a single unit. In multicultural California, the surging immigration combined with revisionist history is requiring elementary teachers to rethink their strategies for teaching about the colorful mid-nineteenth century Gold Rush. This "wholistic approach" is growing in popularity because it involves – aside from meaningful classroom activities – living history field trips to Sutter's Fort, and to Indian Grinding Rock State Historical Park where Native Americans explain their view of the Gold Rush (and demonstrate fire ignition and grinding acorns). A collaborative exchange with local university students to construct computer or hand drawn maps on some aspect of Gold Rush environment or culture complements the field research, and exposes students to higher education. The students research original documents (miners' journals, newspapers, early photographs, etc.), located either on-line, from the State Subject Matter Project Offices, or in adjacent library stacks.

Completing the full suite of activities obligates teamwork and the active learning mode. Students investigate multiple historical perspectives while fully integrating both the California History-Social Science and the National Geography Standards. Many teachers seek mini-grants and community support to meet bus and other field expenses. Although our new learning technologies can enrich and expand this endeavor, the real key is dedicated teachers, parents, and administrators. The possibilities are endless when teachers combine research with bold activities that make students active participants in a learning community.

CONCLUSION

This chapter presented a sampling of contemporary "best practices" for teaching about culture from the geographic perspective. Those who instruct this topic will achieve maximum results only through a sustained menu of professional growth. These include content and pedagogy seminars, travel, attendance and presentation at professional meetings, and an insatiable curiosity about how cultural differences both adorn and complicate our daily lives. A deep desire to explore and experiment with content, techniques, and student assessment is also essential. We are what we know, and we teach what we understand, thus lifelong learning is indispensable to enrich teaching about culture.

ACKNOWLEDGEMENTS

Joseph Stoltman (Western Michigan University) contributed many useful ideas in the early stages of this manuscript. I am also indebted to Judy Walton and Mary Beth Cunha (both of Humboldt State University) for thoughtful editorial critique.

REFERENCES

Alleman, J., & Brophy, J. (1999). The Changing Nature and Purpose of Assessment in the Social Studies Classroom. *Social Education*, *65*, 334–337.

Audet, R., & Paris, J. (1997). GIS Implementation Model for Schools: Assessing the Critical Concerns. *Journal of Geography*, *96*, 29–33.

Bednarz, S. (1997). State Standards and the Faring of Geography. *Oblique*, *17*, 1–3.

Benson, J. S. (2000). Centerville/Centreville: An Exercise in Mental Mapping. *Journal of Geography*, *99*, 32–35.

Bing, L. (1992). *Do or die*. New York: Harper Perennial.

Boone, E. A. (2000). How Many a Million? *Journal of Geography*, *99*, 77–78.

Brophy, G., & Alleman, J. (1991). Activities as instructional tools: A framework for analysis and evaluation. *Educational Researcher*, *20*, 9–23.

Burgmann, A. (1992). The Moral Significance of Material Culture. *Inquiry*, *35*, 3–4.

Burton, I. (1963). The Quantitative Revolution and Theoretical Geography. *Canadian Geographer*, 7.

Byklum, D. (1994). Geography and Music: Making the Connection. *Journal of Geography*, *93*, 274–278.

Carpenter, J. (1996). Bird Bottles, the Fourth Estate, and Mr. Jefferson's Music: Bringing Back History From Your Travels. *Social Education*, *60*, 73–76.

Chiodo, J. J. (1993). Mental Maps: Preservice Teachers' Awareness of the World. *Journal of Geography, 92*, 110–117.

Cleaveland, A., Craven, J., & Danfelser, M. (1979). *What are universals of culture? Global Perspectives in Education*. New York: Intercom 92/93.

Cunha, S. (1998). Using Material Culture to Impart a Sense of Place. *Education About Asia, 3*, 42–45.

Cushner, K. (1992). Creating Cross-Cultural Understanding Through Internationally Cooperative Story Writing. *Social Education, 56*, 36.

Ely, D. (2000). Rural and Urban: Using American Factfinder on the Census Bureau Web Site. *Journal of Geography, 99*, 2.

ESRI – Environmental Systems Research Institute, Inc. (2000). *Community atlas project*. www.communityatlas@esri.com.

Everson, J. (1969). Some Aspects of Teaching Geography through Fieldwork. *Geography, 54*, 63–73.

Fellmann, J., Getis, A., & Getis, J. (1997). *Human geography: landscapes of human activities*. Madison: Brown and Benchmark Publishers.

Gallup (1988). *Geography: an international Gallup survey*. The Gallup Organization for the National Geographic Society.

GeoSphere Project (1994). *Material world: A global family portrait*. Santa Monica CA: The Geosphere Project.

GESP – Geography Education Standards Project (1994). *Geography for life: national geography standards 1994*. Washington, D.C.: National Geographic Research and Exploration.

Gould, P., & White, R. (1974). *Mental maps*. Baltimore: Penguin Books.

Hart, D. (1999). Opening Assessment to Our Students. *Social Education, 65*, 343–345.

Hartman, J., & Vogeler, I. (1993). Where in the World is the U.S. Secretary of State? *Journal of Geography, 92*, 2–12.

Hays, D. (1993). Freehand Maps are for Teachers and Students Alike. *Journal of Geography, 92*, 13–15.

Hurt, D. (1997). Main Street: Teaching Elementary School Students Standards-Based Urban Geography. *Journal of Geography, 96*, 280.

Johnson, D., & Brown, B. (1998). How Big is Africa? *Social Education, 62*, 278–281.

Kern, E., & Carpenter J. (1986). Effect of Field Activities on Student Learning. *Journal of Geological Education, 34*.

Keiper, T. (1999). GIS for Elementary Students: An Inquiry Into a New Approach to Learning Geography. *Journal of Geography, 98*, 47–59.

Kirman, J. (1997). A Teachers Introduction to Remote Sensing. *Journal of Geography, 96*, 171–175.

Knapp, N. (1995). *Teaching for meaning in high-poverty classrooms*. New York: Teachers College Press.

Kunstler, J. (1993). *Geography of nowhere: the rise and decline of America's man-made landscape*. New York: Simon and Schuster.

Marra, D. (1996). Teaching to the National Geography Standards Through Children's Picture Books. *Journal of Geography, 95*, 148–152.

Martin, A. S. (1996). Material Things and Cultural Meanings: Notes on the Study of Early American Material Culture. *William and Mary Quarterly, 53*, 5.

McDowell, L. (1994). The transformation of cultural geography. In: D. Gregory, R. Martin & G. Smith (Eds), *Human Geography: Society, Space, and Social Science* (pp. 146–173). Minneapolis: University Minnesota Press.

Merrett, C. (1997). Research and Teaching in Political Geography: National Standards and the Resurgence of Geography's "Wayward Child". *Journal of Geography, 96*, 50.

Murphy, A. B. (1998). Geography's Expanding Place in American Education. "Rediscovering the Importance of Geography". *Chronicle of Higher Education, 45*(10), A64.

NGS – National Geographic Society (2000). *Education Outreach.* National Geographic Society. www.nationalgeographic.org

Nickell, P. (1999). The Issue of Subjectivity in Authentic Social Studies Assessment. *Social Education, 65*, 353–355.

Nickell, P., & Wilson, A. (1999). Observation as an Assessment Tool. *Social Education, 65*, 351–352.

Payne, H., & Gay, S. (1997). Exploring Cultural Universals. *Journal of Geography, 94*, 220–223.

Perez, R. (1991). *Diary of an undocumented immigrant.* Houston (TX): Arte Publico Press.

Peroco, J. (1998). *A passion for the past: creative teaching of U.S. history.* Portsmount (NH): Heinemann.

Relph, E. (1976). *Place and placelessness.* London: Pion.

Risinger, C. (1998). Global Education and the World Wide Web. *Social Education, 62*, 276–277.

Rodrigue, C. (1996). Imaginary Migration Exercise in Multicultural Geography. *Journal of Geography, 95*, 81–85.

Rong, X. L. (1998). The New Immigration: Challenges Facing Social Studies Professionals. *Social Education, 62*, 393–499.

Saarinen, T. F., & McCabe, C. L. (1995). World Patterns of Geographic Literacy Based on Sketch Map Quality. *Professional Geographer, 47*, 196–204.

Sauer, C. (1925). The morphology of landscape. *University of California Publications in Geography, 2*, 19–54.

Schwartz, D. (1985). *How Much is a Million?* New York: Lothrop, Lee and Shepard Books.

Sharma, M. (2000). The Decinnial Census: More Than Numbers. *Journal of Geography, 99*, 2.

Small, M. F. (1996). An Anthropologist's Attic. *Scientific American, 275*, 1.

Smith, R., & Brown, A. (1996). Developing a Sense of Place. *Journal of Geography, 95*, 86–89.

Social Education (1998). Social Studies and the New Immigration. *62*, 390–450.

Sosniak, L. (1999). Professional and subject matter knowledge for teacher education. In: G. Griffen (Ed.), *The Education of Teachers* (98th Yearbook of the National Society for the Study of Education, Part 1, pp. 185–204). Chicago: University of Chicago Press.

TAGE – Texas Alliance for Geographic Education (2000). *Applying Geography.* San Marcos TX: Southwest Texas State University, Department of Geography.

Tan, A. (1989). *The joy luck club.* New York: Ivy Books.

Thomas, D., Creel, M., & Day, J. (1998). Building a useful elementary social studies website. *Social Education, 62*, 154–157.

Trifonoff, K. (1997). Quilting and Geography: Learning Activities for Elementary and Secondary Levels. *Journal of Geography, 98*, 209–218.

Tucker, M., & Codding, J. (1998). *Standards for our schools.* San Francisco: Jossey-Bass.

Varus, L. (1990). Put Portfolios to the Test. *Instructor, 51.*

Wall, C. (1994). Friends in High Places. *Journal of Geography, 93*, 103–104.

Zant, K. (1977). Field Projects in the High School Earth Science Course. *Journal of Geological Education, 25*, 85–86.

PRINCIPLES OF SUBJECT-SPECIFIC INSTRUCTION IN EDUCATION FOR CITIZENSHIP

Judith Torney-Purta, Carole L. Hahn and Jo-Ann M. Amadeo

INTRODUCTION

What makes education for citizenship special as an area of instruction? First, it includes elements that overlap with the teaching of history and economics within social studies, but is not coincident with either. Some research and examination of practice in those areas can be applied, but gaps remain. A considerable amount of learning about citizenship takes place outside the school. Principles for developing instruction must take account of this. Furthermore, in education for citizenship, teaching knowledge of the foundations of one's country's democracy and a sense of civic responsibility are vital, but some believe that it is even more important to make students aware of national political issues, and to convince them of the importance of voting and having political discussions or of becoming involved in their local communities. These factors combine to create a special situation for education relating to citizenship.

Although a variety of positions are reflected in current discussions of citizenship education, it is clear that within the last decade there has been an acceleration of interest in the area and in related research on political socialization. Among the signs of interest at the national level are the issuing of a set of voluntary educational standards in civic education and the

Subject-Specific Instructional Methods and Activities, Volume 8, pages 373–410.

ISBN: 0-7623-0615-7

implementing of a National Assessment of Educational Progress based on these standards (Center for Civic Education, 1994; Lutkus, Weiss, Campbell, Mazzeo & Lazur, 1999); the establishment of a National Alliance for Civic Education; the renewal of emphasis on the civic purposes of education by the National Council for the Social Studies (Thiesen, 2000); federal and state legislation promoting youth service in the community, accompanied by a ten-fold increase in the number of young people participating in community projects; and various scholarly and pedagogical publications on civil society and the democratic purposes of education (McDonnell, Timpane & Benjamin, 2000).

An indicator of renewed interest at the international level is that during the mid-1990s IEA (the International Association for the Evaluation of Educational Achievement) mounted a two-phased, decade-long study of civic education in nearly thirty countries. Reference will be made to the national case studies that formed Phase 1 of the IEA Civic Education Study (Torney-Purta, Schwille & Amadeo, 1999) and to general insights from the Phase 2 test and survey in which 90,000 14-year-olds in 28 countries participated (Torney-Purta, Lehmann, Oswald & Schulz, 2001).

Many of the theories proposed to account for achievement, especially the principles of effective instructional methods delineated by Brophy (this volume), have relevance. They are not fully adequate to deal with education for citizenship, however. The first section of this chapter describes a conceptual framework to augment these principles; the second section describes three current positions on the focus of citizenship education; the third, fourth, and fifth sections deal with each of these positions in relation to principles of effective instruction and to other conceptual frameworks; the final section presents some overall conclusions in this area and needed research.

WENGER'S THEORY CONCERNING COMMUNITIES OF PRACTICE AND EDUCATION FOR CITIZENSHIP

The area of citizenship education calls for a broad framing, one that includes communities as well as schools and encompasses attitudes, motivation, and participatory behaviors as well as knowledge and skills. This framing builds upon what Alexander (1996) calls the second generation of the cognitive revolution. The model emphasizes the social and cultural context that influences learning. It depicts the learner as "situated" within a world of social affairs and practices that have an impact within designated "sites" of learning, such as the classroom and community.

Wenger (1998) has provided a conceptual framework to elaborate on this process, noting that learning involves the growth of *meaning, practice, community, and identity*. Observations of adult workers formed the empirical basis for this framework, but it has considerable utility in framing citizenship education since it considers both the individual and the broader context of civil society.

Wenger's first point is that learning relates to experience in *constructing meaning* individually and within social groups. Experience both in school and out of school should help young people to understand political concepts in meaningful ways and to connect new civic-related information to what they already know, especially to the powerful ideas important in democracies. When classes are taught in a rote fashion that disconnects governmental structures from their purposes (e.g. representation of potentially conflicting interests) and from the students' experience, meaningful learning is compromised. Describing how students develop meaningful domain-specific cognitive structures that can incorporate new information is central in many approaches, including that of Brophy (this volume) and that of the American Psychological Association's Learner Centered Psychological Principles (Alexander & Murphy, 1998).

To elaborate on Wenger's second point, learning is related to *practice* or mutual engagement in action, including the opportunity to try out knowledge in interpersonal situations, for example, participating in discussion in the classroom or identifying and trying to solve school or neighborhood problems. Cobb and Bower (1999) believe that learning in all types of classes requires both self-engagement and also the opportunity to contribute to the engagement of others and to changes in communal practices. Political scientists Conover and Searing (2000) speak of the democratic purposes of education as intended to "develop in young citizens the motivations, understandings, and skills necessary to engage in a full practice of citizenship once they reach adulthood" (p. 92). The use of terms like "practice" or "practices" relates to an ongoing stream of discussion in social studies education going back to Dewey (1916/1966), who advocated making schools laboratories in which young people were involved in the practice of solving social problems. Since the 1950s, a major strand of social studies curricula has dealt with "reflective inquiry" in which students investigate and deliberate about social issues (Engle & Ochoa, 1988; Evans & Saxe, 1996; Hunt & Metcalf, 1955; Newmann & Oliver, 1970; Oliver & Shaver, 1966/1974). Wenger's second point about action and practice leads to considering the nature of the group with whom the individual engages in practice.

In his third point, Wenger notes that when students learn, they relate themselves to a *community* that has standards of competence and defines certain behavior and goals as worthwhile. These communities are usually face-to-face groups, but they do not always share a positive sense of affiliation.

Negotiations about potential conflict and boundaries are often involved in becoming related to a community of practice. In fact, the novice or young person may be limited to peripheral or potential participation rather than central or full involvement. Young people cannot be centrally involved in some activities because they cannot vote or run for office. They can be involved, however, in discussing the issues which are important in their nation and their community, in formulating strategies for action, and in civic actions such as contacting public officials and writing letters.

Fourth, learning creates *identities*, personal and social histories that link experiences in families with experiences with friends and in school. According to Wenger (1998), a community is an important site for the formation of identity if it presents "life trajectories" that exemplify ways to integrate the past and lead toward the future. A sense of identification with the nation, with a political party, with an ethnic group, or with one's local community – all are traditionally thought of as parts of the citizen's identity (Flanagan & Sherrod, 1998). Conover and Searing (2000) note that although American young people identify themselves as citizens, they seem to lack detailed or creative conceptions of themselves enacting this role in the future.

THREE POSITIONS ON EDUCATION FOR CITIZENSHIP RELATED TO WENGER'S FRAMEWORK AND TO THE CORE INSTRUCTIONAL PRINCIPLES

Wenger's framework and the principles delineated by Brophy (this volume) clearly can be applied in the area of citizenship education. Their contributions can be most clearly illustrated if they are discussed in categories corresponding to three currently debated public positions about citizenship education for young people. Each position is rationalized by reference to a particular body of research. Each also relates to the reflections of practitioners documented by qualitative studies.

First, there is a long tradition of research in political science regarding the knowledge of adult American citizens, usually measured in public opinion surveys with questions about specific government structures, incumbents, and political issues. There is concern about serious deficiencies in adults' knowledge which are "most likely to be found among those who arguably have the most to gain from effective political participation: women, blacks, the poor, and the young" (Delli Carpini & Keeter, 1996, p. 177). In the 1990s the minimal levels of knowledge of civic institutions demonstrated by elementary and high school students were an object of worry (Lutkus et al., 1999), and more rigorous classes to transmit this knowledge were proposed. To address this first set of concerns we have grouped for discussion the instructional principles that deal most directly with the acquisition of content knowledge. This dimension relates to Wenger's concerns about meaningfulness in learning and to his notions about communities of practice where standards of competence are developed.

There is a second current stream of thought about youth citizenship which expresses concern about the declining levels of political trust and efficacy, lack of motivation for reasoned engagement in discussion about issues, and decreasing participation in conventional political activities such as voting among young adults. Those who share this concern would support efforts to address the roots of students' cynicism, and interest them in participating in political dialogue. To address this second set of concerns we have grouped for discussion the instructional principles that deal most directly with reflective inquiry and discussion of issues in the classroom. This is a way to motivate or engage students, give them more confidence in exploring their opinions in relation to important issues, show them how to use evidence in supporting their opinions, and encourage them to become thoughtfully critical about politics. All of Wenger's categories are relevant since discussion often results in more meaningful learning and plans for action, as well as being a part of identity development within communities of practice.

A third group of concerned individuals is looking for ways to more effectively engage students in their local communities, often through "community service" or "service learning" programs. Although appealing to many, a recent review has indicated that not all of these programs provide a satisfactory solution to the current problems of youth civic engagement (Billig, 2000). The approach is related to the communities of practice and trajectories of identity that Wenger has delineated.

It is clear that these three approaches (dealing with content knowledge, inquiry and discussion, and youth service learning) are not mutually exclusive alternatives. It would not be of much value if schools fostered knowledgeable individuals who had no intention of voting or people heavily involved in volunteering who neither knew nor cared about how the social problems they tried to solve were related to government policy. Nevertheless, the grouping of the principles into three categories does correspond to current debates in American society about education for citizenship. It is also clear that similar debates rage within many democracies, new and old, though the particular phraseology often differs, e.g. referring to "project learning" rather than "community service" or "small group discussion" rather than "cooperative learning" (Torney-Purta et al., 1999).

PRINCIPLES, RESEARCH, AND PRACTICE RELATING TO THE ACQUISITION OF CONTENT KNOWLEDGE

Five of the principles laid out by Brophy (this volume) are especially relevant to content knowledge and the development of political concepts: *Opportunity to Learn:* Students learn more when most of the available time is allocated to curriculum-related activities . . .; *Curricular Alignment:* All components of the curriculum are aligned to create a cohesive program; *Coherent Content:* To facilitate meaningful learning and retention, content is explained clearly and developed with emphasis on its structure and connections; *Cooperative Learning:* Students often benefit from working in pairs or small groups to construct understandings; *Achievement Expectations:* The teacher establishes and follows through on appropriate expectations for learning objectives.

First, we acknowledge the importance to citizenship of general excellence in pedagogical practices especially when it leads to high reading literacy and the ability to express knowledge and opinions clearly both orally and in writing. If instruction in line with these principles develops those capacities, it is valuable even if it lacks specific civic content. In fact, Chall and Henry (1991) noted that considerably more than minimal literacy is required for students to read and understand documents such as the Constitution or to locate information in newspapers or election materials. There is also specific civic content to be mastered, however. Research has been conducted on both large and small samples relating to this content knowledge.

RESEARCH ON CONTENT KNOWLEDGE USING LARGE SAMPLES IN THE UNITED STATES

Several surveys of civic knowledge based on representative samples provide clues about the acquisition of content knowledge relating to citizenship. The National Assessment of Educational Progress has regularly tested 4th, 8th, and 12th grade students in civic-related content areas. The NAEP results showed that a relatively small proportion of young people met the requirements set by experts for Proficient or Advanced achievement, but rather were at the Basic level (Lutkus et al., 1999). This generally corroborates the findings of other research, for example, Ravitch and Finn's (1987) multiple choice test covering history administered to a national sample of 17-year-olds.

Three decades ago Langton and Jennings (1968) published research downplaying the value of studying civic topics or taking civics classes in acquiring knowledge. Recently, when the 1988 NAEP data were examined, however, taking classes in which civic topics were included did appear to enhance knowledge of civics (Anderson et al., 1990; Niemi & Junn, 1998). To give some perspective on civics-related classes, the topics on which 8th and 12th graders in 1988 and 1998 reported extensive study were: United States Constitution and Bill of Rights; political parties, elections, and voting; Congress; rights and responsibilities of citizens. National case studies from Phase 1 of the IEA Civic Education Study show parallel general emphases in many countries (Torney-Purta et al., 1999).

Of greatest interest are the predictors of NAEP knowledge scores:

> Given the nature of the NAEP test, it might seem that the best way to raise scores would be for students to memorize facts about the Constitution and be drilled repeatedly about those facts. . . . But the results of the test suggest that this strategy often backfires. Similarly, one might think that frequent tests or quizzes would force students seeking high grades to learn constitutional provisions. But again the results suggest the opposite (Niemi & Junn, 1998, p. 79).

Niemi and Junn's analyses showed that the significant predictors of the 12th graders' knowledge of American government and civics included whether the students planned attendance at a four year college; whether they liked the study of government; and whether they had participated in mock elections, governmental bodies (e.g. legislature) or trials. The amount and recency of taking civics courses was also significant.

Nearly every analysis of NAEP data shows the poorer performance of students from African-American and Hispanic groups and those attending schools where many students are below the poverty line. In the United States college preparatory programs and more motivating types of instruction (including mock elections) are more likely to be offered in neighborhoods where more highly educated parents reside. There is a cluster of school and individual (family) characteristics which are associated with high achievement, and it is hard to separate them.

A program called Kids Voting appears to have somewhat lessened the knowledge gap between high and low socioeconomic status young people and families (McDevitt & Chafee, 2000). A quasi-experimental evaluation was conducted using telephone interviews with 5th through 12th graders and their parents (participants and non-participants in San Jose, California). This school-based program focused on information about how to register to vote, the role of political parties in elections, and how to organize information to make voting decisions (including newspaper reading, analysis of political advertisements, and debates about campaign issues). It concluded with a "Kidsvention" and mock election held on the same day as the general election. Energizing students' interest in elections resulted in more communication with parents about politics, more use of political media such as newspapers in the home and higher knowledge levels. These effects were stronger for low-SES students and families. The authors speculated that the success of young people in stimulating political communication in low SES families was in part because the behaviors fostered were "functional for developmental goals, such as the adolescents' desire to express opinions or the parent's need to keep up with maturing children" (McDevitt & Chafee, 2000, p. 285). The program itself has considerable potential for reducing the SES gaps such as those documented by NAEP. The components of the program, active involvement and attention to news media, could be used by many teachers even if the full program was unavailable.

RESEARCH ON CONTENT KNOWLEDGE USING LARGE SAMPLES INTERNATIONALLY

Another large data base comes from two IEA Civic Education Studies that tested about 30,000 students in 1971 (Torney, Oppenheim & Farnen, 1975) and about 90,000 14-year-olds in twenty eight countries in 1999.

The IEA Civic Education Survey, conducted in 1971 by the International Association for the Evaluation of Educational Achievement, an international consortium of educational research institutes, was the most rigorous study up to that time. Multiple choice tests of knowledge of political institutions were administered to stratified national samples in Western European countries and the United States. The fourteen-year-old students in the United States ranked fourth out of eight industrialized countries (Torney, Oppenheim & Farnen, 1975). Among the characteristics associated with a high score on this test in all countries were coming from a high socio-economic status home with well educated parents. Also predictive of high achievement were being in a classroom where the teacher encouraged students to express opinions, where patriotic rituals were infrequent, and where worksheets were infrequently used. The student's interest in television programs about public affairs was also a predictor. A classroom climate encouraging discussion and expression of opinion was a significant predictor across countries, even when the socio-economic level of the student's home was partialed out. The negative impact of rote learning and the positive impact of involving discussions were clear (Torney-Purta & Schwille, 1986).

For up-to-date comparative data (from the 1999 IEA study) on where United States 14-year-olds stand in achievement on knowledge and skills in civic-related areas in relation to students in more than twenty-five other countries, as well as information about concepts of good citizenship and the responsibilities of government, and a range of attitude measures see Torney-Purta et al. (2001)

RESEARCH ON STUDENTS' CONSTRUCTION OF CONCEPTS OF CITIZENSHIP

Developmental psychologists have conducted interviews with children or adolescents concerning political concepts to study how young people grasp the "powerful ideas" relating to democracy, government and citizenship (Connell, 1972).

In an inventive set of studies employing structured interviews, Helwig investigated young people's conceptions of freedom of speech, freedom of religion, and fair government. Helwig (1995) questioned 7th and 11th graders along with college students in the United States about these topics in the abstract and also in a contextualized situation where granting freedom of speech implied a conflict with harm to others (e.g. a speech advocating physical violence against a rival political party). In the situation of conflict the older groups were much more able to flexibly coordinate the various perspectives to arrive at a coherent answer. In another study, Helwig (1998) interviewed 6–11 year old Canadian students about their view of fairness and their preference for five systems of government: representative democracy, direct democracy, democracy by strict consensus, pure meritocracy, and pure oligarchy. Concepts of fairness were invoked as justifications at all ages, though the eleven-year-olds used a broader range of rationales and integrated a larger number of dimensions. Younger children "somewhat contradictorily maintained both that freedom of speech was a right held independently of existing law (e.g. rules restricting freedom of speech were judged universally wrong) and also that unjust rules prohibiting freedom of speech must be followed" (p. 528). The difficulties faced by many children up to the age of about twelve in dealing with complex situations, contradictions, and conflict suggest the importance of tailoring instruction to children's levels of understanding.

Osborne and Seymour (1988) developed a set of stories in which political concepts were embedded. They organized the stories and activities around a few key ideas that were understandable and interesting to students (such as disagreement and power) and related them to issues that could motivate students to consult media sources such as newspapers. Attitude scales were used in a quasi-experimental evaluation in twenty-seven 6th grade classes. Positive effects were found for increases in sense of political efficacy, acceptance of political conflict, approval of moderate political activity and for decrease in cynicism. Concept development was not measured, although students' written comments indicated that they learned about the pervasiveness of politics and conflict in everyday life and about political decision making.

Another study regarding the construction of concepts was undertaken by Sinatra, Beck and McKeown (1992). They examined the influence of instruction and development on concepts of government representation, building on work concerning conceptual change in science. The same students were interviewed before and after instruction in American history in grades five and eight. Over time the researchers found more consistency than change within students, and characterized most political concepts as vague, fragile, unstable, or showing "nascent" understanding. When change did take place, it was likely to be the inclusion of more references to structures of government. Responses focusing on government structures rather than on the notion of representation characterized the older students, and seemed in some cases to replace rather than build on nascent understandings of representation expressed earlier. The results confirmed previous studies criticizing textbooks for failing to provide information in a way that would lead to the development of more sophisticated concepts.

Personal freedom to do as one wished in everyday life was more important to students than institutions protecting freedom. The authors proposed more attention to students' active part in organizing information and relating it to concepts they already possessed, for example, relating their views of personal freedom to formal features of government (rather than memorizing the structures of government).

Wade (1994) conducted a qualitative study of changes in conceptions of rights in a classroom of 4th graders over two semesters, observing as well as conducting interviews. She wanted to find out how knowledge construction took place and to what extent students might be memorizing isolated facts to use them on a test rather than attempting more in-depth understanding. Many of the students (like those in Sinatra's study) had the misconception that "rights" means the freedom to do whatever you want. Teaching strategies used in this classroom included discussion, cooperative learning, role play and simulation, and class meetings. Students who began the study with less self-oriented concepts were generally more successful in developing in-depth understanding over time. Her major conclusion, however, was that it was an error not "to confront the misinformation that many students possessed, diagnose their personal agendas, and teach for assimilation and accommodation rather than content coverage" (Wade, 1994, p. 91).

These studies all point to the variety of ideas which children have about political concepts, often based on vague or partial views. Instruction, if it is to be meaningful and effective, needs to better deal with those understandings and misunderstandings and with the contradictions that exist within individual students. This is especially important because research with elementary school children suggests that political knowledge and concepts are important in assisting students to understand political discourse such as that found in newspaper articles (Allen, Kirasic & Spilich, 1997).

RESEARCH USING THINK-ALOUD PROBLEM SOLVING TO EVALUATE A COMPUTER-ASSISTED SIMULATION EXPERIENCE IN INTERNATIONAL POLITICS

Computer-assisted instruction is less frequent in citizenship education than in some other subject areas. One project both illustrates this approach and has been evaluated for its impact on conceptual development. The International Communication and Negotiations Project, ICONS, is a program developed by the University of Maryland Department of Government and Politics, organizing simulation exercises in which schools in the United States and abroad participate. Teams of students (age 12 and older) are assigned to role-play diplomats from a given country. A sophisticated computer networking program (based on the Web) allows communication between these teams of student-diplomats. After preparation of a position paper on their country, students begin negotiations centered around issues such as the global environment, immigration, and international debt (Torney-Purta, 1996).

Communication about the messages that will be sent during the negotiation takes place face-to-face within the team; communication between teams takes place through electronic mail and during scheduled on-line conferences in which students participate simultaneously with teams representing other countries. A great deal of debate takes place about the wording of what the team will send over the network. This cognitive conflict among peers is a potent stimulus for individuals to restructure political concepts.

Research strategies derived from cognitive psychology were used to evaluate the ICONS project (Torney-Purta, 1992, 1996). Individual participants were asked to think aloud while solving a hypothetical political problem (e.g. that faced by the finance minister of a developing country which cannot pay its debt). A graphic model representing each individual's underlying image or schema of the political system was developed from these responses. Three types of elements were linked in these schema-maps: *actors* in the political or economic system, *actions* in which these actors could engage and *constraints* upon these actions. Comparing these concept maps from early and late in the session showed that the majority of students went from a limited number of unconnected and unconstrained actions to a larger number of connected and constrained actions.

There were several *characteristics of this program* which seemed to promote the observed restructuring of political concepts. First, the computer screen was an object of highly focused student attention because it was constantly providing new information. After a proposed message was entered and appeared on the screen (but before it was sent), the process of group revision surfaced individuals' misconceptions. Second, the student teams had responsibility for developing criteria for message clarity and content, and they were often more stringent than a teacher would impose. Third, the existence of the messages in written form, as well as repeated pressure to participate as if they were diplomats discussing real policy issues, gave a sense of authenticity of practice.

The Web-based network and team structure have also been used for debating issues involved in governmental changes in new democracies, e.g. how conflicts between freedom of the press and minority rights should be dealt with. Similar discussions promoting conceptual change within the peer group were observed.

Teachers have a choice between joining a program such as this or using elements from it, for example extended role playing exercises that ask students to identify with the role of policy maker or diplomat and involve preparing messages taking positions on civic issues to be sent to other groups using email.

RESEARCH WITH TEACHERS AND RESEARCH DOCUMENTING LESSONS FROM CLASSROOM PRACTICE

Recent research by Davies and his collaborators in England, which has not had a strong tradition of citizenship education, bridges some of these interests in conceptual development and issues important to teachers in the classroom. Davies, Gregory and Riley (1999) formulated models of civic education after administering a questionnaire to nearly 700 teachers from 75 schools in all the major regions of England. Good citizenship could be defined in terms of: (1) social concern and tolerance for diversity; (2) knowledge of government, current events, and (to some extent) the ability to question ideas; and/or (3) subject/obedient role characteristics including acceptance of authority, patriotism, acceptance of responsibility. About 80% of these British teachers rated social concern as the most important. In interviews teachers stressed that knowledge should be seen as a means either to critical examination of issues or to more effective ways of advancing others' well-being or ameliorating social concerns, not as an end in itself.

Based on this study, the authors suggest identifying elements within particular schools with potential for citizenship education: examining unintended ways in which the organization of the school mitigates against participation; involving other staff in looking at ways to pursue fairness and justice in school governance; organizing whole school events such as mock elections; creating councils composed of students and taking their work more seriously; finding ways to stimulate and recognize the professionalism of teachers in generating new ideas; instituting more ways for students to work in the community (with rigorous attention to connections to the school curriculum). Some of these activities go beyond fostering content knowledge, but they suggest that at least in England the emphasis on content knowledge for its own sake does not correspond to teachers' beliefs.

Research by Merryfield (1994, 1998) brings together some of these concerns. She conducted extensive observations of teachers of world studies or global education courses in 12 classrooms in Ohio. She interviewed each teacher about her transcripts with attention to influences on the teacher's decision making process – especially contextual factors. Merryfield presented profiles for each teacher which stressed the guiding principles behind their instruction and the factors influencing it. No matter what the curriculum or textbook or state mandate, the teacher's background, beliefs, and the students' characteristics and experience resulted in enormous diversity in actual instruction. All teachers endorsed the presentation of multiple perspectives on issues. Further, they moderated their day to day decisions about the content to be presented according to their students' skills, previous experience, and interests, while showing sensitivity to racial, ethnic, and religious differences in their classes. Teachers were participating in constructing the meaning of many topics relevant to citizenship education for students on a daily basis embedded in the social structure of the classroom and of current events. The more opportunities provided for teachers to be reflective about how the particular students in their classrooms understand and react to civic concepts and topics, the better. By using examples and topics that would be meaningful to diverse individuals, the master teachers were using culturally sensitive pedagogy that would enable students to connect new knowledge to their existing frameworks.

SUMMARY OF THE SECTION ON CONTENT KNOWLEDGE

In terms of the principles stated at the beginning of the chapter, *opportunity to learn* was addressed primarily in terms of classes taken and students' reports of topics studied. The NAEP analysis by Niemi and Junn is the major source supporting the importance of opportunity to learn. However, its interpretation is a nuanced one. Those who believe that knowledge is a critical prerequisite to citizenship should note that frequent testing is not sufficient and may even be counterproductive in improving achievement. Opportunity to learn is important, but classroom activities need to be personally involving for students and tailored to the concepts they already possess. Researchers, such as Sinatra, criticize much of what is taught in this subject area as causing students to learn terms referring to government structures rather than developing more adequate concepts of how such structures contribute to the ideas behind the structures, such as representation. Wade largely agrees that presenting factual material is unlikely to result in meaningful learning unless there is an effort to address students' misconceptions. Osborne's article and the studies where teachers were the focus showed similar disillusionment with disconnected fact-based instruction and some success in connecting instruction to political narratives and concepts embedded in classroom life.

Curriculum alignment is addressed somewhat indirectly in citizenship education. It may be more useful to look at whether the school offers several complementary routes to acquiring knowledge about citizenship. Support for such an approach can be inferred from the NAEP studies, the 1971 IEA study, Sinatra, Wade, and Osborne (and is also suggested by Davies). Of particular value are activities which actively engage students' interest.

The same studies, with the addition of Kids Voting and ICONS, point to the importance of *cooperative learning* and content-based peer discussions. It appears that discourse within the peer group is an essential context for the development of social and political knowledge because this is the community of practice which matters most to the young person where ideas can be explored.

The principle concerning coherent content as necessary for meaningful learning is addressed most directly by research about concept development. Helwig, Sinatra, and Wade all argue for the centrality of this aspect. Teachers should become more aware of students' misconceptions, discuss them, and help students to arrive at fuller understandings. Young people's self-oriented rather than system or community perspectives need to be addressed. Political concepts can be effectively introduced into the curriculum much earlier than adolescence, especially if stories and conflicts are used to engage children's interest and thinking. The studies of Osborne, Merryfield and Davies all point to the role of sensitive teachers who can recognize and adapt to students' concepts and interests.

High expectations for students, influenced by both home and school, are also valuable. This is clearest in the NAEP study (where students aspiring to college were the most knowledgeable), in the Kids Voting program (where students were expected to carry messages about the importance of politics into their homes) and in the ICONS study (where student-diplomats established expectations that messages would meet high standards of adult-like authentic practice).

PRINCIPLES, RESEARCH, AND PRACTICE RELATING TO ISSUE-CENTERED INQUIRY, DISCUSSION AND DECISION-MAKING IN THE CLASSROOM

The following three principles from Brophy (this volume) are especially relevant in considering the role of issued-centered inquiry and discussion within the classroom: *Supportive Classroom Climate:* Students learn best within cohesive and caring learning communities; *Thoughtful Discourse:* Questions are planned to engage students in sustained discourse structured around powerful ideas; *Cooperative Learning:* Students often benefit from working in pairs or small groups to construct understandings.

It is not fully satisfying to separate content and discourse issues from each other since considerable research shows their close connection. To be concrete, one would not want a classroom discussion devoid of content, no matter how motivating to students. Additionally, the three principles need to be implemented in the context of democratic decision-making. In the area of citizenship education there is a particular value placed on inquiry and discourse about political issues on which there are differing opinions. Indeed the cumulative research in social studies and civic education indicates that it is important to plan instruction where issues-centered content is combined with discourse (pedagogy) in a supportive classroom climate (Hahn, 1996).

In Hahn's (1998) study of citizenship education in five countries the students who reported the most issues discussion in a supportive classroom climate were the most likely to report high levels of political efficacy, political interest and political trust. The ways in which issues were approached differed among countries, as did the frequency of issues discussion. For example, in England where, until recently, citizenship education has not been emphasized, students tended to discuss personal ethical issues rather than public policies. Without a subject such as civics or social studies, discussions were most likely to occur in lessons on religious education, English, or personal and social education. In the United States, students were most likely to discuss issues in their social studies courses in the context of current events, historical events, or public policy debates. In Denmark, elementary school children frequently held class meetings to resolve class and school problems, to advise their school council, and to plan class trips. Older students in Denmark engaged in discussions and decision-making about the topics they would study in many of their classes, as well as conducting inquiry into public policy issues.

Students' opportunities for discussion about public issues vary not only across national contexts, but also within countries, as indicated by a recent study in the United States by Conover and Searing (2000). The researchers purposefully selected four differing communities in which they conducted interviews with high school students, their parents, and community leaders. They found that students from the suburban and rural communities were considerably more likely than students in the urban and immigrant communities to report that they often experienced discussions of political issues in their classes. Conover and Searing argue that discussion in school seems to develop skills and motivation for discussion in other settings, such as in families and with peers. In communities such as those in urban and high-immigrant areas, where there may be less family encouragement for civic participation, it appears especially important that schools provide opportunities for democratic discourse.

RESEARCH ON PARTICIPATION IN THOUGHTFUL DISCOURSE

The research in this section shows that the process dimension of instruction involving sustained discussion in which students actively participate is particularly important in developing dispositions needed for democratic participation. Repeatedly, researchers found courses (content) in civics and government, or social studies, could enhance students' civic knowledge (as detailed in a previous section), but they had negligible effects on student attitudes related to eventual participation (Ehman, 1980; Ferguson, 1991;

Patrick & Hoge, 1991). However, when attention was paid to the process of instruction within classes, variables that make a difference became apparent. An open classroom climate, or classroom atmosphere in which students report frequently discussing issues, hearing and exploring alternative views, and feeling comfortable expressing their own opinions has been found to be associated with participatory attitudes (as well as with knowledge).

In studies of secondary school students in the United States and in other western democracies, researchers found that open climates were associated with comparatively high levels of political interest, political efficacy, and political participation (Blankenship, 1991; Ehman, 1969, 1977; Hahn, 1998; Torney, Oppenheim & Farnen, 1975). However, the mere presence of public issues about which there is controversy is insufficient. Indeed, when they are presented in a "closed" climate in which only one view is presented and/or in which students do not feel comfortable expressing their views, lower levels of political efficacy, participation, and citizen duty are likely to result (Ehman, 1969).

RESEARCH AND EVALUATION OF INSTRUCTIONAL TREATMENTS

Additionally, curriculum projects and instructional treatments that require students to investigate social issues, consider alternative views, deliberate about the consequences of alternative solutions, and make decisions about their preferred policy have yielded benefits. Evaluators of the Harvard Social Studies Project, in which students were taught to use the "jurisprudential approach" to analyze controversial public policy issues, reported that the students who were taught the approach were better able to analyze argumentative dialogues than students who were not taught to use the approach (Levin, Newmann & Oliver, 1969; Oliver & Shaver, 1966/1974).

In a recent application of the public issues instructional approach to discussions about issues raised in segments of the Channel One news programs, evaluators similarly found positive effects. Students who used a public issues discussion format after viewing the news programs scored significantly higher on a written public issues analysis test and on a test of current events knowledge than students who did not view the programs (Johnstone, Anderman, Milne, Klenk & Harris, 1994). Moreover, the evaluators reported that lower-ability students seemed to benefit the most when structured public issues discussions were combined with news viewing.

In another study, researchers looked specifically at the effects of using a public issues centered approach with slow learners (Curtis & Shaver, 1980). In an in-depth study of housing problems in their area, students examined a variety of resources that represented different views on the issue and conducted a community survey. The students showed increased self-esteem, decreased dogmatism, increased interest in housing and other social issues, and increased scores on a test of critical thinking. Furthermore, these students were more likely to express the belief that citizens could make a difference in improving conditions than students who were not exposed to the public issues centered approach. Several of these studies showed positive effects on both knowledge and attitudes.

Other researchers have looked at the effects of instructional units in which students confront issues associated with civil liberties. In one study of a three-week unit, students examined controversial issues related to topics such as freedom of religion, search and seizure, freedom of expression, and due process of law (Goldenson, 1978). Students conducted group research projects and met with community members holding varied views. At the end of the three weeks, substantially more of the students exposed to this unit exhibited attitude changes in the direction of greater support for civil liberties than students not exposed to the instructional unit. The researcher speculated that the methods, which encouraged students to do research about an issue about which people disagreed, affected students more than the particular materials used.

Similar results were obtained in a study of a four-week unit on tolerance for diversity of beliefs (Avery, Bird, Johnstone, Sullivan & Thalhammer, 1992). Junior high school students studied the psychological, sociological, and historical dimensions of intolerance and tolerance. After analyzing case studies, role-playing, conducting mock interviews, and engaging in discussions, students were asked to decide what limits, if any, should be placed on freedom of expression in a democracy. Over the treatment period, students who participated in the unit – unlike students who did not – moved from mild intolerance to mild tolerance of rights for people with whom they disagreed.

Although none of these studies of particular instructional units measured how comfortable students actually felt expressing their views, the units did include varied activities in which students heard and discussed diverse and conflicting views. In addition, all of the units encouraged students to weigh alternatives before making up their minds about public issues. Students were using their knowledge in authentic ways. At the same time, they were developing skills and dispositions that would later enable them as adult citizens to participate in debates about public policy issues. They were becoming connected to communities of practice.

RESEARCH DOCUMENTING LESSONS FROM CLASSROOM PRACTICE

Numerous social studies methods textbooks provide models and suggestions for teachers undertaking public issues instruction (eg. Banks & Banks, 1999; Evans & Saxe, 1996; Parker & Jarolimek, 1997, Parker & Zumeta, 1999). Additionally, there are instructional materials that can be used by secondary students to consider diverse views on controversial public policy issues (National Issues Forum; Opposing Viewpoints, Social Science Education Consortium). These reflect the "wisdom of practice," as do studies of teachers using issues-centered instruction.

Qualitative research that sheds light on the role of public issues instruction to education for citizenship includes recent case studies of classroom practice (Bickmore, 1991, 1993; Hess, 1998; Newmann, 1991; Rossi, 1995, 1998). Researchers have observed classes and interviewed teachers and students to identify the components of issues-centered instruction. They have also identified a number of dilemmas that teachers face in undertaking this type of instruction.

Bickmore (1991, 1993) studied four U.S history and world studies teachers who approached controversy or conflict differently depending upon their conceptions of citizenship. One of the teachers, Tom, believed that citizens must be able to resolve conflicts. For that reason, he used both "conflictual content" (e.g. students studied about conflicts in United States history) and "conflictual pedagogy" (e.g. students debated differing interpretations of an historic event). Another teacher avoided both types of conflict. A third teacher, Ruth, used "conflictual content," in the sense that she acknowledged that different cultures have different ideologies, values, and viewpoints. She treated conflict or controversy as if it were a neutral fact rather than a potentially charged issue, however, and relied on a minority of confident students to answer her questions (which usually had a single factual answer rather than prompting the expression of opinion). For the most part Ruth did not have students confront, explore, or debate the controversies presented.

Because a fourth teacher in the study, Sarah, used a great deal of "conflictual pedagogy," her case is helpful in revealing how teachers might engage students in discussion of controversial public policy issues. She often began a lesson with a divergent question, soliciting views from as many students as she could. Sometimes she would ask for a show of hands to determine how many students approved or disapproved of a particular position and then encourage them to explain their views. She regularly led discussions of current events and asked students to take the perspective of someone living in the society involved and to comment on how they might feel. In contrast to many descriptions of social studies classes, "student voices were heard more than the teacher's and disagreement was common between student and teacher as well as among students" (p. 368). Bickmore concluded that conflictual content can teach lessons about diversity in society but it is through the use of conflictual pedagogy that more students are included in the learning. Yet, this is no panacea. Both Tom and Sarah worried about teaching enough content. And even in their classes, up to about one-third of students had learned a "trick of invisibility" and rarely spoke. Other case studies further elaborate on challenges such as these.

Rossi (1995, 1998) studied classrooms of teachers who used "in depth" instruction. In such classes students study authentic issues or questions that contain ambiguity, doubt, or controversy. Sustained time is given to a single topic or issue. Students use knowledge that is complex and divergent about the topic or issue, consulting sources that range beyond a single textbook and that present varied viewpoints. In one public issues class that Rossi (1995) observed students spent three weeks each investigating issues related to freedom of speech and affirmative action. The units were similarly organized: an introductory activity to engage student interest, whole class and small group discussions of background readings that linked the issue to a conceptual framework, use of previously taught discussion skills to enable students to identify issues and defend positions on them. Each unit concluded with a "scored discussion" in which students demonstrated their ability to state policy questions, identify and explain definitional, value, and factual issues, and defend positions on controversial issues. Each activity required social interaction among students and between the teacher and students. "Discourse was the heart of the process" (p. 100).

Rossi (1995) identified three dilemmas that face many teachers who undertake this instructional approach. First, the teacher wanted students to set the agenda for classroom discourse, in a sense constructing new practices. He wanted students to draw upon their attitudes, experiences, and perspectives and link them to public issues. At the same time, the teacher wanted to provide students with information and concepts that he felt were needed to understand the issue. The second dilemma related to using a systematic model so that the students would acquire skills for discussing public issues, which they could use on authentic questions as members of the class and in their futures as citizens. However, the model seemed to inhibit some students who perceived it to be threatening and judgmental. Finally, in talking with students, Rossi noted that not all students responded in the same way to in-depth public issues study.

Some liked the emphasis on sophisticated discourse and reported that even outside class they found themselves more likely to use evidence; others did not. Yet, most did gain awareness that knowledge could consist of many viewpoints. They expressed tolerance toward points of view other than their own – important understandings for citizens of a pluralistic democracy.

In a study of in-depth issues-centered instruction in classes of low-ability students, Rossi (1998) identified further challenges to using this type of instruction. A preeminent concern of the teachers he studied was the importance of motivating low-ability students. It was difficult to develop questions that would not only capture and sustain student attention, but that would also open doors to meaningful and significant themes to broaden students' worldviews. A second challenge, exacerbated with low-ability students, was developing structures, strategies, and activities that promoted student autonomy, skill acquisition, and conceptual understanding. Too often structures, such as the use of worksheets, generated boredom or passivity. More engaging activities, such as group investigations and presentations, induced student frustration. A third challenge was that more teacher time and energy was needed than with traditional forms of instruction. However, the teachers' efforts seemed to pay off in terms of student learning. Student interviews revealed that although they struggled, many learned to make inferences and decisions from in-depth study of issues. They also valued and learned from expressing their opinion and debating issues as valued members of a classroom community. Further, for some of the usually low achieving students, the in-depth investigation and discussion "opened up their minds to a deeper awareness of the complexity, differing perspectives, and inter-relatedness of the world" (p. 405). Research such as this reveals that teachers need to have a realistic understanding of the challenges. In so doing, they can benefit from the insights of master teachers regarding issues such as whether students should be formally graded on discussion participation and whether teachers should disclose their own personal views (Hess, 1998). Pre-service and in-service programs could provide opportunities for teachers to practice discourse about these issues, hear alternative views, weigh consequences in light of their values, and make decisions in a manner that is consistent with the instruction they will undertake.

Most of the research on democratic discourse and decision-making has been conducted in secondary school classrooms. However, such instruction is also appropriate for elementary students. Indeed, many educators have argued that preparation for democratic life should begin in elementary classes where students discuss issues of importance in the lives of children and in society at large (Angell & Avery, 1992; Banks & Banks, 1999; Skeel, 1996). A recent case study of an elementary class that used class meetings and rudimentary parliamentary procedure as the vehicle for discussion and decision-making is instructive. Angell (1999) found that "children were eager to practice the difficult skills of democratic talk" to solve problems of classroom life. With time, the students developed personal confidence, efficacy, and interest in participation; they also exhibited respect for each other, for justice, and for the law (Angell, 1999, p. 168). In another case study of an elementary class, Bickmore (1999) found that students seemed to develop confidence through the frequent use of discussion. As part of a conflict resolution emphasis, students studied about and discussed global conflicts, as well as interpersonal conflicts confronted in school. Through such an emphasis young children developed skills and understandings that are needed for future civic participation.

For years civic educators have advocated that education for citizenship should be based on inquiry into and discussion about public issues, including those on which there is controversy. By using conflictual content and pedagogy in an open climate, teachers can achieve important goals of education for citizenship (Hahn, 1996). Moreover, the research indicates that such instruction can foster meaningful and authentic learning and the development of conceptual understanding, skills, and dispositions that citizens can use in the future. Yet, there is comparatively little instruction of this type that occurs on a daily basis in American schools, perhaps because teachers are unprepared to overcome the substantial challenges they will face or because of the current emphasis on testing. As Van Sledright and Grant (1994) commented even before the current accelerated focus on standards and assessment, "as long as extensive subject matter coverage is mandated by curricular policy and perceived as an important goal of school learning, constructive citizenship and the layers of learning opportunities added to the curriculum will remain entrenched in a losing battle with traditional academic content unless powerful ways of wedding the two are developed that do not consume additional time" (p. 336).

SUMMARY OF THE SECTION ON DISCOURSE

In terms of the principles stated at the beginning of the chapter, it appears that the notion of supportive classroom climate is valid. In education for citizenship it has been addressed by considering the openness of the classroom in its support for discussion (rather than its more general supportive character).

The value of *thoughtful discourse* structured around powerful ideas is supported by the large majority of the research reviewed. Often this takes place in small cooperative learning groups, but it also occurs on a whole class basis.

In the area of citizenship education there is a particular value placed on inquiry and discourse about political issues on which there are differing opinions. To engage citizens in authentic use of knowledge and to help them develop needed skills and attitudes for participatory citizenship, classes should implement issue-centered instruction together with discourse.

PRINCIPLES, RESEARCH, AND PRACTICE RELATING TO THE APPLICATION OF KNOWLEDGE THROUGH COMMUNITY AND SCHOOL ACTIVITIES

Brophy (this volume) has specified one instructional principle having to do with application: *Practice and Application Activities:* Students need sufficient opportunities to practice and apply what they are learning, and to receive improvement-oriented feedback. In education for citizenship, application is a particularly important matter. In this section we will deal with research on two issues: the students' engagement in community service together with what is sometimes called "service learning;" and students' participation in in-school but outside-class activities.

INTRODUCTION TO LEARNING THROUGH COMMUNITY SERVICE

Community service and service-learning are terms used to describe a wide range of experiences. Although they are sometimes used interchangeably, Niemi (2000) offers the following distinction between community service and service-learning. Broadly speaking, community service refers to the volunteer work done in the community which is not directly linked to an academic course or part of the formal school curriculum. This volunteer experience, however, may be encouraged or even formally arranged by the school. Service-learning typically refers to service that is directly incorporated into a course or the formal curriculum. Here, the service is typically preceded with conceptual and/or political and social information and followed up by classroom discussions and written reflections. Reflection is a vitally important element in service-learning. The emphasis on thinking and writing about the service activity is what distinguishes service-learning from other youth volunteer programs. Although practitioners and researchers make this distinction between service-learning and community service, there is still little consensus about duration of time that should be spent in a service program or type of activity. Service programs can and do take on many forms.

Service programs have a long history in the United States where there is a strong tradition of relying on private and religious organizations for social services. There is also a history in the United States of "advocating the pedagogical value of service participation in youth" (Youniss & Yates, 1999, p. 8). Some trace the pedagogical roots of service programs to educational theorists who advocated the importance of students' active involvement in applying what they learn (Billing, 2000) as well as to Alexis de Tocqueville's writings on the importance of volunteerism in a democracy (Yates, 1999). Within social studies, there has been a tradition of emphasizing that social action should follow study and reflection (Banks & Banks, 1999; National Council for the Social Studies, 1979; Newmann, 1975).

In the past, service programs were implemented primarily by religious and community organizations or colleges and universities. More recently, efforts have been made to engage younger youth in service projects, to link these projects more closely to schools and other formal education settings, and to assess their effectiveness (Killen & Horn, 1999). Especially in the last decade there has been a resurgence of interest in the United States in community service and service-learning with the development of programs designed to link experiences in the real world with academic learning (Stukas, Clary & Snyder, 1999). During the 1990s legislation aimed at increasing young people's service to the community was passed at federal, state, and local levels. For example, there is the National and Community Service Act and the National Service Trust Act which funded AmeriCorps and established the Corporation for National Service. On the state level, Maryland made 75 hours of community service a mandatory requirement for high school graduation, with similar provisions in other states (including South Carolina, Delaware, Kentucky, and Vermont). Local jurisdictions, including cities such as St. Louis and Detroit, passed laws requiring student participation in community service (Stukas et al., 1999); a 1995 survey reported that 15% of the nation's 130 largest school districts required community service (Yates, 1999). In addition to government-supported programs, several major businesses have taken the lead in promoting service activities among youth. Organizations, such as the Constitutional Rights Foundation, specializing in law-related education have also incorporated community service projects in their programs. The result of this resurgence of interest is that community service programs now exist in every state and the number of young people involved in service programs has increased substantially.

Proponents of service programs often argue that with the nation's growth and industrialization, American collective identity is fragmented and that one way to restore participatory democracy and encourage a collective civic identity is to include service programs as part of public education (Wade, 1997). This interest in promoting the civic involvement of youth was also sparked by the perceived need to improve civic education, and to develop leadership skills and a sense of self-efficacy among young people. Critics, however, point to superficiality in some students' involvement.

EXAMPLES OF RESEARCH ABOUT COMMUNITY SERVICE ON LARGE SAMPLES

In the spring of 1992, as part of the National Education Longitudinal Study of 1988 (NELS, 88) follow-up, high school seniors were asked about any community service they had performed during the previous two years. Researchers found that 44% of the cohort had been involved in some type of community service in this time period. In addition, the characteristics of those students most likely to be involved in service activities were identified by the NELS data. Females, students from higher socioeconomic status families, those with higher reading proficiency, and those in private or in urban schools were more likely to perform community service than other students (National Center for Education Statistics, 1995). However, students were not asked about the frequency or duration of their community service, how it related to curricular-based activities, or whether it included the component of discussion or "debriefing" that has been found to be important.

The National Education Household Surveys of 1996 and 1999 (NEHS, 96, 99) also investigated the degree to which American adolescents participate in service activities. The NEHS data indicate that just over half of the high school students in the sample participated in some kind of community service activity during the previous school year (Kleiner & Chapman, 1999).

In summary, there is a fairly high level of service activity among American adolescents. However, the extent to which these service hours and activities contribute to civic knowledge or later civic participation more broadly conceived remains a question. Further, Yates (1999) notes that community service learning has not really made the transition "from theory to widespread practice within the school curriculum" (p. 18). One explanation for the lack of widespread adoption could be the paucity of rigorous evaluations that document the benefits of service programs, making the implementation of often-complex and time-intensive service-learning programs less appealing to educators than more traditional, classroom instruction.

ISSUES IN EVALUATING AND IMPLEMENTING SERVICE-LEARNING PROGRAMS

There have been attempts to define "best practices" for service-learning programs. Initially, much of the service program research was qualitative and largely descriptive, often relying on student journals (Stukas et al., 1999). One clear challenge facing community service researchers is the role that student self-selection plays in determining student outcomes. For example, Johnson, Bebe, Mortimer, and Snyder (1998) found that a service program's effects on high school students' self-esteem, sense of well-being, and grade point average were nullified when preprogram factors were considered.

Another difficulty in assessing the effects of service programs is that the terms are so broadly defined that they encompass a wide range of programs and activities. It is, therefore, important when researching and evaluating service programs to keep in mind the goals associated with different service activities. Based on analysis of data from the National Service-Learning Clearinghouse, Stukas et al. (1999) point out that service-learning coordinators often view their service programs as a way "to enhance personal growth of students, particularly their self-esteem and social responsibility" (p. 2). However, this view may not be shared by the academic instructors who incorporate service-learning into their courses as a way to teach or motivate students to learn content. Service program goals may differ based on the age of the student participant (middle-school student versus high school student) as well as whether there is a "stand alone" program or one that is content-based and linked to a specific academic course (Stukas et al., 1999). The program goals may also depend on the needs of the service recipients, the duration of the service activity, and whether or not the service is mandatory or voluntary. Some programs may be designed to enhance the self-efficacy of the student, while others may be designed to promote civic involvement, improve student grades, or teach substantive academic content.

In a note of caution, Niemi (2000) concludes that much of community participation is "decidedly nonpolitical, possibly even anti-political, in the sense that it conveys to participants that the way to get things done is by direct, personal action by nongovernment groups rather than by any governmental programs" (p. 17). Although such experience could have an impact on a sense of membership in a community of practice, a number of scholars question whether that happens in most programs.

In conclusion, the evaluation of service programs and research related to outcomes (in particular student outcomes) is a complex effort. Researchers must take into consideration the type of program under investigation, the program goals, student characteristics, and the effect of self-selection.

EXAMPLES OF EVALUATIONS OF SERVICE-LEARNING

Research on K-12 service programs has focused primarily on the benefits to the student. Generally, researchers investigate student outcomes in three domains: psychological development, social development, and academic learning. Evidence on the impact of service learning on academic learning or intellectual development is not conclusive, and findings related to the social, moral, and psychological development of students is also mixed (Wade, 1997).

Some empirical evidence suggests a positive influence of service-learning and community service on student outcomes. For example, based on surveys and the student records of 1,000 middle and high school students from 17 Learn and Serve programs in nine states, Melchior (1997) found positive impacts of service-learning on students': (a) civic attitudes (including measures of acceptance of cultural diversity, leadership, and personal and social responsibility); (b) school engagement; (c) school grades (in math, social studies, and science); and (d) students' educational aspirations (measured by their desire to graduate from college). On the other hand, Melchior found no statistically significant impacts for students on measures of social or personal development, including communication skills, work orientation, or risky behavior.

It is important to note that the data from the Learn and Serve program evaluation were collected from only those 17 Learn and Serve programs (established by the 1993 National and Community Service Trust Act and administered by the Corporation for National Service) identified as "high quality." They were selected from a pool of approximately 210 programs through "a purposive sampling process aimed at identifying well-established, fully-implemented service-learning programs" (p. 3). All of the programs in the sample were linked to school-based curricula.

Other researchers have found relationships in the social and/or personal development domains. Youniss, Yates and Su (1997) found a relationship between lower marijuana use and community service among high school students. Shumer (1994), based on his study of 60 high school students involved in a program focusing on youth service and a group not involved in service, found that community service was effective in improving school attendance and school grades. Switzer, Simmons, Dew and Regalski (1995) examined the effects of participation in a school-based helper program for seventh grade students and found that it had a positive effect on boys' self esteem, attitudes, and problem behavior. Switzer's results were gender specific and compared with a group of students not required to participate in the helper program.

Based on an extensive review of the literature, Billig (2000) argues that several program design characteristics enhance the possibilities for positive impact. These program characteristics include attributing responsibility to students, student autonomy, student choice, direct contact with the service recipient over a reasonable period of time, high quality activities, and well prepared teachers. In essence, Billig is arguing for a more complex model for service learning than now exists and calls for more and better research to evaluate the effectiveness of service programs.

In conclusion, community service and service-learning programs have grown rapidly during the last decade. Many educators, researchers, and policy-makers see service-learning as an effective strategy to enhance academic knowledge, foster personal growth, promote civic-mindedness, and improve conditions both at the school and community levels. While some research seems to support the value of youth service programs and activities, more research and evaluation is warranted.

RESEARCH ON PARTICIPATION IN EXTRA-CURRICULAR ACTIVITIES AND ITS IMPACT IN LATER LIFE

In addition to community service and service-learning, many educators and researchers advocate the importance of participation in extracurricular activities as a means of teaching civic knowledge and promoting citizenship. The strategy of teaching democratic values and practices by providing opportunities for students to engage in democratic activities can be seen in many schools and out-of-school programs. Encouraging participation in such activities views young people as already citizens in many respects, not just on the way to becoming citizens. Further, many educators as well as politicians in the United States have pointed to a paradox: While many young people participate in service and community activities, relatively few exercise their right (and responsibility) to vote or undertake other political responsibilities when they reach young adulthood. The need to bridge this gap is widely acknowledged.

Based on a review of the literature, Youniss, McLellan and Yates (1997) contend that high school students involved in school or civic activities are likely to grow up to be involved adults. Activities may range from involvement in a 4H or scout program to work on the school newspaper. The authors argue that organized group participation has a lasting impact because it introduces adolescents to basic group roles and processes, and also because it helps adolescents to incorporate civic involvement into their personal identities at a time when identity issues are salient. The formation of civic identity is the developmental link between active adolescents and involved adults.

Similarly, Smith (1999), using data from the National Education Longitudinal Study (NELS), found that participation in extracurricular activities in youth as well as close family relationships and religious participation, were significant predictors of political and civic participation in young adulthood. The sample, 25,000 eighth grade students, was followed over a six year period.

Also using longitudinal data from the National Center for Education Statistics, Damico, Damico and Conway (1998) examined women's participation in high school activities and their subsequent participation in community and civic life. Based on their analysis of the survey data, Damico and colleagues found that participation in high school extracurricular activities (ranging from membership in honorary societies to student government and athletics) is important for women's level of civic and political engagement later in life.

Verba, Schlozman and Brady (1995) also found a link between participation in civic, social, and volunteer activities during youth and increased volunteer activities during adulthood. They argue that civic and volunteer activities influence political engagement and participation, which they view as essential to democracy. Their data came from a large, two-stage survey of American adults.

Data from the first phase of the IEA Civic Education Study indicate that in some countries opportunities for participatory civic education occur most frequently in special student projects and extra-curricular activities centered around issues (rather than within the formal school curriculum). For example, in Germany, secondary school students have the opportunity to participate in projects ranging from expanding the activities of student government to broader social action. Competitions are held which showcase these efforts (Schwille & Amadeo, forthcoming).

Project Citizen is a notable example of an issue-centered approach developed for young adolescents in the United States and implemented in countries associated with the Center for Civic Education (Vontz & Nixon, 1999). Students choose a problem in their community, develop explanations of why it is important, alternative policies to deal with it, a public policy that the majority of the group agrees to, and an action plan for approaching government. Display materials are developed to describe each step of the process. The culminating activity is a hearing where students role play expert witnesses testifying before community members. Competitions are also part of the process.

In a study of the effects of Project Citizen, students and teachers reported that students developed confidence in their ability to make a difference in their communities, acquired knowledge of public policies and the policy making process, and developed research, communication, and group skills (Tolo, 1998).

The Constitutional Rights Foundation's (1999) program Active Citizenship Today takes students through five units in which they assess the community, select a community problem and conduct research, evaluate policies, examine options, and take action to address the problem. Another program of CRF, CityWorks, combines service learning with lessons in government.

SUMMARY OF THE SECTION ON ACTIVITIES RELATING THE STUDENT TO THE COMMUNITY

The majority of community service and other co-curricular and extra-curricular activities relating to the community provide students with opportunities to *practice and apply their learning* as outlined in the principles. Yet, the pedagogical value depends greatly on the type of activity, the students' preparation and debriefing, and the extent and type of improvement-oriented feedback. While these activities do appear to contribute to citizenship education and may enhance students' self-concept and identity as a member of the community, they probably should not be relied upon as the sole or even primary means of conveying the information and skills necessary for citizenship.

CONCLUSIONS AND SUGGESTIONS FOR ANALYSIS AND NEW RESEARCH

The most general conclusion from this review is that principles of good classroom practice identified across subjects apply in the area of citizenship education. They need to be augmented, however, with emphasis on students' motivation to become engaged, on the application of knowledge, on inquiry and discourse which models democratic process, on opportunities to explore identity, and on perspectives gained in communities of practice outside and inside the school. The materials, teacher training, and public support necessary to realize these aims are largely lacking, however.

The three positions we have identified – those who stress knowledge, those who stress discourse about political issues and participation, and those who stress service in the community – have research to bolster their prescriptions. Collaborative effort would be extremely useful, however. For example, the recommendations coming from research about content knowledge credit opportunity to learn but also note that student engagement in practice is vital, especially when it involves inquiry and discourse about authentic issues. Curriculum integration involves in-class, in-school, and out-of-school activities, such as opportunities for community involvement. Many of these efforts would require teacher collaboration and community support.

A difficulty with this subject area is that there are many perspectives, most of which seem complementary but none of which has been explored in sufficient depth. There is the work of social studies educators some of whom have a focus on civic education, while others work from frameworks such as multicultural education. Their research has been illuminating but has done more to frame and define the issue and to provide materials than to evaluate specific instructional principles. There are those who take an ethnographic point of view that documents the realities of children and classrooms, but such studies exist in only a limited number of schools.

Some political scientists specialize in studying socialization. They do not explore school or classroom processes very thoroughly, but they do call attention to the many sources outside the classroom and to links with adult political culture. Other political scientists are interested in laying out content frameworks describing what young people should know. They provide a content floor upon which much else can be built and ask for a great deal of detailed knowledge, but they sometimes pay little attention to fostering meaningful learning that builds upon students' concepts or to encouraging participation. The work of cognitive, developmental, and educational psychologists is central to understanding meaningful concept learning and the ways community service experience can contribute to political identity, but there has not been sufficiently broad or deep research in diverse groups of students. Many of those who believe that there is a central body of knowledge that is essential for citizens to master are frustrated when developmental psychologists point to the misunderstandings and fragile concepts which some young people hold. It is always the case that more research is called for; here it seems desperately needed on an interdisciplinary basis that brings together some of these disparate strands.

With respect to instruction that focuses on content knowledge, there are several statements by researchers about the need for studies that deal with a broad range of dimensions of young people's political achievement and engagement (their concepts, grasp of factual knowledge, skills in interpreting political communication, attitudes toward civic society, and political activity itself). The IEA Civic Education Study's data base whose first results will be available in 2001 and whose full data base from 90,000 14-year-olds (nationally representative samples from 28 countries) will be made available for secondary analysis in mid-2002, has a wealth of potential data for these analyses. The IEA study's first phase has already shown that many programs of civic education across the world are based on the assumption that content knowledge is essential. This usually means understanding the principles of democracy and also facts about current governmental structures and processes in one's own country. The IEA knowledge/skills test of 38 items is based on an internationally derived content framework dealing with democratic principles and skills and has alpha reliabilities of .80 and higher in all the participating countries. It will allow examination of the association between knowledge and other facets of engagement for students within countries (Torney-Purta et al., 2001). Although researchers will not be able to examine how knowledge scores relate to actual voting behavior, questions have been asked about the intent to vote and engage in other political activities, as well as the extent to which students have confidence in their ability to participate with groups at school. There is also some evidence in this study that both open classroom climate and participation in organizations such as school parliaments have a positive influence (as would be suggested by research).

Understanding how to enhance civic content knowledge also requires research that combines methods – well designed multiple choice tests that include both factual information and skills in interpreting political communications, interviews with students regarding political concepts, in-depth exploration of teachers' concepts and approaches, observations of exemplary teachers, and evaluations of intervention programs in a variety of settings.

With respect to instruction in which students investigate, discuss, and make decisions about public issues, future research is needed along four lines. First, although there are advocates for issues-centered instruction in the elementary grades, there is comparatively little research at that level. More studies are needed of the process and outcomes of such instruction with young children. Second, although discussion and public deliberation are at the heart of democratic citizenship and issues instruction, more research is needed on how to most effectively teach discussion skills to young people, so that they can effectively deal with public issues. Third, although pluralistic democracies require that citizens be able to make decisions with people whose cultures and viewpoints differ from their own, there is only a little research on the multicultural dimensions of issues-centered instruction. Fourth, "structured controversy," a form of cooperative learning has been recommended as one way to foster reflection and develop skills for participation in a pluralistic democracy. Research is needed on the process and outcomes of that form of instruction in social studies classes.

In the area of youth service programs, research focused on the variations in types and goals of programs, the duration of time spent in the service activity, age-related student outcomes, and on the long-term effects of service participation would make a significant contribution. Application of ideas gained in classrooms and schools is clearly a valuable principle, but how this should take place is not very well understood.

Finally, it is important to note that in this subject area it is difficult to implement many of the principles derived from research. Citizenship education has often not been a subject of high status to policy makers, administrators, or teachers. Although many subjects are civic-related (social studies in general, history, social problems) and cross-curricular dimensions are widely recognized, this sometimes seems to make it more difficult to find a manageable focus. The emphasis on covering material in a textbook which has characterized the area for many years and the recent emphasis on standardized testing has in some cases resulted in approaches that are rather far from the kinds of instructional principles dealt with in this chapter.

There are actions suggested by these instructional principles. First is the importance of balance between opportunities to teach content, encouragement of inquiry and discourse about issues, and chances for reflection about practice. Second, is the value of thinking beyond the traditional classroom using a multifaceted approach including whole-school or community based activities. Third, is the vital importance of motivation and engagement, which may be enhanced by authentic activity, high expectations, or computer-assisted instruction. Finally, there are projects which have been successful (Project Citizen, ICONS, Kids Voting) which can be joined with issues-centered engagement or from which elements can be extracted.

Education for citizenship is a challenging area – one that can benefit from further exploration of instructional principles as well as from conscious attention to the communities of practice in which young people are and have the potential to be engaged.

ACKNOWLEDGMENTS

Support at the University of Maryland was provided by the Graduate School through a Research Board Grant to the senior author and from the William T. Grant Foundation.

REFERENCES

Alexander, P. (1996). The past, present, and future of knowledge research. *Educational Psychologist*, *31*(2), 89–92.

Alexander, P., & Murphy, P. K. (1998). The research base for APA's learner-centered educational principles. In: N. M. Lambert & B. McCombs (Eds), *How Students Learn: Reforming Schools Through Learner-Centered Education*. Washington, D.C.: American Psychological Association.

Allen, G., Kirasic, K., & Spilich, G. (1997). Children's political knowledge and memory for political news stories. *Child Study Journal*, *27*, 163–176.

Anderson, L., Jenkins, L. B., Leming, J., MacDonald, W. B., Mullis, I., Turner, M. J., & Wooster, J. (1990). *The civics report card*. Princeton: National Assessment of Educational Progress, Educational Testing Service.

Angell, A., & Avery, P. G. (1992). Examining global issues in the elementary classroom. *The Social Studies*, 113–117.

Angell, A. (1999). Practicing democracy at school. *Theory and Research in Social Education*, *26*, 149–172.

Avery, P. G., Bird, K., Johnstone, S., & Sullivan, J. L. (1992). Exploring political tolerance with adolescents. *Theory and Research in Social Education*, *20*, 386–420.

Banks, J. A., & Banks, C. (1999). *Teaching strategies for the social studies: Decision making and citizen action*. New York: Longman.

Bickmore, K. (1991). *Practicing conflict: Citizenship education in high school social studies*. Unpublished doctoral dissertation, Stanford University.

Bickmore, K. (1993). Learning inclusion/Inclusion in learning: Citizenship education for a pluralistic society. *Theory and Research in Social Education*, *4*, 341–384.

Bickmore, K. (1999). Elementary curriculum about conflict resolution: Can children handle global politics? *Theory and Research in Social Education*, *27*, 45–69.

Billig, S. H. (2000). Research on k-12 school-based service-learning. The evidence builds. *Phi Delta Kappan*, (May).

Blankenship, G. (1991). Classroom climate, global knowledge, global attitudes, political attitudes. *Theory and Research in Social Education*, *18*, 363–384.

Center for Civic Education (1994). *National standards for civics and government*. Calabassas, CA: Center for Civic Education.

Chall, J., & Henry, D. (1991). Reading and civic literacy: Are we literate enough to meet our civic responsibilities? In: S. Stotsky (Ed.), *Connecting Civic Education and Language Education*. New York: Teachers College Press.

Cobb, P., & Bower, J. (1999). Cognitive and situated learning perspectives in theory and practice. *Educational Researcher*, *28*(2), 4–15.

Connell, R. (1972). *The child's construction of politics*. Melbourne: Melbourne University Press.

Conover, P. J., & Searing, D. D. (2000). A political socialization perspective. In: L. M. McDonnell, P. M. Timpane & R. Benjamin (Eds), *Rediscovering the Democratic Purposes of Education*. Lawrence, KS: University Press of Kansas.

Constitutional Rights Foundation (1999), *ACT Handbook for Teachers*. Los Angeles, CA: Author.

Curtis, C. K., & Shaver, J. P. (1980). Slow learners and the study of contemporary problems. *Social Education*, *44*, 302–309.

Damico, A. J., Damico, S. B., & Conway, M. M. (1998). The democratic education of women: High school and beyond. *Women & Politics, 19*(2), 1–31.

Davies, I., Gregory, I., & Riley, S. (1999). *Good citizenship and educational provision.* London: Falmer Press.

Delli Carpini, M., & Keeter, S. (1996). *What Americans know about politics and why it matters.* New Haven: Yale University Press.

Dewey, J. (1916/1966). *Democracy and education.* New York: The Free Press.

Ehman, L. H. (1969). An analysis of the relationships of selected educational variables with the political socialization of high school students. *American Educational Research Journal, 6,* 559–580.

Ehman, L. H. (1977). *Social studies instructional factors causing change in high school students' sociopolitical attitudes over a two-year period.* Paper presented at the annual meeting of the American Educational Research Association, New York.

Ehman, L. H. (1980). The American school in the political socialization process. *Review of Educational Research, 50,* 99–119.

Engle, S. H., & Ochoa, A. S. (1988). *Educating citizens for democracy: Decision making in social studies.* New York: Teachers College Press.

Evans, R. W., & Saxe, D. W. (Eds) (1996). *Handbook on teaching social issues.* Washington, D. C.: National Council for the Social Studies.

Ferguson, P. (1991). Impacts on social and political participation. In: J. P. Shaver (Ed.), *Handbook of Research on Social Studies Teaching and Learning* (pp. 385–399). NY: Macmillan.

Flanagan, C. A., & Sherrod, L. R. (1998). Youth political development: An introduction. *Journal of Social Issues, 54*(3), 447–457.

Goldenson, A. R. (1978). An alternative view about the role of the secondary school in political socialization: A field experimental study of the development of civil liberties attitudes. *Theory and Research in Social Education, 6,* 44–78.

Hahn, C. L. (1996). Research on issues-centered social studies. In: R. W. Evans & D. W. Saxe (Eds), *Handbook on Teaching Social Issues* (pp. 25–41). Washington, D.C.: National Council for the Social Studies.

Hahn, C. L. (1998). *Becoming Political: Comparative Perspectives on Citizenship Education.* Albany, NY: State University of New York Press.

Helwig, C. (1995). Adolescents' and young adults' conceptions of civil liberties: Freedom of speech and religion. *Child Development, 66,* 152–166.

Helwig, C. (1998). Children's conceptions of fair government and freedom of speech. *Child Development, 69*(2), 518–531.

Hess, D. (1998). *Discussing controversial issues in secondary social studies classrooms: Learning from skilled teachers.* Unpublished doctoral dissertation. University of Washington.

Hunt, M. P., & Metcalf, L. E. (1955/1968). *Teaching high school social studies.* New York: Harper & Bros.

Johnson, M. K., Beebe, T., Mortimer, J. T., & Snyder, M. (1998). Volunteerism in adolescence: A process perspective. *Journal of Research on Adolescence, 8,* 309–332.

Johnston, J. E., Anderman, L., Milne, L., Klenk, L., & Harris, D. (1994). *Improving civic discourse in the classroom: Taking the measure of channel one.* Research Report 4. Ann Arbor, MI: Institute for Social Research, University of Michigan.

Killen, M., & Horn, S. S. (1999). Facilitating children's development about morality, community, and autonomy: A case for service-learning experiences. In: W. van Haaften, T. Wren & A. Tellings (Eds), *Moral Sensibilities and Education, Vol.II: The Schoolchild.* Bemmel London & Paris: Concorde.

Kleiner, B., & Chapman, C. (1999). *Youth service-learning and community service among 6th- through 12th-grade students in the United States: 1996 and 1999*. Washington, D.C.: United States Department of Education, National Center for Education Statistics.

Langton, K., & Jennings, M. K. (1968). Political socialization and the high school curriculum in the United States. *American Political Science Review, 62*, 862–867.

Levin, M., Newmann, F. M., & Oliver, D. W. (1969). *A Law and social science curriculum based on the analysis of public issues*. Final report . Washington D.C.: United States Department of Health, Education, and Welfare, Office of Education.

Lutkus, A. D., Weiss, A. R., Campbell, J. R., Mazzeo, J., & Lazer, S. (1999). *The NAEP Civics Report Card for the Nation*. Washington, D.C.: National Center for Education Statistics.

McDevitt, M., & Chaffee, S. (2000). Closing gaps in political communication and knowledge: Effects of a school intervention. *Communication Research, 27*, 259–292.

McDonnell, L. M., Timpane, P. M., & Benjamin, R. (Eds) (2000). *Rediscovering the democratic purposes of education*. Lawrence, KS: University Press of Kansas.

Melchior, A. (1997). *National evaluation of Learn and Serve America school and community based programs: Interim report*. Waltham, MA: Brandeis University, Center for Human Resources and Abt Associates.

Merryfield, M. (1994). Shaping the curriculum in global education: The influence of student characteristics on teacher decision making. *Journal of Curriculum and Supervision, 9*(3), 233–249.

Merryfield, M. (1998). Pedagogy for global perspectives in education: Studies of teachers' thinking and practice. *Theory and Research in Social Education, 26*, 342–379.

National Center for Education Statistics (1995). Community service performed by high school seniors (NCES 95–743). Washington, DC: United States Department of Education, National Center for Education Statistics.

National Council for the Social Studies (1979). Revision of the NCSS Curriculum Guidelines. *Social Education, 43*, 261–273.

National Education Longitudinal Study of 1988. Available www.nces.ed.gov

National Household Education Study. Available www.nces.ed.gov

National Issues Forum pamphlets.(Various). Dayton, OH: National Issues Forums.

Newmann, F. M. (1975). *Education for citizen action: Challenges for secondary curriculum*. Berkeley, CA: McCutchan.

Newmann, F. (1991). Classroom thoughtfulness and students' higher order thinking. Common indicators and diverse social studies courses. *Theory and Research in Social Education, 4*, 410–433.

Niemi, R. G. (2000). *Trends in political science as they relate to pre-college curriculum and teaching*. Paper presented at the annual meeting of the Social Science Education Consortium, Woods Hole, MA.

Niemi, R., & Junn, J. (1998). *Civic education: What makes students learn?* New Haven: Yale University Press.

Oliver, D. W., & Shaver, J. P. (1966/1974). *Teaching public issues in the high school*. Boston: Houghton Mifflin, reprinted, Logan, UT: Utah State University Press.

Opposing Viewpoints series.(Various). San Diego, CA: Greenhaven Press.

Osborne, K., & Seymour, J. (1988), Political education in upper elementary school. *International Journal of Social Education, 3*, 63–77.

Patrick, J. J., & Hoge, J. D. (1991). Teaching government, civics, and law. In; J. P. Shaver (Ed.), *Handbook on Social Studies Teaching and Learning* (pp. 427–436). NY: Macmillan.

Parker, W., & Jarolimek, J. (1997). *Social studies in elementary education.* Upper Saddle River, NJ: Merrill/Prentice Hall.

Parker, W., & Zumeta, W. (1999). Toward an aristocracy of everyone: Policy study in the high school curriculum. *Theory and Research in Social Education, 27,* 9–44.

Ravitch, D., & Finn, C. E. (1987). *What do our 17-year-olds know?* New York: Harper and Row.

Rossi, J. A. (1995). In-depth study in an issues-oriented social studies classroom. *Theory and Research in Social Education, 23,* 88–120.

Rossi, J. A. (1998). Issues-centered instruction with low achieving high school students: The dilemmas of two teachers. *Theory and Research in Social Education, 26,* 380–409.

Schwille, J., & Amadeo, J. (Forthcoming). The paradoxical situation of civic education in school: Ubiquitous and yet elusive. In: G. Steiner-Khamsi, J. Torney-Purta & J. Schwille (Eds), *New Paradigms and Recurring Paradoxes in Education for Citizenship.* Amsterdam: Elsevier Press.

Shumer, R. (1994). Community-based learning: Humanizing education. *Journal of Adolescence, 17,* 357–367.

Sinatra, G., Beck, I., & McKeown, M. (1992). A longitudinal characterization of young students' knowledge of their country's government. *American Educational Research Journal, 29,* 633–661.

Skeel, D. (1996). An issues-centered elementary curriculum. In: R. W. Evans & D. W. Saxe (Eds), *Handbook on Teaching Social Issues.* Washington, D.C.: National Council for the Social Studies.

Smith, E. S. (1999). The effects of investments in the social capital of youth on political and civic behavior in young adulthood: A longitudinal analysis. *Political Psychology, 20,* 553–580.

Social Science Education Consortium (1988–1993). *Public issues series.* Boulder, CO: Author.

Stukas, A. A., Jr, Clary, E. G., & Snyder, M. (1999). Service learning: Who benefits and why. *Social Policy Report, Society for Research in Child Development, XIII*(4).

Switzer, G. E., Simmons, R. G., Dew, M. A., & Regalski, J. E. (1995). The effect of a school-based helper program on adolescent self-image, attitudes, and behavior. *Journal of Early Adolescence, 15,* 429–455.

Thiesen, R. (2000). President's message: Looking ahead with optimism. *The Social Studies Professional (newsletter of the National Council for the Social Studies), 157,* 3.

Tolo, K. W. (1998). *An assessment of We the People . . . Project Citizen: Promoting citizenship in classrooms and communities.* Austin, TX: Lyndon B. Johnson School of Public Affairs, University of Texas.

Torney, J. V., Oppenheim, A. N., & Farnen, R. F. (1975). *Civic education in ten countries: An empirical study.* New York: John Wiley & Sons.

Torney-Purta, J. (1992). Cognitive representations of the international political and economic systems in adolescents. In: H. Haste & J. Torney-Purta (Eds), *The Development of Political Understanding* (pp. 11–25). San Francisco: Jossey Bass (New Directions in Child Development).

Torney-Purta, J. (1996). Conceptual change among adolescents using computer networks in group-mediated international role playing. In: S. Vosniadou, E. DeCorte, R. Glaser & H. Mandl (Eds), *International Perspectives on the Design of Technology Supported Learning Environments.* Hillsdale, NJ: Erlbaum.

Torney-Purta, J., Schwille, J., & Amadeo, J. (1999). *Civic education across countries: Twenty-four national case studies from the IEA Civic Education Project.* Amsterdam: IEA and Washington, D.C.: National Council for the Social Studies. ED 431 705.

Torney-Purta, J., & Schwille, J. (1986). Civic values learned in school: Policy and practice in industrialized countries. *Comparative Education Review, 30*, 30–49.

Torney-Purta, J., Lehmann, R., Oswald, H., & Schulz, W. (2001). *Citizenship and education in twenty-eight countries: Civic knowledge and engagement at age fourteen*. Amsterdam: IEA.

Van Sledright, B., & Grant, S. G. (1994). Citizenship education and the persistent nature of teaching dilemmas. *Theory and Research in Social Education, 22*, 321–339.

Verba, S., Schlozman, K. L., & Brady, H. E. (1995). *Voice and equality. Civic voluntarism in American politics*. Cambridge, MA and London: Harvard University Press.

Vontz, T., & Nixon, W. (1999). Reconsidering issue-centered education among early adolescents: Project Citizen in the United States and abroad. In: C. Bahmueller & J. Patrick (Eds), *Principles and Practices of Education for Democratic Citizenship: International Perspectives and Projects*. Bloomington, IN: ERIC Clearinghouse for International Civic Education.

Wade, R. C. (1994). Conceptual change in elementary social studies: A case study of fourth graders' understanding of human rights. *Theory and Research in Social Education, 22*, 74–95.

Wade, R. C. (1997). Community-service learning: An overview. In: R. C. Wade (Ed.), *Community-Service Learning. A Guide to Including Service in Public School Curriculum*. Albany: State University of New York Press.

Wenger, E. (1998). *Communities of practice: Learning, meaning, and identity*. Cambridge: Cambridge University Press.

Yates, M. (1999). Community service and political-moral discussions among adolescents: A study of a mandatory school-based program in the United States. In: M. Yates & J. Youniss (Eds), *Roots of Civic Identity: International Perspectives on Community Service and Activism in Youth*. Chicago: University of Chicago Press.

Youniss, J., McLellan, J. A., & Yates, M. (1997). What we know about engendering civic identity. *American Behavioral Scientist, 40*, 620–631.

Youniss, J., & Yates, M. (1999). Introduction: International perspectives on the roots of civic identity. In: M. Yates & J. Youniss (Eds), *Roots of Civic Identity: International Perspectives on Community Service and Activism in Youth*. Chicago: University of Chicago Press.

Youniss, J., Yates, M., & Su, (1997). Social integration into peer and adult society: Community service and marijuana use in high school seniors. *Journal of Adolescent Research, 12*, 245–263.

ENHANCING ECONOMIC EDUCATION THROUGH IMPROVED TEACHING METHODS: COMMON SENSE MADE EASY

James D. Laney

INTRODUCTION

Marilyn Kourilsky (1996a), an internationally-known and well-respected economic educator, has described the study of economics as "common sense made difficult" (p. 9). Over the past several decades, professional economic educators have assessed the economic literacy of American school children and the American public in general, and they have lamented the apparent economic illiteracy revealed through these tests. For example, studies by Soper and Walstad (1988) and Walstad and Soper (1988a, b, 1989) indicate that students about to graduate from high school have a relatively limited knowledge of economics. The implication is that inadequate instruction in K-12 economics is to blame.

As evinced by Shaver, Davis and Helburn (1979), social studies teachers, in general, are very textbook-driven in their instruction, and they rely heavily on traditional teaching methods such as lecture and guided discussion. Other methods, including student reports, library work, role-playing, and simulation, are also reported as being used, but are much less common, with secondary teachers tending to use a narrower range of practices than do elementary teachers. Schug and Walstad (1991) assert that economics instruction is

Subject-Specific Instructional Methods and Activities, Volume 8, pages 411–435.

ISBN: 0-7623-0615-7

probably similar to general social studies instruction in the ways described above. Thus, the key to enhancing economic education may depend on the enhancement of methods for teaching economics, especially through the use of "best practices" that make economic learning more meaningful, memorable, student-centered, and applicable to students' own lives and experiences. The school experience needs to be changed so that students are better prepared for their real-world citizenship roles as consumers, job-holders, savers, investors, and political decision makers. The purpose of this chapter is to acquaint the reader with a sampling of successful teaching methods that have the potential for reforming economic education by making the study of economics "common sense made easy" (Kourilsky, 1996a, p. 9). The instructional practices to be discussed include deductive and inductive approaches to concept-based instruction, strategies for correcting economic misconceptions, inquiry-oriented approaches, experience-based instructional systems, role-playing and simulations, vicarious story experiences, strategies for analyzing economic issues and policies, and economic mysteries or case studies.

CONCEPT-BASED INSTRUCTION: DEDUCTIVE AND INDUCTIVE APPROACHES

The K-12 curriculum in economics is organized largely around microeconomic and macroeconomic concepts. Concepts, along with facts and generalizations, constitute the building blocks of the discipline of economics. A concept-based instructional approach is believed to be effective in that it provides students with the knowledge base they need to become better economic decision makers in real life. The assumption is that knowledge automatically leads to wisdom. Typically, deductive and/or inductive instructional sequences are used to teach economic concepts.

The deductive approach to concept instruction involves starting with the general concept definition and then showing students specific examples of the concept. For example, in an elementary school lesson on productive resources described in Weidenaar (1982), the teacher defines the sub-concepts of natural resources, human resources, capital goods resources, and product. Then the teacher gives concrete examples of the three categories of productive resources used in the production of some product (e.g. in the growing, harvesting, and selling of apples: natural = tree, land, water; capital goods = wagon, crates, fruit stand; human = people/labor). Later, students are given opportunities to independently identify productive resources associated with other familiar products. According to Weidenaar (1982), the deductive approach tends to be more appropriate when: (1) students are more at home with concrete rather than abstract ideas, (2) students respond better to teacher-centered rather than student-centered activities, (3) teachers have limited class time to introduce the concept, and (4) teachers want to convey a very precise, traditional definition for a concept.

Inductive concept lessons go from specific examples to general definitions. In a collection of exemplary lesson plans in economics, Weidenaar (1982) includes a middle school activity on voluntary exchange that follows an inductive sequence. In this lesson, students are presented with a set of descriptive situations. Some of the situations involve voluntary exchanges (e.g. a boy cleaning up a sporting goods shop in return for lessons from a tennis professional), while the others feature involuntary exchanges (e.g. a man giving up his wallet at gunpoint in order to avoid being harmed). Students are asked to group those situations that seem to belong together and to come up with their own labels and definitions for each group based on identified common properties. Later, the teacher provides the students with the traditional concept labels and definitions. Weidenaar (1982) states that the inductive approach is most appropriately used when: (1) students have had experience with brainstorming and categorizing, (2) students respond well to small-group or whole-class activities that are student-centered, (3) teachers are as much concerned with the teaching of information-processing skills as they are with the teaching of concepts, and (4) teachers are willing to accept tentatively students' less-than-precise concept definitions. The inductive approach may also be more likely to lead to long-term retention of learning, for students may tend to process the information more deeply when they are required to discover the concept definition for themselves.

Based on the reasons mentioned above, Weidenaar (1982) concludes that a teacher's choice of instructional approach is likely to depend on his/her teaching style, the amount of guidance the students need, the thinking skill level of the students, and the difficulty of the concepts. Indeed, research (e.g. Ryan & Carlson, 1973) indicates the necessity of considering individual learner characteristics when selecting between deductive and inductive sequences.

Some educators contend that a combined approach is best. For example, the "Econ and Me" video program series teaches central economic concepts (i.e. scarcity, opportunity cost, consumption, production, and interdependence) using an inductive/deductive framework and concept development process. The first two programs employ an inductive sequence, with subsequent programs moving to a deductive sequence (Jackson, 1989). Videotape and televised instructional programs have been shown to be an effective way to teach economics at the elementary school level (Chizmar & Halinski, 1983; Morgan,

1991; Walstad, 1980), but concerns about transfer of learning have been expressed with respect to "Econ and Me" (Sweeney & Baines, 1993).

STRATEGIES FOR CORRECTING ECONOMIC MISCONCEPTIONS

Kourilsky (1987, 1991), Laney and Schug (1998), Schug (1994), and Schug and Baumann (1991) describe a number of economic misconceptions held by children in grades K-12. Children simply do not think about the economic world in the same way an economist would. According to Kourilsky (1987, 1991), there are three learning hurdles in economics that account for these misconceptions, with the hurdles categorized as conceptual/analytical, semantic/linguistic, and attitudinal.

Kourilsky (1991) notes that conceptual/analytical hurdles stem from the fact that: (1) a student may be developmentally unready to understand a concept at a given level of complexity or (2) the student may have not yet mastered the prerequisite knowledge and/or skills necessary for comprehending the concept. The first type of conceptual/analytical hurdle is common in young children. For example, converting data from a demand schedule into a demand curve is out of the question for a six-year-old still in Piaget's pre-operational stage of cognitive development. Thus, teachers of economics need to be able to match the complexity of the economic content to be taught to the developmental level of their students. The second type of conceptual/analytical hurdle, on the other hand, is more prevalent among high school students. Because economics is a cumulative discipline, the mastery of advanced concepts, such as those taught in high school, requires the mastery of earlier, prerequisite concepts. For instance, if one does not thoroughly understand the concepts of scarcity and opportunity cost, one may have difficulty catching on to the laws of demand and supply.

As described by Kourilsky (1991), semantic/linguistic hurdles refer to seemingly familiar words that have a totally different meaning in economics than they do in common, ordinary usage. These words include scarcity, demand, profit, and investment among many others. The word demand, for example, is commonly used to mean "to ask urgently," "to adamantly desire," or "to insist upon," but an economist would define it as "wants backed by the ability to pay."

Kourilsky's (1991) last identified hurdle, attitudinal blocks, are students' preconceived ideas about the workings of economics that tend to hamper students' attempts at grasping economic concepts. As an example, students and the general public have a hard time accepting the notion of sunk cost, for they spend time worrying over what should or should not been done in a past decision-making situation rather than salvaging what they can. Once money, time, and/or effort are spent producing or buying something, the decision has been made, and it cannot be reversed. Let's say that one has invested a great deal of money, time, and effort into some project. In this kind of situation, many believe that one should go ahead and finish the project, no matter what it costs. The economist, in contrast, would compare the costs of finishing the project versus the potential earning-power of the finished project, and s/he would abandon the project if the cost of completing it were greater.

In using instructional strategies for correcting economic misconceptions, again the rationale is that accurate knowledge leads to better economic decision making in real life. Laney and Schug (1998) suggest some ways for elementary teachers to correct their students' misconceptions and get them thinking economically. First, teachers need to include time for teaching about economic concepts so that they can assist students in replacing their own intuitive knowledge with knowledge from discipline-based inquiry. Second, teachers need to be aware of the types of misconceptions students are likely to have about the economic concepts to be taught in the curriculum. As an illustration, with respect to the concept of opportunity cost, teachers should realize that many children do not see the cost of choosing in certain situations (e.g. whether there is a cost involved in staying up late to watch television). They need to lead their students to the realization that those who choose always give up their second best alternative (e.g. a good night's sleep), even in the absence of monetary cost. Third, the teacher should focus students' attention on the uselessness of the misconception in explaining much about the real world. This step is accomplished through reviewing the correct concept definition and providing additional practice through numerous new examples (e.g. identifying the opportunity cost in owning a puppy, not wearing your eye glasses to school, and reading a book). Fourth, teachers need to make economic learning more meaningful and memorable to students by using experience-based instructional approaches rather than limiting themselves to book-based, expository approaches. For example, the teacher can have the students experience and identify opportunity cost in a real-life classroom experience, such as an art lesson involving a scarcity of art materials and two possible art projects from which to choose. Fifth, opportunities for teacher-child and child-child conversational exchanges relating to economic ideas are desirable, for learning about language (and, in turn, about the ideas/concepts being communicated) is cemented as children struggle to understand and be understood. These conversational exchanges can be accomplished through: (1) cooperative-mastery learning activities (Laney, 1999; Laney, Frerichs, Frerichs & Pak,

1996) and (2) post-experience debriefing sessions, in which children role-play and discuss economic ideas and problems they have encountered in real-life situations (Laney, 1993).

Research by Schug and Baumann (1991) reveals successful strategies for correcting secondary students' misunderstanding of economics. Based on self-report interviews of outstanding high school economics teachers, the researchers reached three conclusions:

- Successful economics teachers are able to describe the understanding (or misunderstanding) of fundamental economic concepts that their students are likely to have.
- Successful economics teachers consider the following teaching practices to be effective: presenting key examples to illustrate economic concepts, relating economic concepts to the students' lives and/or the local community, employing appropriate visuals, and providing for student practice.
- Successful economics teachers have more confidence in teaching micro-economic concepts (e.g. opportunity cost, supply and demand) than in teaching macroeconomic concepts (e.g. creation of money, GNP); thus, there is a need for more motivating and factually correct means for teaching macroeconomic ideas.

As one specific example of successful practices at work, Schug and Baumann (1991) discuss outstanding high school economics teachers' teaching of supply, a concept that was found to be confusing to high school students. The students seemed to see markets only from the perspective of the consumer rather than the producer. Seventy-five percent of the teachers in the study reported using examples to introduce this concept, and the types of examples used were drawn from the local economy and from students' daily lives. One particular teacher had his students imagine how many hours they would be willing to work at a part-time job across a range of hourly wages, and this data was placed on a chart and later converted to a supply curve. Many teachers also used examples to explain how supply works in different situations (i.e. with farm products, commercial airlines, and high school basketball teams), and they reported drawing supply curves on the chalkboard for illustrative purposes and giving students practice in drawing supply curves on their own.

INQUIRY-ORIENTED STRATEGIES

As noted by Welton and Mallan (1996), inquiry is both a teaching strategy and a learning strategy. All inquiry lessons are organized around a central mystery, question, or problem. The emphasis is on the acquisition of process/thinking skills and the manipulation of data, with students actively engaged in the steps of the scientific method in order to solve the mystery, answer the question, or resolve the problem. The rationale behind inquiry-oriented strategies is that knowledge is subject to change while skills are not. Process/thinking skills provide students with the "tools" they need to function as life-long learners and problem solvers in real-world economic situations. To illustrate an inquiry experience in economics appropriate for the middle school or high school, imagine students: (1) proposing hypotheses about the relationship between age and soft drink consumption, (2) gathering survey data from the local population to test their hypotheses, (3) drawing conclusions about the age-consumption relationship, and (4) generalizing about the effect of changes in the age distribution on product demand (Weidenaar, 1982).

Descriptions of the steps of inquiry vary from source to source. The phases of the inquiry process, based on a compilation of ideas from Weidenaar (1982) and Welton and Mallan (1996), can be summarized generally as follows:

- Describe the mystery situation, unanswered question, or problem to be solved.
- Propose possible explanations, hypothetical answers, or tentative solutions.
- Identify the kinds of evidence needed to support or reject each hypothesis.
- Collect relevant data and evaluate its reliability and validity.
- Accept or reject each hypothesis, using logic and the highest-quality data available.
- State a tentative conclusion.
- Test the tentative conclusion.
- Be aware of what you still don't know.

Inquiry is one of the most highly touted teaching strategies in economics and social studies in general. Unfortunately, compared to more teacher-centered, direct instructional approaches, it appears to be under-utilized in practice. Some of the teaching systems and strategies described in the sections below, including the experience-based Mini-Society and Kinder-Economy programs and the case study method (i.e. economic mysteries), are inquiry-oriented in nature.

EXPERIENCE-BASED INSTRUCTIONAL SYSTEMS

Experience-based instruction, teaching through the use of personal experiences followed by teacher-led debriefings on those experiences, is a proven means of successfully teaching economic ideas. When possible, real-life rather than vicarious experiences are used. Proponents of experience-based instruction point to the authenticity of having students actually go through real-life economic dilemmas as a means of learning economic content and thinking skills. This instructional approach rests on three principles of teaching and learning as follows:

- Active participation promotes learning better than passive participation.
- Personal, real-life experiences promote learning better than vicarious experiences.
- One learns better by actually bearing the bottom-line consequences of one's decisions (Kourilsky, 1996b).

Initial learning and retention of economic ideas is promoted through the application of these principles, for new learnings become more meaningful and thus more memorable.

Marilyn Kourilsky's YESS (Youth Empowerment and Self Sufficiency)!/ Mini-Society™, for grades 3–6, and her Kinder-Economy, for grades K-2, are perhaps the most widely known and researched instructional systems which make use of experience-based methods. As described by Kourilsky (1996b), Mini-Society gives students the opportunity to create their own classroom society and to experience and resolve real-life social and economic problems. The program consists of two integrated components: (1) the experience itself (i.e. a market-day period in which students buy and sell student-produced goods and services using a classroom currency) and (2) a formal, immediate, teacher-led debriefing of economic ideas derived from the students' market-day experiences.

The program is initiated by introducing a classroom scarcity situation (e.g. a scarcity of seating space at the learning center or a scarcity of balls on the playground at recess). After experiencing the problem, the students are organized into an inquiry-oriented, interaction discussion group. Within this discussion or debriefing group, students identify the scarcity problem, brainstorm possible alternative solutions to the problem, discuss the anticipated costs and benefits of each alternative solution, and make and implement a decision on how to allocate the scarce resource. Eventually, with repeated exposure to real or contrived classroom scarcity situations, the students are led to try market mechanism as an allocation strategy for resolving their scarcity problems, thereby necessitating the creation of a classroom currency. This same discussion-debriefing group makes other decisions as well, including the development of student standards of conduct within the society and the selection of activities for which students will be paid in Mini-Society currency (so that the money gets into circulation).

With the first exchange of money for something of value, Mini-Society shifts into high gear. Students begin to buy and sell various goods and services, such as pencils, erasers, and their time. Businesses, such as wallet factories, are developed in response to consumer demand within the society. Some students choose to go into business by themselves or with a partner, while others choose to become salaried workers. In short, students begin to experience adult life in microcosm, responding to real-life personal and societal economic problems within their own small society.

Within any Mini-Society classroom, students tend to encounter twenty-five predictable dilemmas. These dilemmas include a wide range of problem situations including economic shortages, inflated currencies, disagreements among partners, whether or not to go into a business, whether and when to change employment, economic surpluses and unemployment, sunk cost, whether to hire a friend, unacceptable money, the question of a bank, working versus stealing, charity versus compensation, why some people are rich and others poor, and various other economic, entrepreneurial, social, ethical, legal, and political problems. After they are experienced first-hand by students, these dilemmas become the basis of the teacher-led debriefings, which follow each market-day period and take the form of an interaction discussion group.

The interaction discussion group operates using a circular seating arrangement, a listed agenda, a chairman/discussion leader, an accepted discussion procedure, an accepted voting procedure, and a goal orientation. Within each debriefing, the following steps are accomplished: (1) describe, (2) identify, (3) teach or review, and (4) relate. Laney (1997) provides a step-by-step description of the initial scarcity debriefing referred to above. After being confronted with a scarcity situation at school, students are first asked to describe the details of the economic event. The students describe the scarcity situation in their own words and then role-play the situation. In the second step, students identify the central issue, problem, or question associated with the event. The teacher helps students recognize that the existence of scarce resources is a relevant decision-making issue. In the third step, the instructor directly teaches or reviews the economic concept(s) needed for dealing with the central issue, problem, or question. Students supply their own invented definition and label for the concept of scarcity, such as not-enough or too-few-for-so-many, and the teacher provides students with the conventional concept label and definition (making it clear that scarcity refers to the idea of virtually unlimited wants versus limited productive resources). Finally, in the fourth step, the teacher guides the students as they relate the new information to the current problem and to their own knowledge base and past experience. The teacher helps the students resolve the scarcity problem by having them brainstorm possible allocation strategies, discuss and role-play positive and negative consequences of each strategy, select an allocation strategy, implement the selected allocation strategy, and live with the consequences of their decision.

The experience-based Mini-Society instructional system is highly motivating to eight- to twelve-year-old students because they see it as real life, not a simulation. As a result, the system encourages students to engage in independent, creative, and reflective inquiry, with the teacher serving in the role of non-interventionist guide. Research (e.g. Cassuto, 1980; Kourilsky, 1976; Kourilsky & Graff, 1986; Kourilsky & Hirshleifer, 1976) suggests the benefits of Mini-Society participation to be numerous – increased economic understanding, improved economic reasoning (i.e. use of cost-benefit analysis in decision making), increased feelings of autonomy, greater willingness to take moderate risks, enhanced self-concept, improved attitudes toward school and learning, less stereotypical images of entrepreneurs, and increased achievement in mathematics and language arts/reading. Likewise, research on similar, experience-based instructional approaches to economics instruction at the elementary level (e.g. Laney, 1989; Laney, 1993) illustrates the power of real-life experiences and post-experience debriefings in enhancing the initial learning and retention of economic concepts.

The effectiveness of Kinder-Economy, Kourilsky's (1992) experience-based instructional system for teaching economics in grades K-2, has also been well established through research. Studies (e.g. Kourilsky, 1977) have shown that children as young as five who participate in the program can comprehend fundamental economic concepts such as scarcity, opportunity cost, supply, demand, and other microeconomic ideas. Participants go through a carefully sequenced set of experience-based situations that focus on real-world, economic decision making. Unlike the students in Mini-Society, Kinder-Economy students do not create their own society that becomes the core of their reality; instead, they role-play within "pretend" experiences. Kourilsky provides a rationale for this difference by pointing out that Kinder-Economy children are too young and too egocentric to perceive themselves as part of a society.

As in Mini-Society, Kinder-Economy students first experience an economic concept and live with its consequences. The concept of opportunity cost, for example, is introduced only after students have experienced school decision-making situations involving a choice between two alternatives (e.g. which playground activity to participate in or which supplies to use in creating an art project). The students are asked to identify their alternative choices and are guided in the use of cost-benefit analysis to determine whether they made a "good" decision.

In Kinder-Economy, the teacher reinforces the economic concepts learned through the use of games, learning centers, worksheets, and audio-visual presentations. Because of student differences in terms of learning style and learning rate, reinforcement activities may be needed to ensure mastery by firmly entrenching the concepts in students' minds. One typical reinforcement activity is a concentration-like card game in which students match animals (e.g. two mice) to food items (e.g. one piece of cheese) to create scarcity situations (e.g. two mice and one piece of cheese, assuming each mouse wants one piece of cheese to eat).

Employment of experience-based methods is not limited to the elementary school, for the experience-plus-debriefing model has been used to enhance economics instruction at the middle-school and high-school levels as well. For instance, Kourilsky's Maxi-Economy calls for students to participate in a modified market economy and to develop their own businesses. This program has demonstrated its effectiveness in enhancing secondary students' understanding of economic concepts and use of economic reasoning (Kourilsky, 1985).

ROLE-PLAYING AND SIMULATIONS

Role-playing and simulations are two specific applications of experience-based instruction, for both involve student experiences followed by teacher-led debriefings. Much of the attractiveness of these two techniques can be attributed to their ability to interest teachers and motivate students. Students are given the opportunity to vicariously experience economic ideas and dilemmas as preparation for real-world situations they are likely to encounter in the future. In studies by Gretes and associates (1991) and Weiser and Schug (1992), they have demonstrated their effectiveness in promoting elementary and secondary students' understanding of economic concepts and American financial markets.

In role-playing, students assume the character, feelings, and ideas of another person as they experience a hypothetical problem situation drawn from real life. The purpose of the activity is to analyze the human problem. A few students serve as the role-players, while the rest of the class observes the action in order to identify what the characters are feeling, what the characters' underlying motives are, and other possible solutions to the problem.

As described by Kourilsky and Quaranta (1987), teachers should follow the steps outlined below in order to conduct a role-playing activity:

- Select a suitable dilemma based on a familiar problem situation that is important to the learners.
- Warm-up the class using exercises such as charades, pantomime, and/or make-believe group actions.
- Provide an overview of the problem situation to the class and specific instructions to the role-players.
- Provide instructions to all audience members regarding their roles as observers.
- Have the role-players do the enactment while the audience observes.
- Stop the role-play when the main points/behaviors have been observed.
- Lead a discussion of the role-playing situation, focusing on the audience's observations and reactions and ending with a generalization about human behavior and/or a new insight into how to respond to a specific problem situation.
- Evaluate the role-play using students' evaluative comments and one's own notes taken during the role-play.

As mentioned previously, role-playing is often utilized within Mini-Society and Kinder-Economy debriefing sessions. Students have the opportunity to role play economic problems encountered in their classroom or classroom society, and they get to experiment with alternative solutions to these problems. Because role-playing encourages adult-child and child-child conversational exchanges, children also benefit from context-based, oral language experiences in which they apply their knowledge of newly acquired economic concepts and concept labels.

Banaszak (1992) shows how role-playing is also an appropriate technique for use with middle school and high school students. With a simple activity, teachers can demonstrate the benefits of using financial intermediaries (e.g. banks and savings and loan associations). Students assume the roles of borrower (business executive), savers/lenders, and banker, while role cards provide background information on the amount of money needed by the borrower, the amount of money each saver/lender has to loan and/or deposit in the bank, the interest rate of the loan, and the interest rate on bank deposits. In the first round of the role-play, the borrower approaches many savers in order to get the loan. Prior to the second round, the banker accepts deposits from the savers and assures the savers that their money is safe (insured). The second round consists of the borrower approaching the banker for the loan. In the debriefing following the enactment, students compare and contrast the two ways of getting the loan, discovering the relative ease of approaching a single banker to get an answer.

As described by Welton and Mallan (1996), simulations are instructional activities that aim to recreate reality, but they do not attempt to generate perfect, real-world models. Instead, they deliberately simplify reality so that the real world can be better understood without overwhelming students with too many complex factors or considerations. Decision-making and role-playing activities are often employed within the format of a simulation. Typical topics used in designing simulations are decision-making situations involving zoning questions, community priorities, deciding what is newsworthy, and school-policy questions.

Perhaps the most classic illustration of a simulation lesson at the elementary school level involves the recreation of a factory assembly line. Yeargan and Hatcher (1997), for instance, describe a cupcake factory activity that serves as a vehicle for introducing students to the concepts of margin of profit, division of labor, and productivity.

According to Weiser and Schug (1992), many middle school and high school students have experienced financial market simulations such as The Stock Market Game™ (SMG), a simulation administered and distributed by the Securities Industry Association and the National Council on Economic Education (formerly the Joint Council on Economic Education, 1988). In this simulation, teams of three to five students invest hypothetical funds in common stocks on the various national exchanges. During a ten-week period, teams have the opportunity to buy, sell, buy on margin, and sell short. They receive weekly updates on the value of their respective portfolios through printed reports. At the end of the simulation, the teams with the highest equity value in their portfolios receive recognition through certificates, cash prizes, award banquets, or other means. A version of the SMG simulation is now available at an interactive site on the World Wide Web (http://www.smg2000.org/).

As with any simulation, the SMG does not completely reflect reality. Wood, O'Hare and Andrews (1992) and Kagan, Mayo and Stout (1995) note several differences between the SMG and the real world. One noteworthy example is that money not invested in securities does not earn interest in the SMG. Thus, financial market simulations can stimulate student interest in learning about this part of our economic system, but, at best, they can simulate only some of the important aspects of investing.

VICARIOUS STORY EXPERIENCES

Although not as powerful as real-life experiences, vicarious experiences can also be used to successfully introduce and/or reinforce economic concepts and to prepare students for their real-world roles as economic decision makers.

Vicarious experiences permit one to imagine, through one's mind's eye, the images, memories, and feelings of another person. Because they provide vicarious experiences, both economic fables and children's trade books have been used as vehicles to teach economic content.

Kourilsky (1996a) has written a number of economic fables that: (1) allow elementary school children to vicariously experience economic concepts and (2) directly teach the economic concepts by using each economic concept's traditional label and definition within the memorable context of a story. Her fables cover such concepts as scarcity, opportunity cost, supply and demand, price ceilings and price floors, and inflation. When used in conjunction with cooperative-mastery learning elements as in Laney, Frerichs, Frerichs and Pak (1996) and Laney (1999), economic fables can foster the development of economic concepts even without the use of real-life instructional experiences, but Kourilsky prefers to use them for the purpose of reinforcing and enhancing learning within post-experience debriefing sessions.

Kourilsky's (1996a) typical lesson procedure for an economic fable is in four parts as follows: (1) read the fable, (2) discuss it in class, (3) do related follow-up activities or projects, and (4) review and draw generalizations. "The Curse of Inflate-Shun Fable," for example, is the basis of Kourilsky's lesson on the economic concept of inflation. In this fable, the townspeople of Same-O-Same-O are excited to learn that they will each inherit an equal amount of money from the recently deceased Ms. Inflate, the rich town founder. They wrongly assume that they will now have enough money to buy anything they want, for more people with money chasing after goods all over town quickly results in prices going up. Soon after, all of the factory workers in town demand increased wages, and the factory owner complies. The workers are happy until they see prices going up on everything yet again. Judge Shun explains to the people that the factory owner had to raise the prices on the goods he makes in order to pay the increased wages of his workers, and he tries to convince the people not to ask for more, not to demand so much, not to bid up prices, and to save their money rather than spend it. The people conclude that the whole thing is a curse on the town and label the curse as "Inflate-Shun." After the fable is read, it is discussed by the class, with questions focusing on the cause of the price increases, the children's encounters with inflation in their own lives, and possible ways to stop the curse of Inflate-Shun. Discussion is followed by follow-up activities, such as: (1) surveying parents about the cost of items ten years ago versus today and (2) conducting a class auction on two occasions with the money supply available for bidding doubled at the second auction. Lastly, the class reviews what has been learned by writing the definition for scarcity on the chalkboard and by listing possible remedies for inflation (i.e. reducing spending, increasing taxes, making loans more expensive, and increasing production).

Many economic educators (e.g. Coulson & McCorkle, 1993, 1994, 1998; Coulson, Suiter & McCorkle, 1998; Day, Foltz, Heyse, Marksbary & Sturgeon, 1997; Hendricks, Nappi, Dawson & Mattila, 1986; Kehler, 1998; Savage & Savage, 1993) advocate the use of children's literature to teach economic concepts to elementary and middle school students. Their rationale for using children's books, many of which are Newbery Medal winners, Caldecott Medal winners, and/or teacher/student favorites, rests on the following reasons:

- Literacy in reading, writing, and economics can be achieved simultaneously, thereby increasing teaching-learning productivity.
- The pervasive nature of economics and economic decision-making in our lives is imparted through reading many popular children's books.
- The abundance of children's literature in the existing curriculum ensures that teachers have time to devote to economics.
- Children's books can help students develop their "economic imaginations – the ability to place themselves in the story and empathize with a character's economic experiences" (Kehler, 1998, p. 26).
- The wide range of personal and societal problems encountered in children's books motivates students to study about important economic issues and concerns.

The one drawback to the use of children's literature is that teachers, on their own, may have difficulty identifying books that explore economic concepts because the economic concepts imbedded within the stories are not directly named, defined, or described in the story itself.

A typical children's literature lesson involves the students in: (1) reading the story and listening for the answers to certain questions, (2) discussing and/or answering questions about the facts of the story, (3) receiving explanations of economic concepts relevant to events in the story, and (4) relating this new knowledge back to the story and/or to their own life experiences by answering application-oriented questions. For example, as described in Coulson and McCorkle (1993), Lillian Hoban's *Arthur's Funny Money* can be used to teach about the three basic questions answered within any economy (i.e. what to produce, how to produce, and for whom to produce). First, the teacher reads the story to the class, instructing students to listen for Arthur's problem and how he solved it. In the story, Arthur is penniless, and his friend Violet has a problem with numbers. The characters resolve both their problems by going into business together. Second, the class discusses and answers questions about the nature of Arthur's problem (no money), the money-making business alternatives considered (running errands, washing cars, washing bikes), the pros and cons of these alternatives (e.g. too many other people in the car washing business), the final decision reached by the story characters (bike washing), and how Arthur expanded his operation (through advertising and washing additional items for a higher price). Third, the teacher explains the three basic economic questions and the nature and purpose of advertising and advertising slogans. Fourth, the students field questions about how Arthur answered the three basic economic questions (what to produce – a washing service; how to produce – using pail, sponge, soap, etc.; and for whom to produce – anyone with a dirty bike and twenty-five cents) and about how Arthur advertised (or could have advertised) his service. This information is then used by students to create original slogans for Arthur's business.

STRATEGIES FOR ANALYZING ECONOMIC ISSUES AND POLICIES

On a daily basis, people make decisions about the use of scarce resources. One of the major goals of economic education is to produce a citizenry of critical thinkers and rational decision makers, who have the knowledge and skills required to make reasoned judgments about important personal and societal issues in economics and other areas. Values analysis, one form of values education, is particularly well-suited for examining economic issues and policies, for it makes use of economic reasoning (i.e. the cost-benefit-analysis way of thinking) within its procedures. The implication is that students will be better decision-makers and, in turn, make better decisions if they have internalized this decision-making model.

Although there is some variability, most values analysis models generally include some common elements or tasks. These elements or tasks, derived from a blending of ideas from Hartoonian and Laughlin (1997), Weidenaar (1982), and Welton and Mallan (1996), are summarized below:

- Identify the issue or state the problem.
- Clarify the value question by defining the value issues involved.
- Gather and organize evidence regarding actual or anticipated consequences of alternative solutions.
- Assess the accuracy and relevancy of the evidence collected.
- Identify potential, alternative solutions.
- For each solution, identify and appraise anticipated consequences based on specified criteria, and justify the criteria used.
- Make the decision.

• Implement the decision.

The Decision Tree (Laney, 1990b; LaRaus & Remy, 1978) and the Decision Grid (Weidenaar, 1982) are two classic devices used in association with values analysis. In the Decision Tree activity, the teacher introduces a hypothetical or real-life personal or social dilemma. Students write summary sentences (and/or draw simple stick figures) for each step of the cost-benefit analysis process. A sentence describing the problem is written at the base of the tree, and a sentence describing the decision-making goal is written above the topmost branches. On the lowest branches, students compose sentences summarizing their alternatives. The middle branches are reserved for sentences describing positive and negative consequences of each possible choice, and the topmost branches contain a sentence describing the final decision. In contrast to the Decision Tree activity which focuses on the consideration of positive and negative consequences (in light of one's goals/values) to weigh alternative choices, the Decision Grid activity emphasizes the generation of decision-making criteria as a means for evaluating alternatives. Alternatives are listed down the vertical axis of the grid, and criteria are listed across the horizontal axis. If an alternative meets a criterion, a plus sign (+) is placed in the appropriate grid cell. Minus signs (−) are used to indicate that an alternative does not meet a criterion, and question marks (?) are used to represent irrelevance or uncertainty (requiring re-examination and possible revision of the criterion to make it pertinent).

Research on the effect of the Decision Tree activity on elementary school students' economic reasoning (e.g. Laney, 1990a; Laney, Moseley & Crossland, 1991) confirms the usefulness of the device in promoting understanding and application of the cost-benefit-analysis way of thinking. Theoretically, based on brain lateralization research, the Decision Tree activity fosters learning by requiring students to process information through both words and images. The tree itself becomes a useful mnemonic device for remembering the steps of the cost-benefit analysis process.

The Advocacy Instructional Model (AIM), as described by Kourilsky (1972) and Kourilsky and Quaranta (1987), provides students with experience in advocating a particular point of view with respect to some topic, including topics dealing with economic issues and policies. The model, appropriate for use with both upper elementary and secondary school students, resembles a formal debate. The main difference is that the two sides trade arguments and evidence as they prepare their speeches. Students use their research skills, analytical skills, and speaking/listening skills to take a position on a controversial issue and to develop a case to defend their point of view.

The controversial issue is stated as a proposition of policy. This policy statement advocates a change from the status quo (i.e. present system) and contains the words "should." For instance, at the time of my writing of this chapter, a timely and debatable proposition of policy for a unit or course in international economics might be as follows: "China should be granted permanent, normal trade relations with the United States." In choosing a debate topic, the teacher should consider the interests and abilities of the students and the relevance of the topic to the required curriculum.

Students are formed into two teams of two students each. The affirmative team is in favor of the proposition of policy. Its role in the first half of the debate is to argue that there is a need for a change in the status quo by providing contentions backed by evidence in the form of logic, statistics, and/or authoritative opinion. In the second half of the debate, the affirmative team introduces its plan for change, and it presents arguments and supporting evidence that the plan is both practical and desirable. The negative team, in contrast, is against the proposition of policy. It argues that the present system is adequate and effective in the first half of the debate and that the affirmative's plan is not practical and desirable in the second half of the debate. Like the affirmative team, the negative team uses the best available evidence to support its contentions and counter the other team's contentions. To ensure that both teams are arguing the same issues, the two teams have the opportunity to see each other's arguments and evidence prior to the debate itself.

Kourilsky and Quaranta (1987) outline the basic steps in setting up the debate as described below:

- The teacher selects a debate topic.
- The teacher forms the class into debate teams.
- The teacher explains the function of each team to the class.
- The teacher provides guidelines and assistance to students as they prepare for their debate speeches.
- The debate is held, with the audience assigned specific observational functions to be performed during the debate (e.g. listening for exemplary and/or flawed arguments and evidence).
- After the debate, the teacher leads a class discussion/debriefing, featuring critiques of the speeches and a class vote to determine the resolution of the issue.

Research on advocacy learning indicates that the approach: (1) promotes skill development in logic, problem solving, critical thinking, oral communication, and written communication and (2) may promote affective gains in self confidence, self concept, and sense of autonomy (Kourilsky & Quaranta, 1987).

In a study by Kourilsky (1972), there was a statistically insignificant tendency for students participating in the advocacy model to score slightly higher than non-participants on a test of economic understanding. In addition, teachers reported improvements in students' ability to evaluate arguments and evidence.

ECONOMIC MYSTERIES: THE CASE STUDY APPROACH

Economics is often infused into other social studies courses, such as history and geography. At the secondary education level, debate rages over whether an infusion approach is inferior to a separate course in economics, with some studies (e.g. Walstad & Soper, 1988a) offering empirical evidence that high school students gain little in economic achievement from social studies courses even when teachers report the inclusion of economics within the course.

In light of the doubts and concerns expressed about the infusion approach to teaching economics, Schug (1996), Wentworth and Schug (1993, 1994), and Wentworth and Western (1990) advocate the use of economic mysteries in the elementary and secondary school curriculum as a more meaningful, direct way of teaching economics when it is integrated into social studies. Economic mysteries, which constitute a case study approach to teaching economics, can be used within units or courses in United States and world history to help elementary and secondary students analyze puzzling, unexpected historical events through economic reasoning. The promotion of economic reasoning calls for the use of inquiry teaching and deductive reasoning tasks. The emphasis is on applying economic reasoning to a full range of human behaviors, with special attention to the basic intellectual principle of economics: "Human behavior results from the choices people make based on expected costs and expected benefits" (Wentworth & Schug, 1994, p. 10). As with inquiry-oriented approaches, the rationale is that economic mysteries allow students to practice economic thinking skills that will be useful in the real world – now and in later life.

Wentworth and Schug (1993, 1994) point out that economic mysteries work well within the context of a history unit or course for a number of reasons. First, in the minds of economic educators, economic reasoning (i.e. the application of the cost-benefit-analysis way of thinking to problem solving) is arguably one of the most important goals for K-12 schooling, and history can be seen as a collection of puzzles which lends itself to this form of analysis. Second, the assumption that people weigh costs and benefits in making decisions is not a difficult one for students to grasp. Students can comprehend and apply this assumption without having to know a great deal of additional economic content. Third, using economic thinking to reason about past human behavior makes students aware that history is not a chain of events determined by fate and that the economy is not a machine controlled by external forces. Fourth, economic mysteries bring to life the lessons history has to offer, allowing students to vicariously identify with historical figures who faced real-world problems and who experienced both tragedy and triumph. Thus, for all the reasons cited above, this instructional approach seems to be mutually beneficial to both disciplines – economics and history.

Reinke, Schug and Wentworth (1989) have authored the *Handy Dandy Guide for Solving Economic Mysteries*, a handout from *Capstone: The Nation's High School Economics Course*. This guide consists of a set of six statements that delineate different aspects of economic reasoning related to the basic economic principle about expected costs and benefits, and it serves as a practical instructional tool for reasoning about human behavior in the past. Schug (1996) has modified the wording of the principles for use with younger children. The guide's six elaborative statements (with modifications in parenthesis) are as follows:

- People choose. (People *choose* to do things they think are best for them.)
- People's choices involve costs. (People's choices have *costs*.)
- People respond to incentives in predictable ways. (People choose to do things for which they are *rewarded*.)
- People create economic systems that influence individual choices and incentives. (People create *rules* that affect their choices and actions.)
- People gain when they trade voluntarily. (People *gain* when they decide to trade freely with others.)
- People's choices have consequences that lie in the future. (People's immediate choices have *future results*.) (Reinke, Schug & Wentworth, 1989; Schug, 1996, pp. 115–116)

In order to illustrate how the guide works, Schug (1996) analyzes a problem from United States history involving the near extinction of the American buffalo by the late 1800s. Throughout recorded history, people have usually preserved animals upon which they are dependent. Why, then, did Native Americans and European settlers over-hunt the American buffalo when they were dependent on these animals for food and clothing? After identifying this mysterious event, Schug goes on to show how the six principles of economic reasoning can be used to gain new insight into the problem of the decline of the buffalo as follows:

- *People choose to do things they think are best for them.* The buffalo hunters who hunted the buffalo until they almost disappeared actually believed that they were acting in their own best interest.
- *People's choices have costs.* The buffalo hunters could have hunted other game. They somehow concluded that the benefits of hunting the buffalo now outweighed the costs of having the buffalo available for some future use.
- *People choose to do things for which they are rewarded.* Unable to confine the buffalo and establish ownership, Native Americans and the settlers saw no rewards in preserving the buffalo.
- *People create rules that affect their choices and actions.* Because there were no rules regarding property rights over the buffalo, people had little reason to protect the buffalo.
- *People gain when they decide to trade freely with others.* Native Americans traded buffalo hide and bones among themselves and with settlers. Railroad crews were fed using buffalo meat, and settlers purchased the meat for their tables as well. It became popular to wear buffalo-hide robes and to shoot buffalo as a vocation.
- *People's immediate choices have future results.* Over the years, things have changed. Today, the buffalo population is increasing due to more clearly defined property rights and to a demand for "healthy" buffalo meat.

Schug (1996) also suggests a six-step sequence for introducing and teaching the six economic principles described above. First, the teacher presents an economic mystery and has the students guess possible answers using economic ideas. Second, through direct explanation and guided discussion, the teacher demonstrates how to use the six economic principles to solve the mystery. Third, the teacher introduces a new economic mystery and has small groups of students solve the mystery using the principles in the guide and other questions or clues provided by the teacher. Fourth, the students solve the mystery, and their written solutions are discussed in class. Fifth, at regular intervals, new mysteries are presented to the class for analysis, thus providing repeated practice in economic thinking. Sixth, the teacher encourages the students to look for new mysteries and develop their own mysteries for use in the classroom.

Middle school and high school teachers will find an abundance of curriculum materials featuring economic mysteries. These materials focus on economics in United States history (e.g. Schug, Caldwell, Wentworth, Kraig & Highsmith, 1993a, b; Wentworth, Kraig & Schug, 1996) and in world history (e.g. Caldwell, Clark & Herscher, 1996).

Other curriculum materials (e.g. Schug, Morton & Wentworth, 1997) have applied economic reasoning to environmental problems and environmental education. Schug (1997) articulates four principles of "ecodetection" (p.54), modifications of the six principles of economic reasoning described in detail above, which can be used to help solve persistent environmental mysteries (e.g. why people in poor nations tend to have more children than people in rich nations and why people wait in traffic jams rather than ride buses to help the environment). These principles are as follows:

- People's *choices* influence the environment.
- People's choices have *unintended consequences*.
- People's choices are influenced by *rewards*.
- People are more likely to take better care of things they *own and value* (p. 54).

CONCLUSION

The best teaching practices within any discipline, such as the ones in economics described in this chapter, are based on sound learning theory and research-supported principles of learning and instruction. Teachers who reflect on the use of such practices in their classrooms are engaged in the science of teaching, but they are also acting as artful, instructional engineers who select, adapt, blend, and implement these practices for use with specific groups of learners. In order to improve economic literacy in grades K-12 and beyond, teachers of economics need to move beyond totally didactic instruction to include methods that are more student-centered, activity-based, and life-relevant. Numerous publications from the National Council on Economic education, including the *Master Curriculum Guide in Economics* series, provide teachers with easy access to exemplary lessons and units that employ a variety of effective instructional methods.

REFERENCES

Banaszak, R. A. (1992). Teaching about financial markets. *The Social Studies*, *83*(6), 238–240.

Caldwell, J., Clark, J., & Herscher, W. (1996). *World history: Focus on economics*. New York: National Council on Economic Education.

Cassuto, A. (1980). The effectiveness of the elementary school Mini-Society program. *Journal of Economic Education*, *11*(2), 59–61.

Chizmar, J. F., & Halinski, R. S. (1983). Performance in the Basic Economics Test (BET) and Trade-Offs. *Journal of Economic Education*, *14*(1), 18–29.

Coulson, E. C., & McCorkle, S. (Eds) (1993). *Economics and children's literature*. St. Louis, Missouri: SPEC Publishers, Inc.

Coulson, E. C., & McCorkle, S. (Eds) (1994). *Economics and children's literature, 1994 supplement*. St. Louis, Missouri: SPEC Publishers, Inc.

Coulson, E. C., & McCorkle, S. (Eds) (1998). *Economics and children's literature, Supplement 2*. St. Louis, Missouri: SPEC Publishers, Inc.

Coulson, E. C., Suiter, M. C., & McCorkle, S. (Eds) (1998). *Economics and children's literature 3, Special third supplement: Storybooks for primary grades*. St. Louis, Missouri: SPEC Publishers, Inc.

Day, H., Foltz, M., Heyse, K., Marksbary, C., & Sturgeon, M. (1997). *Teaching economics using children's literature*. Indianapolis, Indiana: Indiana Department of Education.

Gretes, J. A. (1991). Teaching economic concepts to fifth graders: The power of simulations. *Social Science Record, 28*(2), 71–83.

Hartoonian, H. M., & Laughlin, M. A. (1997). Decision-making skills. In: M. E. Haas & M. A. Laughlin (Eds), *Meeting the Standards: Social Studies Readings for K-6 Educators* (pp.164–165). Washington, D.C.: National Council for the Social Studies.

Hendricks, R., Nappi, A., Dawson, G., & Mattila, M. (1986). *Learning economics through children's stories*. New York: Joint Council on Economic Education.

Jackson, P. (1989). *A teacher's guide for Econ and Me: An introduction to basic economic concepts*. Bloomington, Indiana: Agency for Instructional Technology, Canadian Foundation on Economic Education & Joint Council on Economic Education.

Joint Council on Economic Education and Securities Industry Association (1988). *Economics and the stock market game*. New York: Joint Council on Economic Education and Securities Industry Foundation for Economic Education.

Kagan, G., Mayo, H., & Stout, R. (1995). Risk-adjusted returns and stock market games. *Journal of Economic Education, 26*(1), 39–50.

Kehler, A. (1998). Capturing the "economic imagination": A treasury of children's books to meet the content standards. *Social Studies & the Young Learner, 11*(2), 26–29.

Kourilsky, M. L. (1972). Learning through advocacy: An experimental evaluation of an Adversary Instructional Model. *Journal of Economic Education, 3*(2), 86–93.

Kourilsky, M. L. (1976). Perceived versus actual risk-taking in Mini-Societies. *The Social Studies, 67*(5), 191–194.

Kourilsky, M. L. (1977). The Kinder-Economy: A case study of kindergarten pupils' acquisition of economic concepts. *The Elementary School Journal, 77*(3), 182–191.

Kourilsky, M. L. (1985). Economic reasoning and decision making by high school students: An empirical investigation. *The Social Studies, 76*(2), 69–75.

Kourilsky, M. L. (1987). Children's learning of economics: The imperative and the hurdles. *Theory Into Practice, 26*(3), 198–205.

Kourilsky, M. L. (1991). Learning hurdles in economics. In: R. Harris & M. L. Kourilsky (Eds), *Handbook for Survival: A Practical Guide for Today's High School Economics Educators* (ch. 3, pp. 1–9). Sacramento California: California Council for Economic Education and California Department of Education. ERIC Database 1992–1999, ED 377092.

Kourilsky, M. L. (1992). *KinderEconomy+: A multidisciplinary learning society for primary grades*. New York: Joint Council on Economic Education.

Kourilsky, M. L. (1996a). *YESS!/Mini-Society, Economics: Debriefing teachable moments*. Kansas City, Missouri: Center for Entrepreneurial Leadership Inc., Ewing Marion Kauffman Foundation.

Kourilsky, M. L. (1996b). *YESS!/Mini-Society, The framework: Experiencing the real world in the classroom*. Kansas City, Missouri: Center for Entrepreneurial Leadership Inc., Ewing Marion Kauffman Foundation.

Kourilsky, M. L., & Graff, E. (1986). Children's use of cost-benefit analysis: Developmental or non-existent. In: S. Hodkinson & D. Whitehead (Eds), *Economic Education: Research and Developmental Issues* (pp. 127–139). Essex, England: Longman.

Kourilsky, M. L., & Hirshleifer, J. (1976). Mini-Society vs. Token Economy: An experimental comparison of the effects on learning and autonomy of socially emergent and imposed behavior modification. *Journal of Educational Research, 69*(4), 376–381.

Kourilsky, M. L., & Quaranta, L. (1987). *Effective teaching: Principles and practice*. Glenview, Illinois: Scott, Foresman and Company.

Laney, J. D. (1989). Experience- and concept-label-type effects on first graders' learning, retention of economic concepts. *Journal of Educational Research, 82*(4), 231–236.

Laney, J. D. (1990a). Generative teaching and learning of cost-benefit analysis: An empirical investigation. *Journal of Research and Development in Education, 23*(3), 136–144.

Laney, J. D. (1990b). Generative teaching and learning of economic concepts: A sample lesson. *Social Studies & the Young Learner, 3*(1), 17–20.

Laney, J. D. (1993). Experiential versus experience-based learning and instruction. *Journal of Educational Research, 86*(4), 228–236.

Laney, J. D. (1997). Economics for elementary school students: Research-supported principles that guide classroom practice. In: M. E. Haas & M. A. Laughlin (Eds), *Meeting the Standards: Social Studies Readings for K-6 Educators* (pp. 176–179). Washington, D.C.: National Council for the Social Studies.

Laney, J. D. (1999). How cooperative and mastery learning methods can enhance social studies teaching: A sample lesson in economics. *The Social Studies, 90*(4), 152–158.

Laney, J. D., Frerichs, D. K., Frerichs, L. P., & Pak, L. K. (1996). The effect of cooperative and mastery learning methods on primary grade students' learning and retention of economic concepts. *Early Education and Development, 7*(3), 253–276.

Laney, J. D., Moseley, P. A., & Crossland, R. B. (1991). *The effect of economics instruction on economic reasoning: A comparison of verbal, imaginal, and integrated teaching-learning strategies*. Paper presented at the annual meeting of the National Council for the Social Studies, Washington, D. C. ERIC Database 1992–1999, ED 355137.

Laney, J. D., & Schug, M. C. (1998). Teach kids economics and they will learn. *Social Studies & the Young Learner, 11*(2), 13–17.

LaRaus, R., & Remy, R. C. (1978). Citizenship decision-making. Menlo Park, California: Addison-Wesley.

Morgan, J. C. (1991). Using Econ and Me to teach economics to children in primary grades. *The Social Studies, 82*(5), 195–197.

Reinke, R. W., Schug, M. C., & Wentworth, D. R. (1989). *Capstone: The nation's high school economics course*. New York: National Council on Economic Education.

Ryan, F. L., & Carlson, M. A. (1973). The relative effectiveness of discovery and expository strategies in teaching toward economic concepts with first grade students. *Journal of Educational Research, 66*(10), 446–450.

Savage, M. K., & Savage, T. V. (1993). Children's literature in middle school social studies. *The Social Studies, 84*(1), 32–36.

Schug, M. C. (1994). How children learn economics. *International Journal of Social Education, 8*(3), 25–34.

Schug, M. C. (1996). Introducing children to economic reasoning: Some beginning lessons. *The Social Studies, 87*(3), 114–118.

Schug, M. C. (1997). Teaching the economics of the environment. *Children's Social and Economics Education, 2*(1), 47–55.

Schug, M. C., & Baumann, E. (1991). Strategies to correct high school students' misunderstanding of economics. *The Social Studies*, *82*(2), 62–66.

Schug, M. C., Caldwell, J., Wentworth, D. R., Kraig, B., & Highsmith, R. J. (1993a). *United States history: Eyes on the economy, Volume One: Through the Civil War*. New York: National Council on Economic Education.

Schug, M. C., Caldwell, J., Wentworth, D. R., Kraig, B., & Highsmith, R. J. (1993b). *United States history: Eyes on the economy, Volume Two: Through the 20th century*. New York: National Council on Economic Education.

Schug, M. C., Morton, J. S., & Wentworth, D. R. (1997). *Economics and the environment: EcoDetectives*. New York: National Council on Economic Education.

Schug, M. C., & Walstad, W. B. (1991). Teaching and learning economics. In: J. P. Shaver (Ed.), *Handbook of Research on Social Studies Teaching and Learning* (pp. 411–419). New York: Macmillan Publishing Company.

Shaver, J. P., Davis, O. L., Jr., & Helburn, S. W. (1979). The status of social studies education: Impressions from three NSF studies. *Social Education*, *43*(2), 150–153.

Soper, J. C., & Walstad, W. B. (1988). What is high school economics? Posttest knowledge, attitudes, and course content. *Journal of Economic Education*, *19*(1), 37–51.

Sweeney, J. A., & Baines, L. (1993). The effects of a video-based economics unit on the learning outcomes of third graders. *Social Science Record*, *30*(1), 43–56.

Walstad, W. B. (1980). The impact of Trade-Offs and teacher training on economic understanding and attitudes. *Journal of Economic Education*, *12*(1), 41–48.

Walstad, W. B., & Soper, J. C. (1988a). High school economics: Implications for college instruction. *American Economic Review*, *78*(2), 251–256.

Walstad, W. B., & Soper, J. C. (1988b). What is high school economics? TEL revision and pretest results. *Journal of Economic Education*, *19*(1), 24–35.

Walstad, W. B., & Soper, J. C. (1989). What is high school economics? Factors contributing to student achievement and attitudes. *Journal of Economic Education*, *20*(1), 53–68.

Weidenaar, D. J. (Ed.) (1982). *Master curriculum guide for the nation's schools, Part II, Strategies for teaching economics: Using economics in social studies methods courses*. New York: Joint Council on Economic Education

Weiser, L. A., & Schug, M. C. (1992). Financial market simulations: Motivating learning and performance. *The Social Studies*, *83*(6), 244–247.

Welton, D. A., & Mallan, J. T. (1996). *Children and their world: Strategies for teaching social studies* (5th ed.). Boston, MA: Houghton Mifflin Company.

Wentworth, D. R., Kraig, B, & Schug, M. C. (1996). *United States history: Focus on economics*. New York: National Council on Economic Education.

Wentworth, D. R., & Schug, M. C. (1993). Fate vs. choices: What economic reasoning can contribute to social studies. *The Social Studies*, *84*(1), 27–31.

Wentworth, D. R., & Schug, M. C. (1994). How to use an economic mystery in your history course. *Social Education*, *58*(1), 10–12.

Wentworth, D. R., & Western, R. D.(1990). High school economics: The new reasoning imperative. *Social Education*, *54*(2), 78–80.

Wood, W. C., O'Hare, S. L., & Andrews, R. L. (1992). The stock market game: Classroom use and strategy. *Journal of Economic Education*, *23*(3), 236–246.

Yeargan, H., & Hatcher, B. (1997). The cupcake factory: Helping elementary students understand economics. In: M. E. Haas & M. A. Laughlin (Eds), Meeting the Standards: Social studies readings for K-6 educators (pp.169–170). Washington, D.C.: National Council for the Social Studies.

DISCUSSION

Jere Brophy

INTRODUCTION

Developing this volume has been an experiment, or perhaps I should say a pilot study. It was designed to see if writers representing relatively narrow slices of the K-12 curriculum could identify and write about instructional methods and activities that are given unique (or at least, characteristic) emphasis in teaching their respective subjects, and if so, whether these could be connected along common dimensions and understood as extensions of or supplements to the set of generic principles outlined in the introductory chapter. I believe that readers will agree with me that the first part of this experiment has been quite successful: The chapter authors have produced scholarly treatments of contemporary thinking about best practices in teaching fourteen school subjects. In the process, they have identified quite a variety of instructional methods and learning activities, ranging in scope from general instructional models to specialized activities used in particular situations. It is my task in this discussion chapter to at least make a beginning on the second part of the experiment: identifying common dimensions that will allow for comparison and contrast of what each chapter had to say with what the other chapters had to say and with the guidelines embedded in the generic principles outlined in the introduction. This is a daunting task, given the range of curriculum areas, instructional issues, teaching methods, and learning activities discussed in these chapters.

When I read through the chapters for the first time, three things struck me immediately. First, all of the authors felt the need to preface their chapters with initial statements about the nature and goals of their respective subjects, often including historical overviews outlining developments over time. My initial

Subject-Specific Instructional Methods and Activities, Volume 8, pages 437–468.
2001 by Elsevier Science Ltd.
ISBN: 0-7623-0615-7

reaction to these introductory sections was one of concern that, given the intended focus of the volume, the space devoted to them might have been devoted to fuller coverage of instructional methods and activities. However, I came to appreciate the value of them, both because they are interesting and informative in their own right and because they help us to understand what the contributors have to say about best practices and why specialists in their respective subjects value the practices that they do.

Also immediately noticeable were certain family resemblances within subsets of the chapters. The three chapters dealing with language arts all had certain features in common, as did the two mathematics chapters, the four science chapters, and the five social studies chapters. I also could see "extended family" resemblances relating to distinctions between "basic skills" subjects (language arts, mathematics) and "content area" subjects (science, social studies). Some of these family resemblances are rooted in the disciplines that inform these respective subject clusters. Others appear related to the nature of the learning goals emphasized (particularly differences in relative emphasis on content vs. skills and on cognitive goals vs. affective goals).

A third immediate impression concerned the perspective taken by chapter authors on the degree to which their subject should be construed as induction into an academic discipline vs. as general preparation for life in the present and future. Here there was considerable variation. Some authors made little or no distinction between the school subject and its underlying discipline: they discussed best practices as methods for orienting students to the discipline and enabling them to begin to function as active members of its community of practitioners. Other authors cast learners more as consumers than as producers of discipline-based knowledge, talking about best practices more in terms of developing appreciations and dispositions that would enrich students' inner lives and equipping them with knowledge and skills that would help them meet the demands of occupational, citizenship, and other roles embedded in life in contemporary society. The authors' positions on this basic issue carried implications for their thinking about the major purposes and goals for their subject, and thus about the kinds of instructional methods and activities that would constitute best practices.

Contrasting positions on this issue sometimes led to differences in the intended meanings associated with commonly used terms. For example, to authors who view school subjects as induction into the disciplines, an authentic learning activity is one that engages students in doing things that disciplinary practitioners do (e.g. engaging in discipline-based inquiry using the modes of thinking and the investigatory tools that are characteristic of the discipline). In contrast, to authors who emphasize school subjects as general preparation for life, an authentic learning activity is one that engages students in applying their subject-matter learning to problem solving or decision making in life outside of school.

Additional impressions emerged as I became more familiar with the chapters. One clear general trend was the influence of social constructivist and sociocultural theories on current notions of best practice, and in particular on ideas about the teacher's role in stimulating and managing productive discourse. Although varying in the particulars of their preferred discourse models, most authors took strong positions favoring thoughtful discourse structured around powerful ideas over scattershot recitation formats.

Most contributors discussed the latest technological innovations in teaching their subject, although with variation in the stances taken. Some were highly enthusiastic about technological innovations, especially web-based activities that allow students to assume the role of disciplinary practitioners (e.g. earth system scientists) as they collect and process data to conduct inquiry or solve problems. Others were more cautious about or even somewhat dismissive of technology, arguing that there is no substitute for a teacher who monitors and scaffolds students' progress and that technology-based simulations, although perhaps useful up to a point, misleadingly simplify the inquiry process and in other respects are not adequate substitutes for more direct inquiry conducted under realistic conditions.

I will return to these issues in subsequent sections focusing on the subject-specific methods and activities discussed in the fourteen chapters. First, however, I turn to the authors' reactions to the principles that were put forth in the introduction to this volume.

REACTIONS TO THE TWELVE GENERIC PRINCIPLES

I was eager to see how the chapter authors would respond to these twelve principles, given that they were described as synthesizing generic aspects of what is involved in teaching school subjects for understanding, and thus as applicable to statements about best practices in teaching any subject. Previously I had received quite positive comments on them, both in correspondence from International Academy of Education members who reviewed successive drafts of the booklet that introduced them (Brophy, 1999a) and during a session organized to discuss this booklet that was held at the 2000 meetings of the Invisible College for Research on Teaching and Teacher Education (an organization for researchers in these fields that meets prior to each year's meetings of the American Educational Research Association). Most comments at the Invisible College meeting were positive, although one participant noted that the principles made no mention of technology and another indicated that more could have been said about motivation (not just about stimulating students' interest in the content but also about helping them to maintain positive self-efficacy perceptions and to self-regulate their learning productively). Although these elaborations were suggested, no one voiced any complaints about what was already included in the principles. However, most of the participants at this meeting were researchers whose work had focused on generic aspects of teaching. Prophetically, one of them said that whereas he personally was happy with the principles, people interested in problem-based learning would see them as too teacher-centered.

Teacher-centeredness was one of several concerns raised by contributors to this volume whose scholarly work has focused on subject-specific rather than generic aspects of teaching. I believe that most of these concerns can be accommodated easily by clarifying or elaborating on what was said about the principles in the introduction, but nevertheless it is worth noting that several of these subject-specific scholars found fault with the principles, whereas generic teaching scholars did not.

There was considerable variation in response to the principles. The authors of six chapters simply accepted them as applying to their subject and made no suggestions for change; the authors of three chapters made only minor suggestions for elaboration on only one principle; and the authors of the remaining five chapters raised significant concerns about several of the principles.

Concerns were expressed about eight of the principles. The authors either did not mention the other four principles or underscored a qualification on their applicability that already had been noted in the introduction. The four principles that were endorsed by all of the authors included those dealing with opportunity to learn (2), curricular alignment (3), scaffolding students' task engagement (8), and achievement expectations (12).

GENERAL CONCERNS

Some contributors raised general concerns about the set of principles as a whole or the assumptions that underlie them. One basic issue was raised in Chapter 4, where Stein communicated acceptance of the validity of most of the principles but argued that teacher education efforts are more likely to be effective and to avoid inducing unintended inferences if they are subject-specific rather than generic. As one example, she suggested that Principle 10 might induce some mathematics teachers to use small group methods inappropriately because they fail to take into account important content issues.

In this regard, she quoted a staff developer who suggested that small groups are not ordinarily the method of choice in teaching mathematics, but they may become necessary "when tasks are so rich, solution strategies so numerous, and eagerness to talk so widespread" that it makes sense to temporarily divide the class into groups rather than continue with a whole-class lesson. As another example, she suggested that Principle 6 might induce some teachers to overprepare lists of questions to the point of being insufficiently flexible in responding to events that develop as lessons progress, or to focus too much on issues such as cognitive level or wait time and not enough on thinking through the mathematical content involved and anticipating probable student responses.

In responding to this concern, I first note that the potential problems illustrated in Stein's examples would result from inappropriate use of the principles rather than any inherent flaw in the principles themselves. I also suggest that both generic and subject-specific aspects of teaching need attention in teacher education. One can focus overly exclusively on content issues without paying sufficient attention to human interaction issues, as well as vice versa. That said, I acknowledge that Stein's concerns are well taken. The list of principles (including the elaboration that follows each of them) is meant as a short summary that needs to be unpacked in considerable detail in teacher education, as well as supplemented by subject-specific principles and principles dealing with instruction to different age levels, different student populations, and so on. It would be inappropriate to treat the principles as the complete scope for a teacher education curriculum, or as the sole basis for an assessment device to measure teachers' effectiveness.

In Chapter Two, Alvermann and Hruby raised another concern that also relates to the distinction between a generic and a subject-specific purview. They suggested that my principles are informed mostly by educational psychology and thus well suited as guidelines for teaching content area reading, but they are less well suited to literature study because it is informed more by literary theory. They characterized the psychological perspective as giving cognition pride of place, viewing it as the serial solving of problems, construing texts as containers of information, emphasizing strategies that will allow students to learn from texts effectively, and assessing this learning using standardized tests. In contrast, they characterized the literary perspective as centering on experience, imagination, and emotion; viewing reading as having experiences to be appreciated and evaluated; construing literary texts as means for generating such experiences; being more concerned with developing open-mindedness, tasteful discrimination, and moral sensibility as responses to text than with the learning of specific content; and assessing using written essays, journals, portfolios, or group projects. Other chapter authors also expressed versions of this same basic concern that the principles are focused on knowledge and skill learning without taking into account aesthetic experiences, positive attitudes toward the subject, or other affective outcomes.

These concerns reflect a misimpression about the intended scope of some of the principles, albeit one that is quite understandable given what was said in the introduction. In developing the principles, I intended to refer to aesthetic and other affective experiences that might be included as instructional goals (e.g. coming to appreciate the elegant simplicity and symmetry of mathematics, experiencing the excitement and generative power of science, and so on). Ironically, I have written at some length about incorporating aesthetic and other appreciation goals into instruction in a textbook on motivating students to learn (Brophy, 1998) and an article on ways that teachers can use modeling, scaffolding, and other instructional strategies to stimulate interest in and aesthetic appreciation for their subject (Brophy, 1999b). Unfortunately, little of this was evident in my elaboration of the principles in the introduction to this book.

A basic assumption underlying the principles is that different types of learning call for different kinds of teaching, and Principle 3 states that all components of the curriculum should be aligned to create a cohesive program for accomplishing intended outcomes. These outcomes are described as knowledge, skills, attitudes, values, and dispositions to action that society wishes to develop in its citizens. Under "attitudes, values, and dispositions" I meant to include the kinds of aesthetic and emotionally tinged experiences that occur when students become intensely engaged in learning activities that have important personal meaning for them. However, I failed to convey this clearly. Future presentations of these principles will need to elaborate on Principle 3 to clarify this point. Elaboration on other principles may be needed as well to establish the points that instructional goals usually include aesthetic or other emotionally tinged learner experiences, and that teachers will need to include attention to these intended outcomes in planning lessons, learning activities, and assessment methods.

The most radical departure from the notions about teaching embodied in the principles appeared in Chapter 9, where Duschl and Smith put forth a radical challenge not only to several individual principles but to the underlying general assumption that teaching will be directed mostly to the whole class and individualized only around the margins. These authors argued that certain unique aspects of earth system science, combined with technological advances that allow students to work with data accessed through the web, have made it feasible and desirable for classes in this subject to oscillate between inquiry groups, individual work, and classs in whole-class settings, with inquiry groups forming the agendas for whole-class lessons. I remain open to this possibility but currently unconvinced.

Pending appearance of evidence to the contrary, I continue to believe that K-12 teachers ordinarily will need to develop both the content bases and the process guidelines needed for carrying out complex learning activities before releasing students to work on those activities. More specifically, I believe that besides incorporating the inquiry and problem-based learning formats that are well suited to the individual- and small-group settings favored by Duschl and Smith, most earth system science instructors often will want to teach more traditional lessons in the subject in whole-class settings, focused on development and discussion of text readings. In any case, whether or not these authors' model of best practices in teaching earth science should prove feasible and desirable, their reaction to the principles underscores the fact that their applicability is limited by the assumptions that underlie them. One of these assumptions is that the student/teacher ratio and other constraints within which most teachers work require them to rely primarily on whole-class instruction. Significant adaptation of the principles, or perhaps even development of a separate set, would be needed to summarize best practices in self-regulated learning situations in which learners work autonomously most of the time.

CONCERNS ABOUT SPECIFIC PRINCIPLES

All of the authors accepted Principle 1 (learning community) as far as it went, but the need to elaborate it in various ways was argued in Chapters 1, 3, 4, and 13. Hoffman and Duffy would extend the principle to include providing students with experiences in seeking meaning: have abundant and varied texts available, read to students, and engage them regularly with texts in ways that help them learn that reading enlightens, enlivens, enhances, and informs. I share these authors' enthusiasm for these teaching practices, but rather than classifying them under Principle 1, I would elaborate Principle 3 to refer to aesthetic and other emotionally tinged experiences and elaborate Principle 2 to clarify that opportunity to learn implies opportunity to attain affective outcomes along with knowledge and skill outcomes.

Citing equity considerations, Freedman and Daiute argued that Principle 1 should be adjusted to indicate that diversity must be addressed, not transcended. I take their point. I meant to include at least some of this when I spoke of the need to connect with students' home cultures and involve parents in their children's learning. However, I agree that more could be said about addressing diversity proactively, and the term "transcends" might be replaced with a phrase such as "honors the individuality and diversity among students who differ in" (gender, race, etc.).

Stein suggested that Principle 1 needs to be elaborated to address the teacher's role in establishing "sociomathematical norms" regarding how mathematics gets done, who has the right to question and propose solutions, and the basis on which claims and counterclaims will be judged. Torney-Purta, Hahn, and Amadeo would elaborate this principle to mention support for honest discussion of controversial issues. Some of what was said in other chapters also implied the need to extend Principle 1 to say something like the following: "Socializing students to function within a learning community might also involve communicating expectations for their participation in certain forms of subject-specific discourse (e.g. sharing responses to literary selections, discussing alternative methods of approaching mathematics problems, assessing evidence and drawing conclusions in scientific inquiry, or discussing controversial social or civic policy issues). Where this is the case, it will be important to teach students how to engage in the desired forms of discourse and so that these expectations are met during subsequent discussions."

The authors of Chapters 4, 5, 9, and 13 raised a set of concerns that apply in varying degrees to Principles 4 (establishing learning orientations), 5 (coherent content), 6 (thoughtful discourse), and 9 (strategy teaching). Their core concern was that these principles appear to imply that teachers routinely would first transmit key ideas and only then move to problem solving or other application opportunities. However, they recommended practices that involve initially engaging students in inquiry, problem solving, or other higher order thinking about the topic, meanwhile withholding more formal statements and elaboration of key ideas until after students have had opportunities to communicate and discuss their own ideas.

Alvermann and Hruby suggested that what readers take from a literary work is personal and basic to what makes reading pleasurable, so it would be poor pedagogy to preempt these personal responses through prior explicit instruction in what a poem means, what an author is trying to say, or the significance of a particular figure of speech or motif. They added that the meanings of literary texts are crafted by readers and can be ambiguous, ironic, or otherwise "beyond the grasp of explicitly taught strategies for simplification and summarization."

Stein characterized the principles as presenting an active teaching model calling for initial development of content through teacher presentation. She contrasted this with her emphasis on first providing opportunities for students to make sense of problems by elaborating their existing networks of understanding, and thus encouraging them to develop identities as makers and users of mathematics (which will not occur if teachers do too much too early). She concluded that project-based pedagogy is the pedagogy of choice for structuring meaningful student learning of mathematics. Similarly, Battista emphasized that attempts to establish appropriate orientations to geometry learning should emphasize sensemaking and understanding rather than rule following as in traditional mathematics teaching; that content must be coherent for students but need not be explained initially by teachers; and that strategy teaching should include attention to metacognitive components, to what is involved in making sense of a mathematical idea, and to problem solving heuristics (not just algorithms).

Duschl and Smith warned against overreliance on "reductionist" schemes and principles for organizing earth science content, an approach which would rob students of opportunities to understand science as a living discipline and experience learning outcomes that balance conceptual, epistemic, and reasoning goals. Torney-Purta, Hahn, and Amadeo noted that meaningfulness (coherent content) can be enhanced through stories and discussion of conflict, and that inquiry and discourse about political issues on which there are differing opinions should be an important part of thoughtful discourse.

Most of these concerns boil down to two related points: (1) students need to develop coherent networks of knowledge structured around big ideas, but in the process they also need opportunities to create and negotiate understandings (so that attempts to teach through direct transmission that treat students as passive receivers rather than active constructors of knowledge are unlikely to be very effective, no matter how elegantly the knowledge may be structured), and (2) often (perhaps most of the time in certain subjects, such as literature or mathematics), inductive/exploratory/experiential/problem-based approaches that initially engage students in inquiry or problem solving and only later move to negotiation and synthesis of what was learned are preferable to traditional approaches that begin by establishing a knowledge base and only then moving to problem solving or other applications.

I have several responses to concerns surrounding these two points. Let me say first that I accept the two points (at least when phrased in the somewhat qualified manner in which I have just rendered them). I had intended to communicate this through phrasing indicating that the teacher might provide "or elicit" coherent explanations of big ideas. However, I see that I failed to do this consistently, and the introduction to this volume does suggest a "provide coherent explanations first, and only then move to applications" model. Clearly, the presentation of the principles (and especially of Principles 4, 5, and 6) needs to be adjusted to incorporate these two points and their implications.

In making these adjustments, however, I would insert some qualifications of my own. One of these concerns limitations on the feasibility and desirability of social constructivist teaching methods. Problems with transmission methods are well known. When these methods are used poorly or simply too much, they tend to lead to student boredom, reliance on rote learning methods, and acquisition of disconnected items of knowledge that are mostly soon forgotten or retained only in inert forms. However, there are predictable difficulties with social constructivist methods as well. Airasian and Walsh (1997), Weinert and Helmke (1995), Windschitl (1999), and several contributors to this volume have observed that these methods are difficult to implement effectively: They require teachers to possess deep and well-connected subject-matter knowledge (and related pedagogical knowledge) that enables them to respond quickly to only partially predictable developments, and they require students to participate more actively and take more personal risks in learning.

I share these concerns and suggest another: heavy reliance on social constructivist discourse models increases the possibility that the discourse will stray from the lesson's intended goals and content, and that even when it stays goal-relevant, progress toward construction of the intended understandings may be erratic and include frequent verbalization of misconceptions. In terms originally introduced by Jacob Kounin (1970) in another context, it might be said that such lessons have a rough rather than a smooth flow, a poor signal (valid content)-to-noise (irrelevant or invalid content) ratio, and frequently interrupted or sidetracked momentum.

These potential dangers of overreliance on social constructivist methods become acute when teachers face either or both of the following conditions: (1) they are working with young learners who as yet have only moderately developed skills for learning through speaking and listening and minimally developed skills for learning through reading and writing, and (2) they are working with learners (of any age) whose prior knowledge of the domain is very limited and poorly articulated, so that questions frequently fail to produce a response or elicit irrelevant or invalid content. One implication is that the social constructivist discourse model that has been developed primarily with middle school and secondary classrooms in mind will need to be adjusted to make it more attuned to the discourse forms and rhythms that commonly occur in the primary grades (e.g. more frequent but shorter exchanges, with more scaffolding to help students express themselves).

Another implication is that the feasibility and desirability of social constructivist methods may vary with student prior knowledge and related signal-to-noise ratio issues. A supportable principle might call for incorporating as much of the social constructivist ideal as is desirable under the circumstances, but in ways that result in smoothly flowing lessons that have acceptable signal-to-noise ratios and include planning for either avoidance of or focused attention to predictable misconceptions (especially if these involve big ideas and are particularly memorable or difficult to eradicate once verbalized). A corollary implication (hypothesis) here is that eliciting certain common misconceptions may promote progress toward instructional goals, but other misconceptions are likely to disrupt such progress and are best left out of the lesson unless students verbalize them spontaneously.

In responding to some of the more specific concerns about Principles 4, 5, and 6, I would refer readers back to what was said earlier in this chapter about instructional purposes and goals subsuming a broad range of learner outcomes and about elaborating Principle 1 to include socializing students into subject-specific forms of discourse. Thus, I would certainly want to include the development of connoisseurship and other aspects of literary appreciation as important goals of literature study, a disposition toward mathematical sensemaking as an important goal of teaching mathematics, a richly developed appreciation of science as an important goal of teaching earth system science, and an interest in and disposition toward productive discussion of controversial issues as an important goal of citizenship education. Following up, I would incorporate these concerns into what I had to say to teachers about establishing learning orientations, coherent content, and thoughtful discourse.

This suggests another clarification of Principles 4, 5, and 6, one that also applies to Principle 9 (strategy teaching). It is understandable that both my treatment of these principles in the introduction and the nature of the research that led to them caused several chapter authors (and probably many readers) to assume that the principles apply primarily if not exclusively to the teaching of knowledge and skills. In contrast, I maintain that the principles are just as applicable to attitudinal, value, and dispositional goals, including aesthetic appreciations. Further, I would argue that teachers promote progress toward such goals by structuring and scaffolding activities that engage students in relevant experiences; that these experiences are most likely to occur when teachers manage the discourse so as to model or elicit certain key ideas or affective responses; and that this may also involve teaching students relevant strategies for making sense of and responding to particular classes of activities.

For example, much of what is involved in helping students learn to take pleasure in reading and to respond to literary works at a variety of cognitive and affective levels involves orienting learners toward the nature and goals of literary activities, managing the discourse with these goals in mind, and teaching literary response strategies. To me, this is what distinguishes literary instruction from merely exposing students to literature. I have written about this at some length (using the teaching of Shakespeare's plays as a running example) in an article that talks about working in the "motivational" zone of proximal development and using modeling, coaching, questions, and feedback to scaffold students' development of appreciation for the aesthetic experiences and other personal satisfactions that may be derived from subject-specific activities (Brophy, 1999b).

One final clarification on this point: Alvermann and Hruby inferred that my conception of appreciation is bound up with the psychological instrumentalist idiom and thus is about utilitarian reasons for learning. In fact, I meant the term "appreciation" to subsume a much broader range of reasons for learning, including the aesthetic appreciations emphasized in their chapter. One of the points that I emphasize in my textbook and other writings on motivation is that a great deal of theory and research has been conducted on the expectancy aspects of motivation (success/failure expectations and attributions, self-efficacy perceptions, etc.), whereas remarkably little has been published on the value aspects (appreciating why a learning activity is worthwhile and experiencing its satisfactions and rewards).

In this regard, I follow the suggestion of Eccles and Wigfield (1985) that subjective task value has three major components: (1) attainment value (the importance of attaining success on a task in order to affirm self-concept or fulfill needs for achievement, power, or prestige), (2) intrinsic or interest value (the enjoyment we get from engaging in the task) and (3) utility value (the role that engaging in the task may play in advancing our career or helping us to reach other larger goals). However, I suggest that classroom applications of this scheme might expand it to include experiencing the satisfaction of achieving understanding or skill mastery under attainment value, aesthetic appreciation of the content or skill under intrinsic value, and awareness of the role of this learning in improving the quality of one's life or making one a better person under utility value (Brophy, 1998).

There were a few comments about other principles. In Chapter 5, Battista acknowledged that students need practice with improvement-oriented feedback (Principle 7), but argued that the practice ought to be accomplished in appropriate contexts. He suggested that this can be done by revisiting core ideas and skills in many different contexts or in games. I agree, although I would repeat my caution about insuring that sufficient practice is provided.

Chapters 4 and 5 suggested interesting qualifications on the applicability of cooperative learning (Principle 10) beyond those noted in the introduction. Stein indicated that small-group contexts ordinarily would not be the method of choice in teaching about number, but conceded that there may be times when breaking the whole class into subgroups makes sense because everyone is eager to talk and this is a way to accommodate "the flood of student thinking and communication." Battista suggested that small-group activities in which all group members work as partners dealing with all aspects of problem solving are appropriate in teaching geometry, but not Jigsaw models that call for individuals to work on separate tasks (because many of these individuals may never come to understand the larger problem or its solution).

Finally, concerns about Principle 11 (assessment) were expressed in two chapters. Alvermann and Hruby inferred that this principle calls for assessing part skills through standardized tests rather than for more holistic and broad-ranged assessment. Battista cautioned that assessment needs to focus not only on behavior (getting correct answers to geometry problems) but also on cognition (making sure that students understand the mathematical concepts and problem-solving procedures involved). These authors read my phrasing of Principle 11 more narrowly than it was intended. As part of the more fundamental principle of curricular alignment, I meant to convey that the forms and content of whatever assessment methods are used need to be designed to assess attainment of the full range of instructional goals (not just the knowledge and skill goals), with particular focus on the major intended outcomes.

COMMONALITIES ACROSS CHAPTERS

In addressing their respective subject-specific issues, the contributors made reference to a great many instructional methods and learning activities that extend our notions of good teaching considerably beyond what is included in my twelve generic principles. Some of these relate to dimensions addressed by those principles but raise considerations not included in them. Others introduce dimensions that the principles do not address at all.

Looking across the chapters, I see far more evidence of congruence and complementarity than of dissonance or contradiction. When different chapters addressed the same teaching methods or learning activities, they typically emphasized the same principles for using them productively, even if they used different terminology and cited different theoretical sources. Furthermore, even if a principle was developed in only one or two chapters, the instructional method or learning activity under discussion often has applicability to most if not all school subjects, and the same basic principles for using these methods effectively would apply. Following are some of the commonalities that I noted in making comparisons across chapters.

GOALS: TEACHING FOR UNDERSTANDING, APPRECIATION, AND LIFE APPLICATION

As expected, all authors emphasized teaching for understanding that would equip students with networks of connected knowledge that are structured around big ideas and retained in forms that make them available for life application, in contrast to disconnected knowledge learned primarily by rote and retained (if at all) only in inert forms. In addition, however, the authors tended to emphasize appreciation goals. Those who made few distinctions between their school subject and its underlying discipline often expressed these goals in terms of developing appreciation for the discipline as a valuable and rewarding human enterprise. Stein stated that the development of mathematical understandings and the socialization of students into the practice of making and using mathematics are equally important and complementary goals of mathematics instruction. Duschl and Smith called for a shift of perspective from "final form science" to "science-in-the-making," allowing students to do science authentically and experience its excitements and rewards.

Other authors omitted or placed less emphasis on equipping students with disciplinary skills to use to generate knowledge, but depicted students as consumers of subject-matter learning who would apply the learning in their lives outside of school. Laney spoke of the need to make economic learning meaningful, memorable, and applicable to students' own lives and experiences. Hoffman and Duffy emphasized helping students to experience what it feels like when reading enlightens, enlivens, enhances, or informs, and Alvermann and Hruby emphasized cultivating students' aesthetic responses to texts. I suspect that specialists in all school subjects would endorse instructional goals that construe students as consumers of subject-matter learning who will apply it in their lives outside of school, and that many would endorse goals that construe students as novices learning to use disciplinary tools to generate knowledge (where the school subject is closely associated with a discipline).

Freedman and Daiute elaborated on these themes in talking about how writing instruction has evolved from relatively sterile models emphasized in the past: We now teach the processes for creating various forms of text, not just the processes for rendering them into script. Along with structural aspects of paragraphs, we teach about contextual factors such as the purpose of the paragraph in the larger text, the situation in which it will be read, the nature of the readers, and the intended reaction to be produced in them. Teaching written genres now means making visible the less visible but formative rhetorical demands of purpose, audience, and context as the motivation and processes for using specific text forms. Teachers might organize curricula around classical text forms such as essays, stories, or letters; around more basic elements such as descriptive or cause-and-effect paragraph types; or around theses or central ideas. However, such forms would be taught as moments in socially situated activities such as identifying and solving a social problem via writing letters to influential people in the community, summarizing existing knowledge about the problem, or writing reports that include facts and proposals. In this way, writing is embedded in broader social issues. Forms are not taught as ends in themselves but as tools used to accomplish social purposes.

As part of emphasizing life applications, several authors talked about personalizing the instruction around students' interests or home backgrounds, or about applying the learning to local issues. Many of the examples cited by Duschl and Smith, for example, involved applying earth system science to the local geographical area and ecosystem. Similarly, Cunha spoke of including research on one's own family migration histories as part of study of the geographic theme of movement.

ADDRESSING MULTIPLE GOALS SIMULTANEOUSLY

Authors commonly spoke of attending in a balanced way, or even simultaneously, to multiple types of goals or multiple levels of analysis or understanding. They typically advocated mixtures of knowledge, skill, attitude, value, and dispositional goals, and in particular, stressed addressing content and process goals simultaneously by engaging students in authentic activities. Battista outlined a curricular organizational framework that interlaces three levels of structuring (spatial, geometric, and axiomatic) with four modes of processing geometric information (describing, representing, analyzing, and justifying), as well as a four-phase instructional approach designed to help students progress through the stages in sophistication of geometric thought identified by van Hiele. Wandersee spoke of moving students up through four levels of biological literacy using instruction that connects learning science (understanding important concepts and principles in depth), doing science (learning the kinds of scientific thinking that lead to biological knowledge), and learning about science (the nature of biology and how biologists develop knowledge). Later he stated that biology courses should integrate: (1) cognitive development activities, (2) investigative activities that involve posing and solving problems and persuading peers, (3) issue-driven activities, and (4) interactive presentation activities.

Minstrell and Kraus identified four aspects that need to be interlaced into a framework for teaching physics: identifying and monitoring students' thinking, testing and modifying their understanding (promoting change), teaching for deep conceptual understanding, and applying learning to build functional understanding. Treagust and Chittleborough spoke of interrelating three levels of chemical phenomena: the symbolic level (representations), the microscopic level (movements of electrons, particles, etc.), and the macroscopic level (relating chemical phenomena to everyday experiences). Such emphasis on simultaneously processing content at several cognitive levels was common in subjects that reflect hierarchically organized knowledge bases, whereas emphasis on inclusion of a full range of goals (affective and dispositional as well as cognitive and procedural) occurred in all subjects.

INQUIRY MODELS

Inquiry models, discourse management, and authentic activities were emphasized in most of the chapters, often as parts of a larger approach advocated as ideal. These substantive aspects of curriculum and interaction usually were considered more fundamentally important than aspects of the formats or settings of instruction (e.g. group size) or the instructional technology used. Also, the contributors focused on methods and activities that would be used with all students, making little or no mention of adjustments for struggling students (except in the language arts chapters) or attempts to accommodate supposed learning styles or other individual differences in students.

Traditional notions of instruction typically call for proceeding relatively linearly from lower to higher levels of cognitive objectives, and in particular, for beginning by establishing a knowledge base before moving to application or other higher order activities. In contrast, many of the chapters argued for frequent (and in some cases, primary) use of inquiry models that begin by placing students into application (or other higher order) contexts immediately. Stein called this approach using problem contexts for developing concepts and procedures, and noted that others have called it problem-centered teaching, investigations, inquiry-based teaching, or project-based pedagogy. In Stein's version, students are asked to solve problems and in the process observe patterns and relationships, conjecture, test, and discuss their ideas. Only after developing some understanding of the situation are they encouraged to connect their work with mathematical symbols and conventional procedures. The idea is to develop and build on the personal and collective sensemaking of the class before "bridging" to more formal discipline-based concepts.

Treagust and Chittleborough spoke of teaching chemistry using a problem-solving model that subsumes the following steps: identify the problem and the important issues, evaluate relevant information, plan a solution, execute it, validate it, and assess understanding of what has been learned from the experience. Thornton also spoke of a problem-solving model, this time for use in teaching history: pose the problem, determine what is known and what information must be obtained, develop tentative models or solutions, conduct research to test these, then evaluate the assembled data to draw conclusions.

Wandersee identified the following as key elements of an inquiry approach for teaching biology: active student orientation to the learning task; engagement of students with a scientific question; direct, hands-on experiences with relevant natural phenomena; eliciting students' existing ideas; exchange of, dialogue about, clarification of, and testing of these ideas; gathering contemporary scientific input and negotiating restructuring of existing ideas; evaluating alternative explanations; testing the application of current explanations in new situations; continuous feedback to learners with subsequent modifications of explanations as needed; and culmination in teacher-student review and assessment of what was learned and how it was learned.

Laney spoke of inquiry-oriented strategies for teaching economics. Inquiry lessons are organized around a central mystery, question, or problem. Students are engaged in using scientific methods to solve the mystery, answer the question, or resolve the problem, and in the process acquire skills for processing and thinking about data. Commonly emphasized steps in the inquiry method include; present the mystery, question, or problem; propose possible explanations or solutions; identify needed evidence; collect relevant data and evaluate its quality; accept or reject each hypothesis using logic and available data; state tentative conclusions; test them; and evaluate what has been learned and what still is unknown.

The general approach and the steps common to these inquiry models reflect theorizing that goes back at least to John Dewey. Over the years it has been described not only using the labels mentioned by Stein but also others such as guided discovery or guided exploration.

Laney's chapter also described a decision making model that social studies scholars frequently put forth as better suited to their subjects than the problem-solving model emphasized in mathematics and science. Steps in this decision making model include: identify the issue, clarify its value aspects, gather and organize evidence on the consequences of policy alternatives, assess the accuracy and relevance of this evidence, identify potential alternative decisions, appraise the anticipated consequences of implementing the decisions and justify the criteria used, make a decision, and implement the decision.

Writing in the context of teaching earth system science, Duschl and Smith made several comments about inquiry approaches that probably apply to the teaching of other subjects as well. They argued that inquiry methods stimulate students to develop general inquiry abilities as well as subject-specific conceptual and process knowledge because they: (1) cause students to confront the limits of their knowledge, to respond to unexpected outcomes with curiosity, and produce motivation to learn; (2) place demands on students to gain knowledge needed to complete their inquiry; and (3) provide opportunities for students to apply what they are learning (either to apply science understandings to answer scientific questions or to generate new knowledge by discovering new principles or refining existing knowledge claims). However, Duschl and Smith also acknowledged that inquiry approaches have proven difficult to implement effectively. Citing Edelson, Gordin, and Pea (1999), they suggested that basic strategies need to be augmented with the following elaborations: (1) use meaningful problems to establish a motivational context for inquiry, (2) use staging activities to introduce background knowledge and investigative techniques, (3) use bridging activities to bridge the gap between the practices of students and the more formally disciplined practices of scientists, (4) use scaffolds that embed tacit knowledge of experts in supportive user interfaces, (5) provide a library of resources, and (6) provide record keeping tools to help students record procedures and store generated products. Much of this might be summarized by suggesting that teachers consider not only disciplinary authenticity but appreciation and life application goals in selecting inquiry activities, and that they provide whatever orientation, structuring, and scaffolding may be needed to enable their students to engage in these activities productively.

DISCOURSE MANAGEMENT

Contributors frequently advanced principles for managing discourse, often embedding them within inquiry models of teaching. Respectively, the chapters emphasized the following aspects of discourse:

(1) Discussion of the meanings of and of individuals' reactions to text (especially narratives).
(2) Discussion to enrich students' understanding of content and appreciation of literature.
(3) Analyzing examples of writing genres to develop understanding of the nature of their features and how these might be included in one's own writing.
(4) Mathematical argumentation: using problem contexts for developing mathematical ideas, representations, and procedures.
(5) Geometric inquiry, sensemaking, and problem solving within a classroom discourse that establishes ideas and truths collaboratively.

(6) Biological inquiry framed within the constructivist model of learning and ideas about biology as a domain.
(7) Analysis of physics problems and negotiation of potential approaches and solutions.
(8) Discussion to stimulate students to construct understandings, concept maps, and mental models of chemical phenomena in problem solving or laboratory debriefing contexts.
(9) Engaging in the dialectic between evidence and explanation or observation and theory that occurs in extended inquiry activities in earth system science (more in small-group than in whole-class settings).
(10) Analyzing and responding to historical events, with focus not just on establishing what happened but on considering connections to other events and implications for personal or civic policy decisions.
(11) Negotiation of learning accomplished through field studies, map-based instruction, and simulation work in geography (during subsequent debriefings).
(12) Thoughtful discourse around big ideas in culture studies.
(13) Value-based discussion of social and civic issues, especially controversial issues.
(14) Economics-based discussion occurring in the context of concept teaching, inquiry, or debriefing following experiential learning or simulations.

Some chapters offered extended models of desired discourse, focusing on concerns that have special relevance to their respective subjects. For example, citing the Vygotskian notion that people develop intellectual skills and dispositions by first using them in social interactions with others, Stein urged mathematics teachers to gradually socialize their students into mathematically accepted ways of thinking, reasoning, and valuing – that is, to help them become competent conjecturers, evidence providers, and critics of others' conjectures and evidence. She argued that students must learn how to make a claim and present a well-grounded case for it, as well as how to listen and respond to claims made by others. Teachers establish such discourse by selecting tasks that are open to different solutions or allow different positions to be taken and defended, such as how to determine the area of an irregular shape. Ideally, two or more different answers will be advanced and defended. The teacher elicits these, legitimizes the diverging positions, calls for collaborative reasoning and problem solving, and focuses students' attention on the rationales underlying the answers. A key strategy is revoicing: reuttering a student's contribution by aligning it with important academic content and opening slots for others to agree or disagree. This allows the teacher to steer the conversation in productive directions, yet credits the contents of the reformulation to the student. This makes the experience more meaningful to the class and helps them develop ownership over their ideas and the self-confidence to see themselves as capable of making sense of tasks and contributing to developing conversations. It also models the process of integrating the discussion and moving forward. The teacher might have to do some conventional telling at various points, but for the most part will act as discussion leader.

Torney-Purta, Hahn, and Amadeo cited a case study illustrating desirable handling of discourse in citizenship education. The teacher began with a divergent question, solicited views from students, sometimes asked for a show of hands to see how many students approved or disapproved of a particular position, encouraged students to explain their views, and helped them make connections by asking about how what was being studied related to current events or having students take the perspective of someone living in the society under study and asking how they might feel, etc. These authors noted that sustained time is needed to focus on a single issue that is authentic because it involves ambiguity, doubt, or controversy. This may create worries about failure to cover the prescribed curriculum or about certain students remaining uninvolved while the others participate in the discussion. A continuing dilemma pits the desire to teach a particular process or model of inquiry or decision making against the desire to "go with the flow" by getting good discussions started and not trying to control them any more than necessary. Other issues include deciding whether students should be formally graded on discussion participation and whether the teacher should disclose his or her own personal views on controversial issues.

AUTHENTIC ACTIVITIES

Along with ideas about structuring discourse, the authors usually had something to say about learning activities as opportunities for students to develop and apply subject-specific learning. There was a considerable range in relative emphasis on whole-class discourse vs. small-group and individual activities. The chapter on number placed heavy emphasis on the former and the chapter on earth system science placed heavy emphasis on the latter, with other chapters in between.

Authors frequently used the word "authentic" in speaking about desired activities, although the primary meanings they attached to the term varied according to their emphasis on students as generators vs. consumers of subject-specific knowledge. Those who emphasized generation of knowledge tended to speak of authentic activities as activities that engage students in doing what disciplinary practitioners do, whereas those who emphasized consumption of the knowledge tended to speak of activities that allowed students to apply what they were learning to their lives outside of school.

Brophy and Alleman (1991) reviewed the scholarly literature on learning activities and evaluated the activities suggested by major elementary social studies textbook series. Then they identified criteria for the design and selection of individual learning activities and for evaluating a set of learning activities when looking across an instructional unit as a whole. They suggested that any activity included in the curriculum ought to meet all three of these basic criteria:

(1) Goal relevance: The learning opportunities provided by the activity promote progress toward the unit's major goals.
(2) Feasibility: The activity is feasible within the time schedule, physical space, and other constraints within which the teacher must work.
(3) Cost effectiveness: The anticipated learning benefits from the activity justify its costs in time and trouble for teacher and students.

In selecting from among activities that meet these primary criteria, teachers might consider several secondary criteria:

(1) Students are likely to find the activity interesting or enjoyable.
(2) It provides opportunities for interaction and reflective discourse, not just solitary seatwork.
(3) If it involves writing, students will compose prose, not just fill in blanks.
(4) If it involves discourse, students will engage in critical or creative thinking, articulate and defend problem-solving or decision-making approaches, and so on, not just regurgitate facts and definitions.
(5) The activity is targeted for students' zones of proximal development; it is not merely an occasion for exercising overlearned skills.
(6) The activity focuses on application of important ideas, not incidental details or interesting but ultimately trivial information.

Criteria for thinking about the set of activities as a whole include:

(1) As a set, the activities offer variety and in other ways appeal to student motivation, to the extent that this is consistent with curriculum goals.
(2) As a set, the activities include many ties to current events or local and family examples or applications.

Brophy and Alleman's guidelines went on to indicate that such activities need to be structured and scaffolded effectively if they are to have their maximal impact. This means preparing students for the activity in advance, providing guidance and feedback as they work on it, and structuring post-activity debriefing or reflection afterwards. These guidelines seem congruent with what contributors to the present volume had to say about activities. In fact, many chapters mentioned one or more of the same guidelines. Some also put forth additional criteria that seem worth noting here.

All authors emphasized the importance of postactivity debriefing to assess and reflect on what was learned. However, there was some disagreement about the degree to which teachers should thoroughly orient students to an activity before allowing them to engage in it. Alvermann and Hruby, even while noting that preteaching of new vocabulary is a research-supported component of content-area reading instruction, stated that there may be times when one might want to expose students to the material first. Or, instead of more traditional and formal new vocabulary instruction, one might want to embed such instruction in pre-reading activities, such as getting students ready to read *Romeo and Juliet* by having a week's worth of study in Elizabethan history and language, sonnet forms, and thinking activities structured around themes such as star-crossed love, gang rivalries, or obedience to parents. These authors also warned against preempting students' personal aesthetic responses to reading through too much introductory instruction in what a poem means, what an author is trying to say, etc. Authors of the mathematics chapters and several of the science and social studies chapters, in the process of putting forth inquiry or problem-solving models of teaching, also emphasized allowing students to engage in activities and construct understandings relying primarily on their past experience and current vocabularies before "bridging" to syntheses expressed within disciplinary structures using discipline-specific terminology.

Minstrell and Kraus even suggested that up to half of a unit should be spent generating understanding and moving from students' prior conceptions to a class set of ideas that organize the phenomena being taught. The focus would be on knowing how we know and why we believe the ideas. Then, the rest of the unit would be spent organizing experiences around powerful ideas of physics and coming to see the utility of these ideas and the scientific reasoning processes through which they were derived.

Stein featured instructional tasks (along with discourse and norms) as one of three key elements to effective teaching about number. She suggested that good tasks: (1) focus on important mathematical ideas, (2) use multiple representations, (3) can be solved using multiple strategies, and (4) require high-level thinking and reasoning. Elaborating, she suggested that good tasks embody foundational ideas that enable students to understand other mathematical ideas and connect different areas of mathematics; possess longitudinal coherence in that they build on prior understandings and leave behind "residue" that can be accessed and used productively in the future (insights into the structure of mathematics, strategies for solving problems, etc.); and are attuned to how student thinking develops and thus allow teachers to monitor those developments, especially at big transition points (e.g. moving from additive to multiplicative forms of reasoning). Multiple representations (supplementing language with manipulatives, diagrams, sketches, models, etc.) help students to make personal meaning of quantitative situations, think in diverse ways about mathematical concepts, and develop insight into the structure of mathematics. When tasks can be addressed using multiple strategies, teachers can encourage students to invent their own procedures. This is especially useful when students are learning new concepts or procedures. It encourages students to develop flexibility in their thinking about mathematical ideas, problem settings, and solution methods. By requiring high-level thinking and reasoning, good instructional tasks require engagement with mathematical ideas, stimulate students to make purposeful connections, and more generally, provide contexts for the class to work as a learning community constructing mathematical knowledge.

Freedman and Daiute indicated that helping students learn to write requires activities that shift their perspectives from speaker to listener, writer to reader, creator to critic, skeptic to persuader, and so on. Productive writing activities employ varied types of classroom language that support the development of varied cognitive strategies necessary to writing development; take into account the different demands of different types of written language and the fact that students must learn to write in a variety of ways, for a variety of purposes, and for a variety of audiences; and include reflection and assessments by teachers and students to monitor students' development and achievement in writing.

Battista suggested that ideal tasks for promoting students' geometry learning are problematic tasks that can guide the direction of their theory building by productively focusing their attention, encouraging them to reflect on their thoughts and actions, and promoting "perturbations" that require them to reorganize their current theories. Teachers scaffold students' progress on these tasks through phases that begin with initial exploration and proceed through elaboration on intuitive knowledge, introduction of relevant mathematical terminology and formulations, additional problem solving that requires synthesis and use of what is being learned, and eventual summary and integration of the learning.

Treagust and Chittleborough argued that algorithmic tasks that can be completed on a formula basis are less useful for teaching chemistry than conceptual problems that require students to rely on their reasoning skills and construct understandings. Tasks that require students to transfer across levels of representation and reconstruct problems in terms of their own understanding are especially desirable.

These guidelines about tasks made by various authors in talking about teaching their respective subjects appear to have generality across subjects. In addition, some of the authors offered guidelines for carrying out particular types of tasks, which are addressed in the following sections.

WORK WITH THE FIELD'S ARTIFACTS

Even though a daunting number and variety of learning activities are mentioned in the fourteen chapters, a great many of them are comparable in that they share a key characteristic: They engage students in working with artifacts used in applying the school subject or practicing its underlying discipline (e.g. prose and poetry selections, geometric figures, biological specimens, chemical compounds, laboratory equipment, historical source material, maps, cultural artifacts, etc.). Many of the authors emphasized the importance of engaging students in authentic activities using these subject-specific artifacts. Hoffman and Duffy emphasized the importance of keeping a rich variety of reading materials available in the classroom, and of engaging students in reading and discussing these texts in ways that help them realize that reading enlightens, enlivens, enhances, and informs. Similar comments were made by the authors of the other language arts chapters. Battista emphasized the value of allowing students to explore the properties of geometric figures using either physical shape makers or computer-based tools such as *Logo* or *The Geometer's Sketchpad*. Most of the other authors also spoke of emphasizing inquiry or problem-solving activities in which students would examine the kinds of objects or data addressed in the field (preferably data collected themselves), using the field's specialized instruments or data analysis techniques to develop and test models, seek answers to questions or problems, and so on.

For example, Fraser and Stoltman, in talking about teaching physical geography, emphasized the importance of field study in which students work with information collected in the field to address a larger problem. Data collection is followed by processing of the information, discussion, analysis, and negotiation of explanations for observed relationships and construction of follow-up plans. Field studies include literal field work in which students collect data directly through observations and recordings in the local geographical area; field trips to museums, farms, wildlife refuges, and other sites away from the school; and virtual field trips experienced via technology. A benefit of field study is the opportunity to use instrumentation in applying the tools of field research and scientific inquiry to an issue or problem (e.g. measuring tapes, string, maps, thermometers, geographic positioning systems, etc.). Other benefits include the opportunities for students to engage in experiential learning through direct, concrete experiences; to develop observation skills, analytical skills, teamwork skills, and environmental ethics; and to take responsibility for their own learning as they carry out the tasks involved in collecting, verifying, analyzing, and reporting information. Fraser and Stoltman acknowledged that the effectiveness of field work is not yet well documented because very little research has been done on it, but they argued that established principles of teaching and conventional wisdom suggest that hands-on, minds-on study outside the classroom and in natural or human landscapes has compelling benefits, especially when conducted in the context of inquiry or hypothesis testing as a method for teaching and learning.

Other authors made similar assumptions about the value of experiential learning with field-specific artifacts in their respective subjects. Cunha even included the collection of pedagogically useful artifacts on one's own travels as something that effective teachers of culture studies ought to be doing routinely.

Battista distinguished between technological tools for doing mathematics and technological tools for teaching mathematics. The former were developed outside the field of education but can be used for doing mathematics more easily or powerfully (calculators, statistics packages, graphing programs, etc.). The latter were developed specifically for enhancing students' mathematics learning (shapemakers, *Logo, The Geometer's Sketchpad*, etc.). It strikes me that this distinction between tools used in conducting field-related activities and tools specifically designed for teaching in the field is a fruitful one that has applications in all subject areas.

One place that authors disagreed was on the relative value of simulations or virtual experiences compared to direct experiences with real objects and events. Some authors were especially enthusiastic about computer-based learning opportunities, most notably Battista in talking about *The Geometer's Sketchpad* and other tools for teaching geometry and Duschl and Smith in talking about web-based opportunities for students to collect and analyze earth system science data.

Most authors were more restrained in their enthusiasm for simulations and virtual learning experiences, warning that these experiences can oversimplify what occurs in reality and perhaps induce misconceptions in students (which would need to be addressed during teacher-led debriefing of the experiences). Such caution was most obvious in Chapter Six, where Wandersee argued that laboratory experiences are important for letting students see a phenomenon or effect for themselves, decide what to test and then attempt to do so, and carry out open-ended investigations. Even here, he warned that quickie demonstrations have little value, suggesting that any inquiry-based laboratory work that is worth doing will take time to carry out and involve problem posing, problem solving, persuasion of peers, pre-laboratory preparatory orientation, and post-laboratory analysis. He indicated that laboratory activities can be supplemented by demonstrations or computer programs, but not replaced by them, so that the best rule of thumb is to teach with actual objects and natural phenomena unless this is impossible due to economic, time, or safety constraints.

Cunha sounded similar concerns in talking about teaching cultural studies. He cautioned that even the best interactive multimedia limit students to visual and auditory sense perception, so that we should not abandon the "hands-on" and "boot-wet" pedagogy that has served us well in the past.

Treagust and Chittleborough addressed these issues at some length in Chapter Eight. They portrayed laboratory experiences as essential for teaching chemistry, although they acknowledged that many such experiences are not closely connected to the conceptual material being taught or are mere "recipe tasks" in which students simply follow directions without understanding what they are doing. They also noted that cost, safety, and time concerns have led to a reduction of direct laboratory experience in recent years in favor of micro-scale substitutes (featuring simple, plastic, and often improvised equipment instead of large-scale glass-and metalware) for virtual laboratory experiences. They favored adaptation of experimental methods to promote inquiry through challenges, predictions, and experimental designs, especially in group activities that require students to make public predictions and then assess them. The teacher prepares the students based on knowledge of what they know now and where they are headed, bridging the chasms between different levels of representation of chemical knowledge, presenting tasks that connect to the theory being studied, providing clear expectations about how to do the tasks, and preparing them to deal with likely misconceptions. Then, the teacher mostly observes and listens, giving students thinking time and accepting their ideas without judging them. Instead of recipe-driven exercises in which the expected results are known to the students in advance, the teacher emphasizes activities that promote intellectual honesty and the use of authentic scientific processes in the laboratory: open-ended experiments that do not have predetermined or expected results but require students to explain their results in a scientific manner.

Minstrell and Kraus made similar comments in talking about laboratory activities in physics. They added that when engaging students in investigations that do not have an outcome known in advance, teachers should push students' thinking by posing questions to challenge and extend their ideas as they analyze results. The goal is not to arrive at a "right" answer quickly, but to build deep understandings. Teachers facilitate discussion by encouraging other groups to question the presenting group's findings in comparison with their own and by asking questions designed to challenge likely misconceptions.

SIMULATION AND ROLE PLAY

Simulation and role play methods were emphasized in several chapters, especially those dealing with aspects of social studies that do not involve much work with physical artifacts or tools. Traditionally, these methods have been portrayed simply as substitutes for direct experiential learning. However, several contributors noted that simulation and role play also can be used to demystify what is involved in generating disciplinary knowledge (e.g. to give students a taste of science-in-the-making instead of experiencing science only as a body of cut-and-dried knowledge), to help students see life applications of school learning (e.g. to simulate a forensic problem or hold a mock trial of polluters to stimulate application of chemical knowledge), or to promote empathy with people being studied in history or culture courses. Some authors' ideas about the demystification aspects of learning through simulation or role play seem connected to other authors' ideas about promoting the metacognitive and dispositional aspects of learning through engagement in aesthetic literary experiences, compositional writing, or scientific inquiry.

TEXTBOOKS

The authors had remarkably little to say about textbooks, perhaps in part because the generic principles put forth in the introduction covered much of what might be said about textbook-based teaching. Nevertheless, despite their emphasis on inquiry activities, most authors implied that students would be using textbooks at least part of the time. Treagust and Chittleborough noted that students need a user-friendly text to provide good explanations to supplement class work, including photos, diagrams, graphs, problems with worked solutions, and summaries. Cunha even noted that excellent workbooks exist for building skills in the social sciences, although he was referring to workbooks that call for research and analysis rather than filling in blanks. In contrast, some chapters (notably those in mathematics) implied little or no use of textbooks, and Wandersee rejected textbooks as very poor bases for teaching biology for understanding (because they are overly laden with unnecessarily long Greek and Latin terms and bulging with coverage of unnecessary detail).

FOSTERING METACOGNITION AND SELF-REGULATED LEARNING

Many of the authors emphasized teaching in ways that help students to become metacognitively aware of their learning goals and strategies and better able to self-regulate their learning. Frequently this was part of the rationale for the recommended inquiry approaches and discourse management principles. Hoffman and Duffy included methods that support self-monitoring among the three major sets of methods recommended in their chapter. They spoke of teaching students to monitor reading itself (through explicit strategy teaching, thinking aloud to model processes involved in strategy use, questioning to stimulate reader response to texts, etc.) as well as teaching them to monitor their progress as readers by negotiating goals, building portfolios, etc. Freedman and Daiute sounded similar themes in talking about teaching writing. They included detailed guidance about the construction and use of portfolios to monitor and promote reflection about students' development as writers. Much of what many of the other authors had to say about assessment also focused on planning and using assessment data to build metacognition and self-regulation, especially as teachers and students review assessment data together.

Authors often spoke of helping students not only to acquire individual concepts but to organize their learning within well connected networks structured around big ideas (although some warned against presenting formally organized disciplinary knowledge too early, before students have sufficient opportunity to draw on their prior knowledge to address problems or engage in inquiry using their existing ideas and terminology). Both Wandersee and Treagust and Chittleborough argued the value of teaching students to use concept mapping and other methods for graphic portrayal of knowledge networks, claiming that these methods are cost effective in the long run even though it takes time for students to learn to use them efficiently.

TRAJECTORIES, MISCONCEPTIONS, AND REPRESENTATIONS

Many authors emphasized the notion of trajectories in the development of understanding or levels of skill in the subject area. This came out frequently in discussions of using knowledge of trajectories in selecting learning activities and scaffolding students' progress through them, as well as using assessment data (mostly informal) to keep close track of the progress of individual students through those trajectories.

Subsumed within the notion of trajectories are the notions of false starts and dead-ends that occur when students develop or proceed on the basis of misconceptions (sometimes called alternative conceptions by authors who prefer not to use the term "misconceptions"). Laney suggested that misconceptions occur for three main reasons: (1) conceptual-analytic hurdles that occur when students are developmentally unready to understand a concept or have not yet mastered prerequisite knowledge or skills, (2) semantic/linguistic hurdles that occur when familiar words are given specialized meanings in the subject (e.g. scarcity, demand, or profit in economics), and (3) attitudinal blocks that occur when preconceived ideas based on one's experience hamper understanding of seemingly contradictory new ideas (e.g. most students initially have a hard time accepting the notion of sunk cost because it seems counterintuitive). Treagust and Chittleborough also noted that special meanings attached to everyday terms are a source of misconceptions in chemistry learning.

Another source of misconceptions is an overly restricted or otherwise distorted set of representations used to develop basic understandings. In geometry, for example, this can lead students to believe that squares are not rectangles or that a three-sided figure is not a triangle unless its "bottom" side is oriented on the horizontal plane. In history, some students come to believe that accounts of the past are not "history" unless they are about famous and important people.

It seems clear that in all subjects it is important for teachers to be aware of common trajectories of knowledge development (including misconceptions) and to draw on this knowledge to keep their teaching within the zone of proximal development. However, even in subjects for which there is a rich literature about common trajectories and misconceptions, there is not yet much theory or research speaking to issues involved in applying this knowledge (e.g. determining which common misconceptions to address directly and which to ignore unless students bring them up, when and how to refer to or elicit a misconception, or whether to address it immediately when it comes up or to let students struggle with it for awhile). Work also is needed on when and how to use techniques designed to temporarily induce dissonance or other confusion in students, such as when introducing historical paradoxes or cultural anomalies as ways to help students to identify with and see the rationality in behavior that may seem irrational or bizarre at first.

TEACHING WITH STORIES

Much has been written about narrative forms of discourse in recent years as well as about narrative structures as ways in which learners (young learners in particular) may organize and retain information. The language arts chapters in this volume naturally had much to say about the use of stories for instruction, but stories also were emphasized as instructional vehicles in Chapter Thirteen (stories as a basis for communicating political concepts and citizenship values) and Chapter Fourteen (fables as ways to communicate basic economic principles). In addition, Freedman and Daiute spoke about using writing as a way to help students learn to think about conflict or other social issues. I believe that the potential for using narrative structures (not only in written or printed matter but in videos and other multimedia formats) as bases for teaching methods and learning activities needs theoretical and empirical attention.

SOCIAL ASPECTS OF LEARNING

Most authors cited notions of learning community, social construction of knowledge, or sociocultural learning to argue that learning is most likely to be meaningful and accessible for use when it is socially negotiated through classroom discourse. In this regard, some authors hypothesized that special benefits accrue from interactions with peers. The language arts chapters suggested that students are stimulated by and in other ways profit from exposure to their peers' reactions to literature selections or their own writings. Torney-Purta, Hahn, and Amadeo emphasized that the peer group is an essential context for developing social and political knowledge in the "community of practice and discourse" that matters most to young people who are exploring ideas and identities.

Authors who emphasized extended inquiry activities usually assumed that students would carry these out in pairs or small groups and emphasized that the discourse that occurs during these activities is crucial to students' construction of understandings. Collaborative learning in small-group contexts was widely recommended, although usually with cautions that students need to be adequately prepared for the task and for the processes involved in working together, the teacher needs to circulate to monitor and scaffold progress, and the work should culminate with whole-class debriefings in which the activities and findings of individual groups are presented and discussed. The only major reservations were advanced in the mathematics chapters. Battista accepted forms of collaborative geometry learning that call for each student to complete all aspects of the task but rejected Jigsaw methods that require them to work on only one aspect of the task (and presumably learn the other aspects from their peers). Stein stated that small-group instruction is not ordinarily the method of choice for teaching about number, but allowed that breaking the class into groups might be appropriate when eagerness to talk was especially widespread.

CONCLUSION

Given the number and variety of teaching methods and learning activities addressed in this chapter, I will not attempt a summary. Instead, I will offer two thoughts by way of conclusion.

First, I view this chapter as just a beginning, for myself and I hope for readers as well, in identifying some of the commonalities involved in teaching different subjects effectively. The contributors raised many issues that struck me as important but were not addressed in this discussion because they appear specific to just one or two subjects (e.g. service learning activities as ways to develop values and dispositions relating to earth stewardship or citizenship, literature circles and book clubs in literacy instruction, writers' workshops in writing instruction, map work in geography, some of the latest in computer-based technology, chemistry's special problem of dealing with unseeable things, and the use of primary sources in teaching history). However, further analysis may reveal connections between these and other aspects of subject-specific teaching that are currently mentioned in only one of the chapters. Even now, for example, it strikes me that maps are representations, so there are probably many commonalities between what might be said about using maps in geography teaching and about using geometric and other representations in mathematics teaching. Similarly, much of what might be said about using primary sources in teaching history also might be said about using artifacts in teaching archeology or anthropology. I hope that this book will stimulate the making of such connections and thus inform our attempts to elaborate our theories and models of teaching and to help teacher educators prepare people to teach a range of school subjects.

Finally, it seems fitting to conclude this volume that appears early in the twenty-first century by offering some historical perspective. For this I borrow from Wandersee, who concluded his chapter with a list of suggestions for improving high school biology teaching that appeared early in the twentieth century. Based on a survey of teachers, the list featured the following ideas, which seem to apply to other subjects as well:

(1) More emphasis on "reasoning out" rather than memorization

(2) More attention to developing a "problem-solving attitude" and a "problem-raising attitude" in students
(3) More applications of the subject to everyday life and the community
(4) More emphasis on the incompleteness of the subject and glimpses into the "great questions yet to be solved by investigators"
(5) Less content coverage so that the course will progress no faster than the students can learn with understanding.

Sound familiar?

REFERENCES

Airasian, P., & Walsh, M. (1997). Constructivist cautions. *Phi Delta Kappan, 78*, 444–449.

Brophy, J. (1998). *Motivating students to learn*. Boston: McGraw-Hill.

Brophy, J. (1999a). *Teaching* (Educational Practices Series No. 1). Geneva: International Bureau of Education.

Brophy, J. (1999b). Toward a model of the value aspects of motivation in education: Developing appreciation for particular learning domains and activities. *Educational Psychologist, 34*, 75–85.

Brophy, J., & Alleman, J. (1991). Activities as instructional tools: A framework for analysis and evaluation. *Educational Researcher, 20*(4), 9–23.

Eccles, J., & Wigfield, A. (1985). Teacher expectations and student motivation. In: J. Dusek (Ed.), *Teacher expectancies* (pp. 185–226). Hillsdale, NJ: Erlbaum.

Edelson, D., Gordin, D., & Pea, R. (1999). Addressing the challenges of inquiry-based learning through technology and curriculum design. *Journal of the Learning Sciences, 8*, 391–450.

Kounin, J. (1970). *Discipline and group management in classrooms*. New York: Holt, Rinehart & Winston.

Weinert, F., & Helmke, E. (1995). Learning from wise Mother Nature or Big Brother Instruction: The wrong choice as seen from an educational perspective. *Educational Psychologist, 30*, 135–142.

Windschitl, M. (1999). The challenges of sustaining a constructivist classroom culture. *Phi Delta Kappan, 80*, 751–755.